IDEAS, QUALITIES AND CORPUSCLES

LOCKE AND BOYLE ON THE EXTERNAL WORLD

This study presents a substantial and often radical reinterpretation of some of the central themes of Locke's thought. Professor Alexander concentrates on the *Essay Concerning Human Understanding* and aims to restore that to its proper historical context. In Part I he gives a clear exposition of some of the scientific theories of Robert Boyle, which, he argues, heavily influenced Locke in employing similar concepts and a similar world-picture. Against this background, he goes on in Part II to provide an account of Locke's views on the external world and our knowledge of it. He shows those views to be more consistent and plausible than is generally allowed, demonstrating how they make sense of and enable scientific explanations of nature. In examining the views of Locke and Boyle together, the book throws new light both on the development of philosophy and the beginnings of modern science, and in particular it makes a considerable and original contribution to our understanding of Locke's philosophy.

IDEAS, QUALITIES AND CORPUSCLES

LOCKE AND BOYLE ON THE EXTERNAL WORLD

PETER ALEXANDER

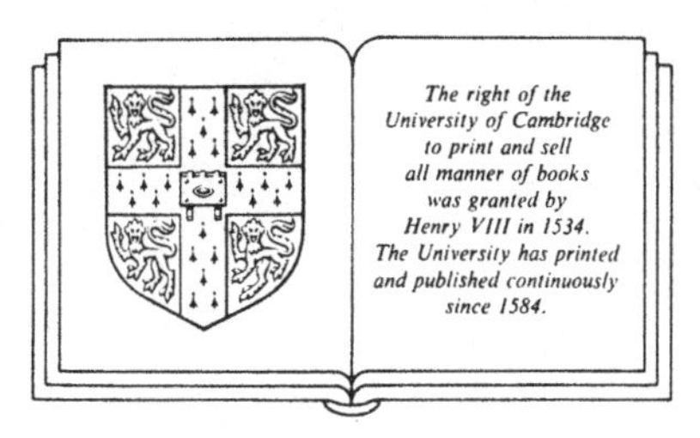

CAMBRIDGE UNIVERSITY PRESS
CAMBRIDGE
LONDON NEW YORK NEW ROCHELLE
MELBOURNE SYDNEY

CAMBRIDGE UNIVERSITY PRESS
Cambridge, New York, Melbourne, Madrid, Cape Town, Singapore, São Paulo, Delhi

Cambridge University Press
The Edinburgh Building, Cambridge CB2 8RU, UK

Published in the United States of America by Cambridge University Press, New York

www.cambridge.org
Information on this title: www.cambridge.org/9780521107341

© Cambridge University Press 1985

This publication is in copyright. Subject to statutory exception
and to the provisions of relevant collective licensing agreements,
no reproduction of any part may take place without the written
permission of Cambridge University Press.

First published 1985
This digitally printed version 2009

A catalogue record for this publication is available from the British Library

Library of Congress Catalogue Card Number: 84–15457

ISBN 978-0-521-26707-6 hardback
ISBN 978-0-521-10734-1 paperback

FOR
PETER ANTON

CONTENTS

ACKNOWLEDGMENTS

The author and the publisher would like to thank the editors of the respective journals for their permission to incorporate material from the following articles in this book:

'Curley on Locke and Boyle', *The Philosophical Review*, LXXXIII, 1974;
'The names of secondary qualities', *Proceedings of the Aristotelian Society*, LXXVII, 1977;
'Boyle and Locke on primary and secondary qualities', *Ratio*, XVI, 1974; and
'Locke on substance-in-general, I and II', *Ratio*, XXII, 1980, and *Ratio*, XXIII, 1981.

PREFACE

In the course of preparing this book I have had helpful criticisms and suggestions from colleagues and students, too numerous to mention individually, in British, Irish and American universities. I have been particularly helped by my colleagues at the University of Bristol, who have remained unfailingly cheerful and stimulating through my many assaults on their time and patience and my frequent insistence on turning the conversation to Locke. I am grateful to Yvonne Kaye and Doreen Harding for typing, advice, moral support and equanimity in the face of my many changes of mind; to my wife for putting up with it all for so long; and to two anonymous publisher's readers who helped me to improve the final version, even if less than they hoped. My thanks to all these.

P.A.

Bristol, March 1984

LIST OF ABBREVIATIONS

E: Everyman Edition of *S.C.* (see below), ed. Pattison Muir. p.310, note 8.

E.C.T.C.: Boyle, *Experiments and Considerations Touching Colours* in *Works* 1772. p.309, note 5.

E.G.: Boyle, *About the Excellency & Grounds of the Mechanical Hypothesis*, in *Works* 1772. p.312, note 3.

F: Facsimile of *first* edition of *E.C.T.C.* p.309, note 5.

L.S.: Locke, Letters to Stillingfleet in *Works* 1823. p.315, note 1.

O.F.Q.: Boyle, *The Origin of Forms & Qualities* in *Works* 1772. p.310, note 1.

S.: Extracts from Boyle's works in Stewart, *Philosophical Papers*.

S.C.: Boyle, *The Sceptical Chymist* in *Works* 1772. p.310, note 8.

INTRODUCTION

This book has been written in the belief that John Locke's *An Essay Concerning Human Understanding* has been sadly misinterpreted ever since it was published in 1690. It was something of a best-seller in the late seventeenth and early eighteenth centuries[1] and immediately gave rise to controversy, in which Locke took part, on most of his fundamental views. Attitudes to it in the present century, at least until about 1960, have largely consisted of a mixture of apathy and derision; it was admitted that Locke had dealt with many problems that interest modern philosophers but it was widely thought that his successors, especially Berkeley and Hume, had developed his 'empiricist' approach in such a way as to render it clearer, more plausible and more consistent than anything he had achieved. It was quite usual for students to be told that Locke's *Essay* was long, tedious and confused and that as Berkeley and Hume had discussed his most interesting views more shortly and more consistently a careful reading of Berkeley and Hume would afford a sufficient acquaintance with those views of Locke that were important. His own views were regarded as providing useful objects for philosophical criticism within the powers of the beginning student.

For the really assiduous student who wished to read Locke, there were, and still are, numerous abridged editions which became the natural source of reference for the student who was good enough to go on to teach philosophy. Even these students, however, had been so corrupted by the idea that Berkeley and Hume had got things more nearly right and by the use of abridgements that they tended to read Locke with a ready-made interpretation in their minds. So the conventional view of his philosophy was perpetuated.

A factor that contributed to this, in our century, was the relative unimportance attached to the history of philosophy among Anglo-American philosophers during the immediate pre- and post-Second World War decades. This had the consequence that many philosophers saw the works of the great philosophers of the past, if they took any notice of them at all, not as subjects for serious, detailed study, but rather as mines from which could be extracted what appeared to be problems currently under discussion together with instructively unsatisfactory attempts at solving those problems. These lumps of ore were hastily torn from their contexts with little consideration of the question whether they had been intended as the problems and solutions that, on a superficial reading, they seemed to be. Locke has suffered particularly from this procedure; even today he is more frequently cited than read.

All this involved, I believe, many mistakes which are gradually being exposed.[2] One mistake was to regard the empiricist basis and aim of Locke's view as identical with those of Berkeley and Hume; I hope to show that although he was, in some sense, an empiricist, his empiricism had more in common with that of the contemporary natural scientists than with that of Berkeley or Hume and that in some respects he was closer to the rationalism of Descartes and Leibniz. I think that in spite of his rejection of Aristotelianism, in which he was much influenced by Robert Boyle, there were elements of the scholastic philosophy that he did not see how to reject, even if he wished to do so. He may have preferred to avoid such sceptical conclusions as Berkeley and Hume were led into by their more wholesale rejection of Aristotelianism. A close study of these matters led me to think that the general descriptive labels 'empiricist' 'rationalist' applied to philosophers may too easily mislead if they are unqualified and used without care.[3]

A related mistake is to suppose that Locke was more inconsistent and less plausible than he really was and that Berkeley was right about this. I shall argue that when Berkeley was clearly referring to Locke he often misunderstood him. However, recent studies of Berkeley[4] have suggested that he was often not referring to Locke when he has been traditionally supposed to be. This raises a further problem: justification for the view that the truth about one philosopher can be obtained by reading another philosopher's comments on him must rest on a correct interpretation of both philosophers. This, in turn must rest upon a careful and independent reading of both.

The abridgement of a philosopher's work also raises problems that are often unrecognized by the student; it involves an interpretation which determines the separation of important passages, which must be included, from unimportant passages, which can be omitted. Different abridgements of a work may represent *radically* different interpretations. I believe that all the abridgements of Locke's *Essay* with which I am familiar are misleading to a greater or lesser extent. One defence that is often mounted is that Locke's writing is very repetitious and the student can do without repeated statements of the same point. There are two flaws in that argument. The first is that the importance a person attaches to a view may be indicated by the frequency with which he adverts to it; the second is that it is often likely that Locke is not just repeating himself when he appears to be but is either saying something different or saying the same thing in a different context in order to shed a new light on it. I am inclined to think that the best available abridgement is the Everyman edition by John W. Yolton;[5] but it is dangerous for me to say even that since it is a personal opinion relying heavily on the fact that it is the abridgement that least conflicts with my own interpretation of the *Essay*.

It may be perfectly legitimate for a philosopher not to take an interest in the history of philosophy but if one does adopt this attitude then it behoves one not to refer to 'the views' of historical figures for criticism and rejection in the course of developing one's own views. Anyone who is important enough to be referred to as often as Locke is worthy of study enough to ensure a reasonable chance of getting him right. I believe, of course, that there is more to be said for the pursuit of the history of philosophy than that it provides a source for outmoded views and instructively bad or usefully good arguments for use in current controversies. A better understanding of contemporary philosophical arguments may be contributed to by an understanding of earlier arguments or conclusions out of which, or in reaction to which, contemporary views have been developed. Of course, not every philosopher can be his own historian of philosophy; for most of us there is just not enough time. But for this very reason sound and scholarly history of philosophy should be encouraged.

My central aim is to achieve a better understanding of the most important features of Locke's philosophy, mainly as contained in the *Essay*. I do not claim, of course, that Locke was correct in all his conclusions but I have not the space here to engage in a full-scale criticism of them from our modern point of view. An essential pre-requisite for such criticism, however, is that we understand Locke's meaning and most of my energies here are directed towards this end. In the course of it I believe it emerges that, if I am right, much of his work is more plausible, consistent and worthy of discussion than has usually been supposed. My project arose largely from my thinking that such a well-read, intelligent and intellectually curious[6] person as Locke was unlikely to have been guilty of the many inconsistencies, muddles and blindnesses of which he has been accused. I thought it probable that he had often been misinterpreted and was not such a dolt as seems to be suggested by many of the criticisms directed against his work. So I looked for a different interpretation.

The difference between the empiricism of Berkeley and Hume, on the one hand, and that of Locke, on the other, can, I believe, be traced largely to the fact that he was much involved with, and influenced by, the work of contemporary natural philosophers, or scientists, while they were much less so. It has frequently been said by commentators that he was influenced by his friend and contemporary the Honourable Robert Boyle, and, less plausibly, by Newton but there has been little exploration of the extent and nature of this influence. It seems strange that critics should not have realized that such influences might provide a key to the understanding of Locke's ideas; perhaps this is another consequence of the belief that Berkeley and Hume had understood him. I wish to stress the importance of this influence and to suggest some consequences for the interpretation of Locke. I do not think that the tracing of influences is important for its own sake; I am not interested simply in where Locke got his ideas from unless this throws light upon the nature of his ideas and so helps us to understand what he was saying.

Locke's *Essay* covers a great deal of ground, raising and attempting to deal with a wide variety of problems. The seventeenth century was a period of great intellectual ferment and any philosopher of that period was likely to have been influenced by many scientists and philosophers holding very different views. Locke is no exception; among those mentioned in the literature as having influenced him are Aristotle, Descartes, Gassendi,[7] Leibniz, Bacon, Boyle and Newton as well as lesser known near-contemporary scholastics and alchemists. Some of Locke's ideas may well have come from, or through, any of these; his criticisms of individuals from this list usually take the form of a rejection of specific ideas and should not be taken as implying that he had no use for other ideas held by them. This makes it dangerous to refer to Locke, or any important figure from this period, as a 'Cartesian', a 'Gassendist', a 'Baconian' or even an 'anti-Aristotelian'. Locke rejected certain Aristotelian views, and that is a central feature of his philosophy as it is of Boyle's, but both of them accepted and built upon certain other Aristotelian views. Locke may have derived his use of '*Ideas*' largely from Descartes or largely from Gassendi; he may have been encouraged in his atomistic views by Boyle, by Gassendi, or by both.

There have been attempts recently to establish that while Boyle was strongly influenced by Descartes, Locke was strongly influenced by Gassendi and no doubt there are respects in which both these statements are true.[8] However, it does not follow, even if we bear in mind the strong opposition between Descartes and Gassendi, that Locke and Boyle are farther apart than I suppose. Much detailed work on ways in which those influences were incorporated into Locke's work would be necessary to establish that. In the meantime, I present a great deal of positive evidence for the thesis that Locke acquired many important ideas from Boyle or, at the very least, used many ideas that Boyle also used. The tracing of influences must be a detailed and specialized matter if it is to carry weight and it would be impossible, I believe, to discuss all the possible influences I have mentioned in one manageable volume. Moreover, my primary aim is to understand Locke's *Essay* and to present a fairly comprehensive interpretation that will make sense of it; in this my thesis is of considerable help. Within the bounds of that interpretation there remains much to be done in the way of elucidating details and a study of many other figures besides Boyle would no doubt contribute to that.

So I concentrate on the influence of Robert Boyle in the belief that many of Locke's central ideas can be explained by considering Boyle's views, without wishing to suggest that Locke was not importantly influenced by other people. It is worth saying that there is good reason to think that the influence of Newton upon him was much less important. Mere dates suggest this. Locke was working on the *Essay* for twenty years before its first publication in 1690. Newton's *Principia* was first published in 1687 and there is little sign of contact between them before 1689 when Locke first met him. The earliest extant letter from Newton to Locke is dated 1690.[9] In later editions of the *Essay* Locke makes a few changes in the text as a result of Newton's work but he makes them grudgingly and with a confession that Newton's ideas are not fully intelligible to him.[10] Moreover, it has recently been argued on the basis of the contents of their works, as well as dates, that it is more likely that Locke influenced Newton than the other way about.[11]

I believe that misunderstanding of Locke's *Essay* frequently begins very early with the passage in the *Epistle to the Reader* about master-builders and under-labourers in which Locke indicates something about his relation to the contemporary natural philosophers. He says

> The Commonwealth of Learning, is not at this time without Master-Builders, whose mighty Designs, in advancing the Sciences, will leave lasting Monuments to the Admiration of Posterity; But every one must not hope to be a *Boyle*, or a *Sydenham*; and in an Age that produces such Masters, as the great *Huygenius*, and the incomparable Mr. *Newton*, with some other of that Strain; 'tis Ambition enough to be employed as an Under-Labourer in clearing Ground a little, and removing some of the Rubbish, that lies in the way to Knowledge; which certainly had been very much more advanced in the World, if the Endeavours of ingenious and industrious Men had not been much cumbred with the learned but frivolous use of uncouth, affected, or unintelligible Terms, introduced into the Sciences, and there made an Art of, to that Degree, that Philosophy, which is nothing but the true Knowledge of Things, was thought unfit, or uncapable to be brought into well-bred Company, and polite Conversation.[12]

I take it that when Locke here talks of 'advancing the Sciences' he is using 'Sciences' in the old sense to mean 'knowledge' and that when he talks of 'Philosophy' he means both natural and metaphysical philosophy, that is, what *we* now call 'science' and 'philosophy', respectively.

First, it is worth commenting on the names mentioned. Locke had in his library 62 items by Boyle, 8 by Sydenham, 3 by Huygens and 2 by Newton of which one, the *Optics*, was not published until 1704, the year in which Locke died. Thomas Sydenham was a distinguished medical man who was one of Locke's mentors. One may suppose that Locke mentioned these men in descending order of their direct importance to him but, perhaps, in ascending order of their current reputations. At any rate, he separates Boyle and Sydenham, both of whom he knew and worked with, from 'the Masters' Huygens and Newton, the first of whom he dubs 'great' and the second 'incomparable'.

The usual view of this passage nowadays is that when Locke talks of himself as an 'under-labourer' he is stressing that he is a philosopher, in one modern sense, that is, that he engages in the exposure and removal of conceptual muddles from various areas of discourse, such as scientific and moral discourse. I believe that to be only part of what Locke had in mind. In the first place, he worked in a more literal sense as an under-labourer, helping Boyle with relatively routine tasks[13] and he read and commented upon some of Boyle's manuscripts before publication.[14] In the second place, and probably much more importantly, the passage quoted suggests that Locke saw himself as helping to make natural philosophy, the views of scientists, more accessible to the intelligent layman, to make it fit and capable 'to be brought into well-bred Company and polite Conversation'. That is, he saw himself as popularizer, in the best sense, of current scientific ideas and controversies. Since the work of the 'new philosophers' frequently involved them in controversies with those who clung to the allegedly Aristotelian ideas of the scholastics and since these ideas were often couched in obscure and ill-understood jargon there was much standing in the way of the layman who wished to know what was going on. Locke was helping the new philosophers to clear the ground of this jargon. This was among the ideals of the newly-formed Royal Society of which both Boyle and Locke were early Fellows. As we shall see, Locke frequently uses the language of Boyle's corpuscular philosophy and may be seen sometimes as reporting, explaining and illustrating that very important hypothesis and recommending its form of explanation.

I would not have it thought, however, that Locke was *merely* reporting the work of natural philosophers or merely trying to remove their, or their predecessors', conceptual tangles. His more specific philosophical task, and that which mainly constitutes his original contribution, was to explore the implications of recent scientific work for our view of the world and of knowledge and belief. This took him into a consideration of what constitutes knowledge and what sorts of things we can or cannot hope to know as well as a reasonably detailed discussion of the nature of language.

There is a related but perhaps even more important feature of Locke's *Essay*, the first clue to which lies, I think, in his treatment of innate ideas. Having put forward, in Book I, direct arguments against the view that we have some ideas innately Locke then says, in effect, that if the reader is unconvinced by these he should read the rest of the *Essay*; it is based on the hypothesis that there are *no* innate ideas and will, he hopes, show that a plausible and comprehensive account of our knowledge and understanding of the world can be given without them, thus showing that it is unnecessary to postulate them. This will give indirect support to the hypothesis. He says

> I know it is a received Doctrine, That Men have native *Ideas*, and original Characters stamped upon their Minds, in their very first Being. This Opinion I have at large examined already [i.e. in Book I]; and, I suppose, what I have said in the fore-going Book, will be much more easily admitted, when I have shewn, whence the Understanding may get all the *Ideas* it has, and by what ways and degrees they may come into the Mind; for which I shall appeal to every one's own Observation and Experience. (II.i.1)

What he in fact claims to show is that it is sufficient for an adequate account of our knowledge of the world to suppose our knowledge to be based upon ideas originating in experience; thus there is no *need* to suppose as well that we are provided at birth with ideas that are independent of our experience.

I take this argument to be of the hypothetico-deductive form of modern scientific argument, to the development of which Boyle and others at this time were so significantly contributing. It is worth noting, in passing, that it has been claimed that Newton was less favourably disposed towards such arguments.[15] I believe this form of argument figures largely in Locke's *Essay*, in particular, in the way in which it uses the corpuscular hypothesis: if an account of our everyday experience and description of the world based on the best available scientific hypothesis were adequate and plausible then this would provide powerful indirect support for that hypothesis. Locke may be regarded as, quite deliberately, helping Boyle and others in supporting it.

A sympathetic reading of the *Essay* as a whole, in conjunction with the relevant works of Boyle, seems to me to put it beyond doubt that the 'lasting monument' of the master-builders that most impressed Locke was the corpuscular philosophy; it is used and referred to over and over again especially in the most central passages. Moreover, Locke uses many examples that figure in Boyle's experiments specifically designed to support his hypothesis.

Locke's view of hypotheses in general is far from uncontroversial. Thus

John W. Yolton, though he admits that Locke was much taken with the corpuscular hypothesis, says[16] that 'the aspect of science which influenced Locke was not the formation of hypotheses' (p.58) but rather it was the Royal Society's 'programme of natural histories which caught Locke's eye' (p.63, n.3). I am inclined to think that Locke attached importance to both these things and therefore to favour Laurens Laudan's account[17] of his attitude to hypotheses in general. I do not wish to enter here into direct controversy but I hope that what I have to say later will help to make my view more plausible.

It must not be thought that either Boyle or Locke regarded the corpuscular hypothesis as established or that they regarded its basic concepts as the only possible ones for scientific explanation. Boyle says

> that, which I need to prove, is, not that mechanical principles are the necessary and only things, whereby qualities may be explained, but that probably they will be found sufficient for their explication.[18]

Locke is similarly cautious when he says

> I have here instanced in the corpuscularian Hypothesis, as that which is thought to go farthest in an intelligible Explication of the Qualities of Bodies; and I fear the Weakness of humane Understanding is scarce able to substitute another, which will afford us a fuller and clearer discovery of the necessary Connexion, and *Co-existence*, of the Powers, which are to be observed united in several sorts of them. This at least is certain, that which ever Hypothesis be clearest and truest, (for that it is not my business to determine,) our Knowledge concerning corporeal Substances, will be very little advanced by any of them, till we are made see, what Qualities and Powers of Bodies have a *necessary Connexion or Repugnancy* one with another; which in the present State of Philosophy, I think, we know but to a very small degree: And, I doubt, whether with those Faculties we have, we shall ever be able to carry our general Knowledge (I say not particular Experience) in this part much farther. (IV.iii.16)

In the *Essay* Locke is, I believe, partly exploring the implications of this hypothesis and partly considering, in general, the nature and place of hypotheses in natural philosophy.

Boyle sought to establish or refute his hypothesis by observation and experiment. If Locke is regarded as attempting to give indirect support to it by considering how broadly it could be applied to familiar everyday phenomena then many of his arguments, especially those concerning primary and secondary qualities, become more intelligible and plausible than they have usually been supposed to be. Of course, Locke was interested also in 'natural histories', as were both Boyle and Bacon, because these were to form the basis for the construction of hypotheses which were not drawn out of thin air, or based upon the authority of Aristotle, or upon *a priori* conclusions from unexamined metaphysical principles as, they alleged, were those of many scholastic philosophers whom they attacked.

It is important to recognize that the context of Locke's *Essay* was the so-called 'Revolution in Science' of the seventeenth century. The nature of this 'revolution' is an exceedingly complex matter but one central feature of it was a rejection of many ideas and attitudes allegedly drawn and developed from Aristotle by the medieval schoolmen. Another important feature was the growing perception that theology, natural philosophy and metaphysical philosophy could and should be pursued independently of one another and that philosophy need no longer be regarded as the handmaid of religion. Part of this was the recognition that philosophical thinking and scientific investigation constituted no threat to religious belief or much theological thinking. The natural philosophers attacked by Boyle were largely influenced by scholastic views or by certain alchemical ideas or both. Boyle was strongly influenced by early and recent atomism, which had never become generally accepted largely because it was regarded as atheistical, it was attacked by Aristotle and was difficult to make consistent with well established Aristotelian views. Much of Aristotelianism had become part of the official philosophy of the Church, so these various points are closely connected. Boyle explicitly mentions two main targets, peripatetics or schoolmen and spagyrists or alchemists, mainly those with medical interests.

Locke, I believe, was joining Boyle in this attack, although partly from a different point of view. He was more directly interested in epistemology, Boyle in empirical investigation. It is difficult to understand either of them without an awareness of the context in which they were working. This is more often said of Boyle than of Locke for the obvious reason that Boyle was much more explicit in stating the views against which he was reacting. One of my aims is to show how a recognition of this helps to clarify what Locke was saying. I have accordingly divided the book into two parts, the first dealing largely with Boyle and his background, the second dealing with selected central features of Locke's *Essay*.

Boyle deserves a book, or many books, on his own account. I have been able to do no more than give a sufficient picture for my purpose of the place of his theorizing in its context by indicating briefly how he saw the views against which he was reacting. His work was highly influential and constitutes a landmark in the history of chemistry. He feared criticism from fellow natural philosophers because of his interest in chemistry which was unfashionable at the time partly, I think, because much alchemy had brought chemistry into disrepute among those involved in the Royal Society. Boyle recognized, however, that many alchemists, especially those with a medical interest, had made valuable empirical discoveries in chemistry. In fact Boyle did not desert physics for chemistry but applied physical ideas to it and so developed out of atomistic views going back to ancient Greece what is probably his major contribution, the corpuscular hypothesis. The aim of this was to allow the development of mechanical explanations of as many natural phenomena as possible. He produced at least a programme for this and some discussion of what its limits might be.

I have tried to give an accurate account of this hypothesis and to extract the epistemological views underlying his discussions of it. This part of my book is obviously inadequate to do more than indicate a few of the complexities of the period and expose some of the ideas that were centrally under discussion. I claim only that it helps us to a clearer understanding of Locke's *Essay* and I shall be satisfied if it does that. I do not claim to make any original contribution to the history of science; I have had to rely on much excellent historical scholarship that has appeared during this century. Neither have I been able to give a detailed account of the views of those Boyle attacked as their proponents saw them; what is important for my purpose is how Boyle saw them.

The second part of the book is not intended to be a commentary on the whole of Locke's *Essay*; I have chosen those topics that seem to me to be central to Locke's thought, explored their relations to one another and argued that they may be susceptible to unorthodox interpretations when they are considered in the light of Boyle's work. One effect of this is to show the *Essay* to be a more coherent and unified work than it would otherwise appear to be. Another is to show it as a developing argument in which new technical terms are introduced gradually in a significant order and ideas are adjusted and refined as the work proceeds.

I have tried in my arguments to rely as little as possible on Locke's celebrated carelessness and inattention and to assume that he is not being inconsistent until I am driven into a contrary view. This assumption is not, however, gratuitous because it is part of my contention that the consideration of the *Essay* in the light of Boyle's work and the regarding of it as a developing argument helps one to remove some apparent inconsistencies. I have not often engaged in piecemeal controversy with other commentators on Locke, partly for reasons of space, partly in an attempt to avoid tediousness but more importantly because my aim has been to present my interpretation whole and to ask the reader whether it does not make more coherent sense of what Locke says in the *Essay* than some more familiar interpretations. I hope that the detailed arguments will be considered not only in isolation but also in relation to the whole. I believe that this is an approach of which Locke would have approved.

I am aware that one consequence of my method is that I have had to ignore many important arguments to be found in the literature about what Locke meant in particular passages but I hope that the effect will be to provide a framework within which these arguments may be considered and, perhaps, more easily assessed. I believe that a philosopher's detailed arguments may be more readily understood and evaluated if the overall direction of his argumentation can be established, with at least some probability, in advance of their consideration.

I have incorporated and developed material from my earlier articles listed in the bibliography. This book goes beyond those articles in attempting to show that the approach I adopted in them has implications for the understanding of much of the rest of Locke's *Essay*. Interpretations not clearly foreshadowed in my articles appear especially in the chapters on ideas, powers, patterns and resemblance, language and knowledge. For example, I propose an unorthodox interpretation of Locke's '*Ideas*' which allows a more plausible account of his views on language and meaning and of the extent of his scepticism about knowledge. No doubt my account of his views about knowledge and natural kinds is more orthodox than the rest but that is because I think that he has been less seriously misinterpreted on these topics by other commentators. However, I hope that these two chapters help to draw together many of my earlier interpretations and to show the *Essay* to be a relatively consistent and well-organized whole.

QUOTATIONS AND REFERENCES

There are difficulties, when one writes about Boyle, in deciding from which editions of his works to quote. The first editions are not easily accessible and, besides, are not very reliable in reproducing what he wrote. As M. A. Stewart says in the Preface to his useful selection from Boyle, 'It is not clear that any replication of Boyle's original editions would be authoritative ... for the simple reason that they were not often well-prepared either for or by the printer.'[19] Stewart's own modernized text, though the most accessible, does not contain all the works to which I wished to refer. There are two editions of *Collected Works*, frequently used by scholars, published in 1744 and 1772, of which Stewart says that the second is the more corrupt. In spite of this I have quoted from the 1772 edition, having checked the quoted passages against the first editions and found no differences that would affect my interpretation. This edition has the advantage of greater accessibility. I have given references to it, followed by references to any modern edition that I consider may be helpful. I have also given a list of corresponding page references in the 1772 edition and the first editions of *The Origin of Forms and Qualities* and *The Sceptical Chymist*, the two works from which I quote most frequently, for the main passages quoted.

Quotations from Locke presented fewer problems. I quote mainly from his *Essay* and have used Nidditch's excellent modern edition. There is, as yet, no authentic modern edition of his letters to Stillingfleet so I have quoted from the 1823 edition of his works, which is familiar to scholars and which I have found to be reasonably reliable by comparison with the first edition. An occasional word in square brackets indicates a word inserted from the first edition.

PART I

1

BOYLE ON EMPIRICAL INVESTIGATION

I

Robert Boyle is perhaps best known to the layman for 'Boyle's Law' and, for some reason that is not clear to me, his book *The Sceptical Chymist* (1661). This book is largely a direct attack on certain doctrines of the 'chymists' or 'spagyrists' and an indirect attack on certain doctrines of the Aristotelians or 'peripatetics'. The context in which he was writing was highly complex. He had little use for alchemy in its most extreme and 'mystical' forms but he had a considerable respect for some of those he called 'chymists' or 'spagyrists' largely because of their empirical interests and discoveries. Spagyrists were iatrochemists or medical chemists some of whom had the theoretical aim of giving accounts of illness and health in chemical terms and the very practical aim of discovering chemical cures for illnesses. They were in certain respects close to alchemy, out of which their practices and doctrines had largely developed. Paracelsus, who rejoiced in the name of Philippus Aureolus Theophrastus Bombastus von Hohenheim (1493–1541), was an important figure whom Boyle both admired and criticized. He regarded the universe as fundamentally chemical, sought to give a chemical account of the creation and held a theory, attacked by Boyle, involving both the 'elements' of the Aristotelians and the 'chemical principles' of the alchemists. The origin of the name 'spagyrist' has been traced to him by at least some scholars.[1] The most plausible account is that it is derived from the Greek words σπάω, I tear (apart), and ἀγείρω, I put together. Thus, in modern terms, a spagyrist is one who works by analysis and synthesis, suggesting methods of which Boyle approved.

Boyle mainly spends time only on those for whom he has some respect. He often uses the terms 'chymists' or 'vulgar chymists' for one of the groups he attacks, probably referring to those who had freed themselves from alchemy at least to the extent of carrying out fruitful observations and experiments while eschewing the secrecy and the wilder imaginative flights of the most extreme alchemists.

Alchemy is exceedingly complex partly because of the several streams of thought that contributed to it, partly because of the secrecy practised by many of its adherents and partly because it attracted charlatans and dolts as well as sincere and intelligent seekers after truth. By the time of Boyle it had a very long history and that history has still only partly been written. Here I can do no more than outline some of the doctrines and practices that Boyle attributed to the spagyrists and attacked. Fortunately it is not necessary for my limited purpose to do more; what matters is Boyle's understanding of some of these doctrines and practices, whoever held or engaged in them, and the arguments he used against them. I shall avoid scholarly controversy about who should be called alchemists or iatrochemists or spagyrists and use the term 'chemists' for those whom Boyle attacked.[2]

In the hands of its best practitioners, such as Paracelsus and J. B. van Helmont (1577–1644), alchemy was regarded as a new science based on observation and experiment, reacting against blind adherence to authority, especially that of Aristotle, and finding new applications of chemistry to medicine as well as to the explanation of phenomena in general. Some practitioners relied to some extent upon Hermetic, neo-Platonic and neo-Pythagorean philosophy and even accepted a kind of 'magic', thought to be compatible with Christianity, and regarded as an empirical study of the occult forces of nature. The well known aim of making gold artificially was of only minor importance to many of them. Besides this they sought recipes for health and longevity and the separation of the 'pure essences' of substances from unwanted impurities. This last goal was of great importance for the development of modern chemistry at the hands of Boyle and his successors. In spite of the anti-Aristotelianism avowed by many of them chemistry continued to rest on the Aristotelian doctrines, often misinterpreted, of, for example, the four elements earth, water, air and fire, and forms and qualities. This is one reason why Boyle thought he could attack spagyrists and peripatetics at the same time and with the same sorts of argument; another reason was that their explanations had a common form which Boyle regarded as misconceived.

Paracelsus was known mainly on the continent until the seventeenth century when his work was taken up and developed by a number of English alchemists.[3] During this period also the works of van Helmont and others were becoming more widely known in England through translations. It was against the background of a renewed interest in alchemy in England that Boyle criticized the chemists.[4] In his published works Boyle is not always as explicit as one would wish about whom he is attacking or, indeed, relying on but apart from Paracelsus and van Helmont he does frequently mention Daniel Sennert (1572–1637), a German medical chemist who attempted the difficult feat of combining Aristotelianism and Atomism. In his unpublished works Boyle mentions some other lesser known figures such as George Starkey, a London 'chemical physician' and 'philosopher by fire' and Frederick Clodius, a somewhat suspect dabbler in the arcane.

Boyle's greatest importance perhaps lay in the fact that he attempted to develop chemical theory on a basis of empirical investigation. He found, especially among the chemists, empirical methods that were confused and imprecise and from which large unjustified inferences were made and, especially among the peripatetics, a great interest in theories but little interest in their empirical confirmation. He also found in both camps, 'explanations' that were unintelligible or circular and so failed to be explanatory. He was primarily interested in giving clear and confirmable accounts of natural phenomena and he saw that this must begin with accurate descriptions of them. In this he was following Francis Bacon and espousing a fundamental aim of the nascent Royal Society (founded 1662–3). Explanations of the phenomena must make clear how they happened. Explanations require general theories; theories to be plausible must be based upon and checked against the results of observation and experiment; and it must be clearly stated how the theory is linked to the observed phenomena in every particular case. Boyle respected the chemists for their empirical interests and the peripatetics for their theoretical interests and attempted to draw a lesson from both while pointing out their complementary weaknesses.

One weakness of both chemical and peripatetic theories was that they relied upon 'occult' qualities and entities and, throughout his works, Boyle inveighs against occultness in natural philosophy. The chemists tend

> to content themselves to tell us, in what ingredient of a mixt body, the quality enquired after, does reside, instead of explicating the nature of it, which ... is much as if in an inquiry after the cause of salivation, they should think it enough to tell us, that the several kinds of precipitates of gold and mercury, as likewise of quick-silver and silver ... do salivate upon the account of the mercury, which though disguised abounds in them; whereas the difficulty is as much to know upon what account mercury itself, rather than other bodies, has that power of working by salivation. (*E.C.T.C.* 724, *F.* 178)[5]

For him something is occult if, first, it is unobservable and the only ground for postulating it is that it 'explains' a particular phenomenon or property and there are no alternative methods of confirming its presence and, second, if no intelligible description of it which is independent of what it is claimed to explain can be given. Suppose I say that a substance has a bitter taste because of the presence in it of *X*. If the only evidence for the presence of *X* is the bitter taste and no other 'explanation' depends upon the presence of *X* and if, further, *X* cannot be described or can be described only as either bitter or as *the cause or principle of bitterness* then X is occult, even if there seem to be *a priori* reasons allegedly supporting its presence. The use of occult conceptions discourages the description of mechanisms that explain how things happen.

Explanations were habitually given by chemists in terms of *principles* and by peripatetics in terms of *elements, forms* or *real qualities* and these were all, according to Boyle, occult. We shall consider these terms and what Boyle says about them in more detail later. His aim was to replace principles, elements, forms and real qualities by inner structures which could be clearly described in terms drawn from experience and which would allow the precise description of mechanisms. These descriptions, even though they would be of unobservable structures, would have empirical content and would allow real explanations. That is the function of the 'corpuscular philosophy' of which Boyle has something to say in most of his works but which he develops most fully in his *The Origin of Forms and Qualities* (1666). According to this view, all material substances are composed of minute particles or corpuscles whose properties are of a kind we meet with in ordinary sense-experience, such as shape, size and motion or rest. The structures of large collections of these corpuscles would, it was hoped, explain the observable properties of substances and their interactions with one another. The mechanism of changes would involve the rearrangement of corpuscles and collisions between them. There would be no need for any occult relationships between corpuscles, such as 'sympathies' or 'antipathies' or, indeed, for forces of any kind. It was hoped that inferences would be possible from the nature of what was observed, if it was observed in sufficient detail, to the unobserved underlying structures and mechanisms and that descriptions of them in terms we understand could be given. This would allow the intelligible description of the connections between the phenomena and these structures which would have real explanatory power. It also implied that the more careful and detailed the observations and experiments, the more likely they were to reveal and distinguish between the hidden mechanisms.

The originality of this approach, at least in the context of seventeenth-century science, was that Boyle was attempting to give an account of chemical properties and reactions in physical terms, that is, in terms acceptable to the natural philosophers, mostly physicists, who were the prime movers in the 'Revolution in Science'. They were hostile to chemistry and Boyle feared that they thought he was wasting his time dabbling in chemistry which they associated with the obscurantism of alchemy.[6] However, he was a brilliant experimentalist, he was utterly wedded to his corpuscular hypothesis and he was able to bring the two together in many ingenious ways and so win the respect of his critics for his general approach. He seldom loses sight of his corpuscular hypothesis and the possibility that his experiments will support and shed new light on it.

Boyle was still a long way from the concepts of modern chemistry such as those of a *chemical* reaction and of an element as a *chemical* entity that could remain unchanged through a reaction, but he was even farther from the mystical and arcane ideas of alchemy so his work marks an enormous step in the direction of modern chemistry.

Throughout his critical writings Boyle is basically attacking two current kinds of view about what can loosely be called 'elements', that is, fundamental units in nature in terms of which natural phenomena were thought to be explainable. Although the peripatetics used the word 'element' while the spagyrists used the word 'principle' their conceptions and the forms of explanation based on them were similar. Boyle defines these words in at least two places in *The Sceptical Chymist.* It is of the utmost importance to realize that these are not definitions of 'element' in the sense of modern chemistry.[7] In the opening section, entitled 'Physiological Considerations' he defines them as

> those primitive and simple bodies, of which the mixt ones are said to be composed, and into which they are ultimately resolved. (*S.C.* 468, *E.* 18)

and, in the Sixth Part, as

> certain primitive or simple, or perfectly unmingled bodies; which not being made of any other bodies, or of one another, are the ingredients, of which all those called perfectly mixt bodies are immediately compounded, and into which they are ultimately resolved... (*S.C.* 562, *E.* 187)[8]

He immediately goes on to cast doubt on the existence of such elements and eventually to deny it (*S.C.* 583, *E.* 224). What he particularly objects to is the idea, to be found in both the peripatetics and the spagyrists, that every substance in nature is compounded out of some proportion of *all* the elements or principles.

The Aristotelian elements, earth, water, air and fire were thus considered by the peripatetics to be unanalysable components, in various combination, of observable bodies and substances whose observable qualities could be explained in terms of properties attributed to these elements. It was hoped that they would be obtainable in nearly pure form from observable substances as the ultimate products of analysis. According to the spagyrists the ultimate products of analysis were the chemical (or physiological or hypostatical) principles. Most spagyrists, among those Boyle was attacking, held that there were three principles, salt, sulphur and mercury, called the '*tria prima*'. They too were regarded as being responsible for the observable properties of observable substances and the basis of all explanation of chemical phenomena.

There was considerable controversy about details both between the spagyrists and the peripatetics and between the members of each group but the two groups had views in common besides their conception of elements or principles. The favoured method for the investigation of the composition of substances was 'analysis by fire' seen as the breaking up of those substances into their constituents by heat, usually by exposure to naked flames. They also both held that their principles or elements were not the same as the substances which, in common parlance, go by the same names; they were the *pure forms* or *essences* of ordinary salt, sulphur and mercury or of ordinary earth, water, air and fire, respectively. For example, Boyle in one place says

I know it may be said, that chymists in the opinion above recited mean the principle of sulphur, and not common sulphur which receives its name not from being all perfectly of a sulphureous nature, but for that plenty and predominancy of the sulphureous principle in it. (*E.C.T.C.* 722, *F.* 178)

The aim of analysis by fire was, if possible, to separate out the pure forms or essences from one another and from useless dross. The situation is complicated even farther by the fact that some chemists (*e.g.* Paracelsus) held that the Aristotelian elements were analysable into the principles and others (*e.g.* Sennert) that the principles were analysable into the elements. Yet others held that there were five principles rather than three (*e.g.* N. Lefebvre, 1615–69) or that there was just one element (*e.g.* van Helmont).

However, such details are not relevant here since Boyle believed that he could throw serious doubts on all these views by questioning the alleged empirical evidence for the elementary nature of anything so far claimed to be an element or a principle, especially if they were thought to be obtainable by analysis by fire. Most of his empirical arguments depend upon experiments designed to show that the theories he was attacking were at least no better and usually less well supported than other possible theories, especially his corpuscular theory. This allowed him to press the claims of his own theory to greater intelligibility and plausibility. *The Sceptical Chymist* is largely concerned with the question of empirical support. However, as I have suggested, Boyle also had abstract logical and theoretical objections to the views he was attacking, especially concerning the concept of explanation involved, and he glances at these at the end of *The Sceptical Chymist.* He developed these arguments more fully in his *The Origin of Forms and Qualities* to which I shall return later. The main burden of this work is that what is allegedly done by theories involving mysterious elements and principles could be actually done by a theory replacing them with perspicuous and describable corpuscular structures.

I now turn to *The Sceptical Chymist* in a little more detail. The book is rambling and repetitive but I propose to pick out only the main lines of argument that are relevant to my central purpose. It is in the form of a dialogue between a peripatetic (Themistius), a spagyrist (Philoponus), a corpuscularian (Carneades) and a common man (Eleutherius) with Boyle himself acting as narrator and occasional commentator. On the whole, of course, he is sympathetic to Carneades but he warns us that he does not always agree with him. This is probably partly because Boyle wishes to appeal to all those sympathetic to corpuscular theories even if they disagree in some of the details and partly because his own view on many of the details awaits further empirical investigation. The work becomes substantially a monologue by Carneades, with short interjections by the others, as it moves towards the exposition of the corpuscular hypothesis.

II

In the Preface and 'Physiological Considerations' with which Boyle opens *The Sceptical Chymist* he sketches the general background and makes some methodological remarks. His two groups of opponents are worthy of attention because they have both made contributions to chemistry. However, these are embedded in a mass of empty explanations, unsupported speculation, unintelligible statements of positions and unjustified conclusions from insufficiently controlled observations and experiments.

The remarks in the Preface (*S.C.* 458–63, *E.* 1–10) appear to be largely theoretical but it soon becomes evident that they are leading to considerations concerning the empirical confirmation of hypotheses. The 'naturalists and physicians' have performed useful experiments and proposed efficacious chemical remedies but 'their notions about the causes of things and their manner of generation' are weak because they rely for explanations on the three principles, salt, sulphur and mercury, whose nature is far from clear. The advantage of paying attention, on the contrary, to 'the motions and figures, of the small parts of matter and the other more catholick and fruitful affections of bodies' is that a theory based on these gives a much more flexible basis for explanations and a more comprehensive explanatory power. This will be explained in Chapter 3.

Carneades' arguments for the corpuscular hypothesis, it is admitted, may not be conclusive; he merely aims to 'propose doubts and scruples' which will show that his opponents' arguments are not conclusive either. His objections need not even be consistent with one another; if he proposes two incompatible hypotheses and each is 'but as probable as that he calls in question' he has shown that the theory he is opposing requires further support. If he proposes several hypotheses, each of which is probable on the available evidence, any hypothesis with which they all disagree is open to doubt unless that evidence shows it to be more probable than they.

If a counter-instance can be brought to any universal hypothesis and is not open to dispute then that hypothesis is overthrown. The hypothesis that all 'mixed bodies' are composed of a determinate number of fixed ingredients, whether elements or principles, would be refuted by the discovery of just one such body that is not so composed. Boyle also points out that it is easier to find objections to an hypothesis than to construct an alternative one that is not open to objections, a statement that may be construed as either modest or boastful.

He protests against the secrecy surrounding the chemists' theories and the obscurity with which they are expressed and calls upon 'the more knowing artists' to 'either explicate or prove the chymical theory better than ordinary chymists have done'. In spite of their having produced valuable empirical results, he is 'no admirer of the theorical part of their art' although its obscurity has allowed him to learn much from the vulgar chemists' experiments while leaving him free to theorize independently. His arguments may be regarded as largely requests for clarification and, then, further support for their theories. The result of his independent theorizing is that he is strongly inclined to the corpuscular hypothesis but is willing to accept any alternative that is 'intelligibly explicated and duly proved'; he loves 'fluctuation of judgment little enough to be willing to be eased of it by anything but error'.

In 'Physiological Considerations' (*S.C.* 464–73, *E.* 11–28) Carneades stresses that the peripatetics rely heavily on books while the chemists rely heavily on laboratories and says that the peripatetics' strong point is that they value reasoning, general theories and consistency while their weak point is that they neglect experimental testing. He has already regretted that their attacks on the chemists are based on 'little experimental knowledge in chymical matters'. With the chemists it is the other way about; their strength lies in their experimentation, their weakness in their theorizing. Boyle appears to respect Aristotle, and at least some peripatetics, for the systematic nature of their thinking, although it is with some irony that he allows Themistius, his Aristotelian, to say

> it is much more high and philosophical to discover things *à priori*, than *à posteriori*. And therefore the Peripatetics have not been very sollicitous to gather experiments to prove their doctrines, contenting themselves with a few only, to satisfy those, that are not capable of a nobler conviction. (*S.C.* 469, *E.* 20)

They employ experiments, Themistius says, to illustrate rather than to establish or, he might have added, to attempt to refute. When we see how harmonious *Aristotle*'s doctrine of the elements is with his other principles of philosophy; and how rationally he has deduced their number from that of the combinations of the four first qualities from the kinds of simple motion belonging to simple bodies (*ibid.*)

we see that one experiment will suffice to illustrate the theory. The favoured experiment is that in which green wood is burnt and is seen to produce earth, water, air and fire. I shall return to this.

I think it is clear that Boyle has some sympathy with these peripatetic criticisms of the chemists even if Themistius' rhetoric is misplaced. Themistius points out that the Aristotelian view survived unopposed for centuries until

in the last century Paracelsus and some few other sooty empirics, rather than (as they are fain to call themselves) philosophers, having their eyes darkened, and their brains troubled with the smoke of their own furnaces, began to rail at the Peripatetick doctrine, which they were too illiterate to understand, and to tell the credulous world, that they could see but three ingredients in mixed bodies; which, to gain themselves the repute of inventors, they endeavoured to disguise, by calling them, instead of earth, and fire, and vapour, salt, sulphur and mercury; to which they give the canting title of hypostatical principles. But when they came to describe them, they shewed how little they understood what they meant by them, by disagreeing as much from one another, as from the truth they agreed in opposing: for they deliver their hypotheses as darkly as their processes... (*S.C.* 470–1, *E.* 22)

Later, Carneades says

when I acknowledge the usefulness of the labours of Spagyrists to natural philosophy, I do it upon the score of their experiments, not upon that of their speculations; for it seems to me that their writings, as their furnaces, afford as well smoke as light.... (*S.C.* 550, *E.* 166)

Boyle's sympathy, however, is two-edged. It emerges that what he admires about the peripatetics is their *attitude* to theorizing and that what he admires about the chemists is their *attitude* to experimentation, though here he admits that they have made real contributions; it is evident that he thinks little of the content of the theories of both groups. The peripatetics, as much as the chemists, show how little they understand what they mean by their own explanatory concepts. He wants a theory of matter whose concepts are intelligible and have empirical content and whose statements are based upon and supported by accurate observation and experiment. He wants, above all, a theory which will allow explanations of the phenomena in which the relations between theoretical statements and descriptions of phenomena are perspicuous and detailed enough to enable us to see how things happen and why the phenomena are as they are. It becomes clearer in *The Origin of Forms and Qualities* that Boyle sees in the peripatetics and the chemists a common tendency to give explanations that are unintelligible or empty because circular and that do not exhibit precise ways in which events are brought about. Not surprisingly, as it is the other side of the same coin, their theories must remain unconfirmed; since it is not clear what requires confirmation, it is not clear what would confirm them. In *The Sceptical Chymist*, however, Boyle is, as a step in that direction, arguing the need for clearer descriptions of the phenomena and exposing confusions by showing how ambiguous the results of the favoured experiments are.

Boyle's enthusiasm for detailed descriptions of the phenomena before theorizing begins may be regarded as Baconian and so it avowedly is. However, Bacon is often pictured as rejecting hypotheses from science and accepting only descriptions of phenomena and generalizations from them.[9] Boyle was clearly not a Baconian in that sense but then, I think, neither was Bacon. That view will not survive a careful reading of Bacon; what he rejected was over-hasty hypothesizing based on too little careful observation. Boyle regards hypotheses, of the kind Bacon approved, as of central importance in scientific investigation and as being useless unless they could be tested by observation and experiment.

III

The peripatetics and the chemists, in basing their theories on elements and principles, respectively, were at least conforming to a general view of explanation that seems, on the face of it, the most plausible. In the context of natural science it seems reasonable to suppose that the similar observable behaviour of different samples of the same substance or of different substances depends upon some similarity of their inner constitutions and that different behaviour depends upon some difference of inner constitution. The explanation of the behaviour of many kinds of thing in terms of few kinds of inner constitution seems a powerful way of reaching general accounts of phenomena. This much Boyle could accept; what he could not accept was the 'occultness' of the particular conceptions of inner 'constitution' favoured by the peripatetics and the chemists.

Underlying the criticisms that I have already mentioned was a more general methodological or explanatory principle to which Boyle attached considerable importance. The peripatetics and chemists conceived their inner constitutions in terms of ingredients having radically different sorts of properties while Boyle conceived them in terms of differing structures of ingredients (i.e. particles or corpuscles) having the same sort of properties. The difference between *ordinary* salt and sulphur, for the chemist, depended upon the predominance in them of *elementary* salt and sulphur, respectively, these elements being regarded as fundamental and as having radically different properties; the difference for Boyle, depended upon their being composed of corpuscles, all of which had merely such properties as shape and size, arranged in different configurations. Two different substances might even be simply different structures of particles having exactly the same shapes and sizes. Boyle's fundamental ingredients are neutral as between different observable properties; his opponents' were not. That is why he is able to claim a greater flexibility and generality for his theory. It is also why he is able to claim that it allows real explanations.

The chemists held that their three principles, salt, sulphur and mercury, were each responsible for specific observable properties. There were, however, disagreements between them about which element was responsible for which property but if one held, for instance, that mercury was responsible for the colours of substances then different colours would be explained by the presence of different proportions of mercury in the substances in question. Treatment by fire was thought to be capable of isolating these principles so there was at least thought to be some possibility of empirical confirmation. There were two aspects to the regarding of salt, sulphur and mercury as *elementary*. They were regarded as, respectively, the principal ingredients in ordinary salt, sulphur and mercury but not identical with them; they were the 'pure essences' of these substances. However, an added complication was that some chemists regarded the principles as spiritual rather than physical or as hovering uneasily between spirituality and physicality, which might be thought to have prejudiced the possibility of empirical confirmation. Another aspect was that they were regarded by some as elementary in the sense that they were not disintegrable by fire or any other method of analysis.

The idea of chemical principles was much older than Paracelsus, who was one of Boyle's main targets, but it appears that he was responsible for adding salt to an earlier two-principle theory.[10] He appears to have been undecided or inconsistent about their elementary nature since he also used the four Aristotelian elements and spoke sometimes as if they were composed of the principles but sometimes as if the principles were composed of the elements.[11] He was also inclined to think that different observable substances contained different *varieties* of the three principles; at one point he says 'There are as many sulphurs, salts and mercuries as there are objects.'[12] Daniel Sennert held firmly that each chemical principle was made up of the four Aristotelian elements and van Helmont held that they could all be reduced to one Aristotelian element, water. Boyle is able to make much of these disagreements.

He attacks the theory of chemical principles on various grounds but in

The Sceptical Chymist mainly on empirical ones. In the first place, he points out in Part 5 that there was serious disagreement about what properties each principle was responsible for; salt was variously said to be responsible for solidity, permanence, taste and colour, sulphur for toughness, odour and inflammability, and mercury for colour, volatility and the proper mixing of ingredients. In the next place, even when there was agreement about this, there was little attempt to be explicit about the properties of the principles themselves or the criteria for their identification. This had two unacceptable consequences, at least: the only description of, say, elementary salt tended to be of the form 'the principle of solidity', which pointed the way to empty explanations; and there could be no agreement about when a particular principle had been obtained as a product of analysis. In the third place, Boyle questioned the elementary nature of the products of analysis claimed to be principles and showed that many alleged principles were in fact further disintegrable. In the fourth place, he held that some products of analysis eluded classification under the heading of any of the principles, even when they were reasonably clear bases for classification, and so he questioned the *number* of the principles. It is true that numbers other than three were held by some to be correct but he was dubious about the empirical support for all these claims. Finally, he questioned the favoured method of analysis by fire and cited many observations and experiments in support of his doubts.

All these points suggested, in one way or another, the difficulty or even impossibility of empirical confirmation of the theory of chemical principles and the vagueness of the assignment of properties to the principles made it difficult to see how it could lead to the description of mechanisms to which Boyle attached so much importance. The failure to do this was also likely to lead to contradictory explanations. Given only three principles it was necessary to account for more than one observable property in terms of each. Suppose, then, that salt is taken to account for solidity and taste. A hard substance with a strong taste would then contain a large proportion of salt but how could we differentiate between a hard body with a weak taste and a soft body with a strong taste, in terms of this ingredient? A further empirical problem is that if the principles were homogeneous and each had the properties for which it was responsible in the highest degree then this conflicted with experimental results since the ultimate products of analysis which were claimed to be principles were not always found to have the specific properties that had been attributed to the principles. Moreover there was a logical problem about assigning such properties to the principles and then basing explanations on them:

the attempt to explain observable property *X* in terms of a principle having that same property could not provide an explanation of property *X* as such. On the other hand, some properties different from those they were to explain must be assigned to the principles or there would not be enough principles to go round and no one seems to have been clear what these properties should be.

In spite of the fact that some chemists used the Aristotelian elements, Boyle sees the peripatetics as a separate group because they accepted earth, water, air and fire but not the chemical principles. He also regards the peripatetics' theories as largely *a priori* in contrast to those of the chemists. However, they are open to attack by some of the arguments levelled against the chemists because they regard their elements as fundamental ingredients in all observable substances and probably separable by analysis by fire, even though they attached little importance to any empirical confirmation that might be based on this. In one respect, however, their theory went a little further in the direction of Boyle's ideal than the chemists' theories: they made some attempt to give some specification of the properties of their elements rather than distinguishing them only in terms of what was to be explained. They held that there were four fundamental 'first qualities', namely, hot, cold, moist and dry. Each element had two of these, thus:

earth	cold and dry
water	cold and moist
air	hot and moist
fire	hot and dry.

So explanations of observable qualities by proportions of the elements ultimately rested on these qualities. Such qualities as colour or flavour were presumably to be explained in terms of properties other than colour or flavour. That is one essential feature of a satisfactory explanation. However, it was still very difficult to see what kind of mechanism could be proposed for the explanation of, say, colour in terms of 'first qualities' and any other basis for explanation would detract from the possibility of a unified fundamental theory for all physical phenomena. The elements also had further properties consequent on their having 'natural places' in the universe. In terms of these it was relatively easy to develop at least a crude theory of motion for terrestrial bodies; it was enormously more difficult to base a theory of chemical action upon them and this may account for the use of chemical principles along with the Aristotelian elements by some who were interested in both physics and chemistry.

It is perhaps because Boyle thought that the peripatetics' theories were even less clearly related to the phenomena than those of the chemists that he reserves his main attack on them for the more theoretical consideration of their concepts and explanations in *The Origin of Forms and Qualities*. There he considers mainly explanations in terms of other Aristotelian concepts, namely, substantial forms and real qualities, while explanations in terms of the four elements drop out of sight. In so far as he discusses explanations in terms of the elements in *The Sceptical Chymist* the type of explanation is extremely limited saying, for example, that quicksilver is fluid because 'it participates much of the nature of water' and heavy because of the amount of earth in it (*S.C.* 551, *E.* 167–8). Since these are similar in form to explanations given by the chemists they face the same criticisms.

It should be clear that the situation Boyle faced was confused. A further complication, as we shall see, is that both chemists and peripatetics frequently and disastrously misinterpreted Aristotle's views, both theoretical and empirical, without realizing it. Boyle, to his credit, recognized this and often goes out of his way to stress that he is attacking latter-day Aristotelians rather than Aristotle, whom he greatly respects.

IV

Boyle spends a good deal of space questioning the value of analysis by fire. It was regarded by both his groups of opponents as the method of reducing substances to their principles or elements. The chemists regarded it as a method of discovering the constitutions of substances and confirming hypotheses about them while the peripatetics regarded it merely as a way of illustrating their *a priori* thesis that all substances are mixtures of the elements in various proportions. Themistius, as I have mentioned, gives only one example, the burning of green wood in a chimney. When this is done 'you will readily discern in the disbanded parts of it the four elements, of which we teach it and other mixt bodies to be composed' (*S.C.* 470, *E.* 21). *Fire* is there in the visible flames, the smoke coming out of the chimney vanishes into *air* and so 'manifests to what element it belongs and gladly returnes'; *water* boils and hisses at the ends of the burning wood; the ashes show themselves by their weight, firiness and dryness to be *earth*. So perfectly does this illustrate the peripatetic doctrine that no other experiment is necessary; it should not be thought of as intended to establish the doctrine, which is, allegedly, independently established *a priori*.

Themistius says

If I spoke ... to less knowing persons, I would perhaps make some excuse for building upon such an obvious and easy analysis; but it would be, I fear, injurious, not to think such an apology needless to you, who are too judicious either to think it necessary, that experiments to prove obvious truth should be far-fetched, or to wonder, that among so many mixt bodies, that are compounded of the four elements, some of them should, upon a slight analysis, manifestly exhibit the ingredients they consist of. (*S.C.* 470, *E.* 21)

Boyle here, wittily and cunningly, makes Themistius both flatter his opponents and castigate them for superficiality; it is only the ignorant that need empirical confirmation and that because their theories are too crude and implausible to be established by respectable *a priori* reasoning. Every intelligent man recognizes that the Aristotelian doctrine is the true one because it has been so established. Moreover, Themistius continues with a fine disregard for the power of counter-instances, the attempt to establish conclusions by experiment is unreliable since hypotheses built on a few experiments are likely to be knocked down by a few more. The method, which he alleges is Aristotle's, of considering the whole body of earlier hypotheses and, by reasoning about them and modifying them so that they are finally consistent, is much surer and safer. No doubt the thought underlying these extraordinary statements is that truths established by logic, in contrast to those drawn from observation, are universal and eternal and so are not susceptible to empirical refutation and the problem of induction.

Boyle, in the person of Carneades, argues that this experiment and others devised by the chemists cannot be taken as separating substances into their elements or principles and so as either illustrating or confirming the respective theories. He adds many experiments of his own to support this. It is important to remember, when considering these, that analysis by fire involved subjecting them to great heat in a furnace or over an open flame, rather than gentle heat in an oil, water or sand bath or in a retort, and that there was no awareness on either side that air or other contaminants enter into the reaction.

Boyle puts forward a large number of objections and it is worth extracting the most important of them.

1. Fire appears to be powerless to break up some admittedly complex substances into their constituents; prolonged intense heating of gold, for example, shows no signs of producing either elements or principles. (*S.C.* 487, *E.* 53)
2. It is agreed that the elements and principles must be homogeneous but there are other methods of treating substances to give products which are homogeneous and not further reducible. These methods include distillation, freezing and the use of solvents. Alcohol can be separated from beer by freezing it; alcohol is homogeneous and not

further reducible so what is the warrant for *not* regarding it as an element or principle? (*S.C.* 491–2, *E.* 61)

3. Fire may change a substance into another or change its ingredients rather than merely separate them; there is no guarantee that several homogeneous products of heating a substance were original ingredients of it. By heating many vegetables without any addition we may obtain glass. (*S.C.* 501, *E.* 78)
   Heating the salt and earth remaining in the ashes of burnt plants produces glass which resists intense heat without disintegration. (*S.C.* 481, *E.* 42)
4. The products of analysis by fire may sometimes be further broken down and shown to be compound by other methods so even if they were original ingredients of the substance in question they cannot be principles or elements. (*S.C.* 525, *E.* 121)
5. Substances that are admitted by all parties to be compound can often be sublimed, that is, vaporized and condensed again into solids without going through a liquid phase, over and over again, even using considerable heat, without any sign of their disintegrating. Quicksilver can be distilled from cinnabar, a mercury ore, and sulphur from iron pyrites, an iron-sulphur ore, but the products are too 'compounded' to be considered elementary. (*S.C.* 483–4, *E.* 46–7)
6. Heating sometimes gives more than three, or even four homogeneous products. Which of these are to be taken as elementary if we hold a three-element or a four-element theory? Distilling raisins yields an 'alcali', phlegm[13] and earth, an oil and a spirit very different from that of wine. (*S.C.* 516, *E.* 105)
7. Fire sometimes unites substances rather than separates them, as in the making of metallic alloys, glass and soap. (*S.C.* 489–90, *E.* 56–7; *S.C.* 500–1, *E.* 77)
8. The alleged principles or elements may be produced *de novo* whereas it ought to be impossible to make true elements or principles on the theories under examination. Boyle here dwells on van Helmont's famous experiment with a willow tree which he repeated himself with a vegetable squash. (*S.C.* 493–7, *E.* 64–70)

This last item is worth considering in a little detail. Van Helmont claimed that his experiment showed that there was only one true element, water; Boyle drew other inferences from it. His account of van Helmont's experiment goes, briefly, as follows. Van Helmont dried 200 pounds of earth in an oven and planted a small willow tree weighing 5 pounds in it. He watered it for five years with rain water or distilled water and then took the tree out and weighed it again. He calculated the weight of the leaves that had fallen in four autumns. The total weight of the tree and its four crops of leaves was 169 pounds, 3 ounces. He dried the earth and found it to weigh only about 2 ounces less than the original 200 pounds. His conclusion was that 164 pounds of roots, wood and bark (Boyle omits leaves) sprang from the water alone and that the other alleged 'elements' composing the tree were transmuted water (*S.C.* 495–6, *E.* 67–8).[14]

Boyle obtained similar results. It should be noted that this was an ingenious and careful experiment and that, in the context, van Helmont's conclusion was by no means ridiculous. He had taken precautions to eliminate impurities in the water and, allowing for inherent errors in weighing devices, the results were probably fairly accurately recorded. What neither he nor Boyle knew about was the action of the air and no precautions relating to this were taken.

As we should expect, Boyle interpreted the results differently without questioning their accuracy. Water is corpuscular and changes in the tree could be explained not by the mere absorption of water which then remained unchanged but by the rearrangements of the corpuscles of the water to produce other substances. Some other substance might have done as well; the corpuscles of air could be similarly rearranged. Boyle connected this with his experiments on the burning of plants to obtain what were alleged to be elements or principles so van Helmont's experiment appeared to show that they were all capable of being transmuted into water, or *vice versa*, which, according to many peripatetics and chemists, was impossible. Boyle's hypothesis allowed him to say that if this did indeed happen it could be given an intelligible mechanical explanation; transmutation of one substance into another, whether or not it was claimed to be elementary, could be explained by the rearrangement of the corpuscles of the first substance into the structure peculiar to the second. What are truly elementary are not substances but individual corpuscles. Carneades sheds further doubt on the inference drawn from the burning of plants by claiming that the products are a mixed bag: salt, spirit, earth and oil. This list contains one chemical principle, one element and two substances falling into neither category. On Boyle's corpuscular hypothesis that presents no problem in principle.

The enthusiasm for analysis by fire is, Boyle says, based on the Aristotelian account of heat according to which it is a property of, or even part of the definition of, heat 'to assemble things of a resembling and disjoin things of a differing nature' (*S.C.* 488, *E.* 54).[15] However, he takes some of his experiments to show that heat sometimes unites things of a different nature, as in the making of glass, and that cold, which we should expect to have the opposite effect, is capable of separating like parts, as in the separation of alcohol from beer by freezing it.

Boyle is well aware that some chemists hold that although heat alone may be insufficient for the analysis of some substances into principles it *is* sufficient when those substances are mixed with water or some other substance. But, he replies, even with the help of other substances, fire will not separate the three principles from gold, silver or quicksilver. Moreover, water or any other additive may *react* with the original substance to give products that were not its ingredients. Besides, different additives have been shown to give different products with the same original substance so it is necessary to decide which of these products are principles (*S.C.* 487, *E.* 52–3).

The many experiments considered suggest that his opponents have not sufficiently considered what they mean by 'analysis by fire'. The effects of heat appear to differ according to the conditions under which it is applied. It is important to specify

> by what degree of fire, and in what manner of application of it, they would have us judge a division made by the fire to be a true analysis into their principles, and the productions of it to deserve the name of elementary bodies. (*S.C.* 479, *E.* 39)

In other words, experiments, especially at such a fundamental level, are useless unless they are carefully controlled and their results accurately described.

He later emphasizes that the chemists are insufficiently precise in deciding that products of analysis are elementary and in specifying criteria for their identification. He says, for example,

> qualities slight enough may serve to denominate a chymical principle. For, when they anatomize a compound body by the fire, if they get a substance inflammable, and that will not mingle with water, that they presently call sulphur; what is sapid and dissoluble in water, that must pass for salt; whatsoever is fixed and indissoluble in water, that they name earth. And I was going to add, that whatever volatile substance they know not what to make of, not to say, whatsoever they please, that they call mercury. (*S.C.* 528, *E.* 126)

As he remarks, shortly after this, there are easily observable differences between salts so it is urgent to decide which is elementary salt (*S.C.* 534, *E.* 137–8). Boyle was by no means the first to argue in this way; Thomas Erastus (1524–83), for example, had argued in 1572 against Paracelsus that a substance may not consist of substances from which it was generated, and that the intensity with which a body is heated may determine the extent of its decomposition.[16]

I shall deal directly with Boyle's corpuscular hypothesis later but it is worth mentioning here that throughout *The Sceptical Chymist* he says that particular experiments intended to show the production of the three chemical principles may be explained in terms of his hypothesis and none of the experiments cited is sufficient to rule out some form of corpuscular hypothesis.

Against Aristotle's principle concerning heat he says

> the true and genuine property of heat is, to set a moving, and thereby to dissociate the parts of bodies, and subdivide them into minute particles, without regard to their being homogeneous or heterogeneous; as is apparent in the boiling of water, the distillation of quicksilver, or the exposing of bodies to the action of the fire, whose parts either are not (at least in that degree of heat appear not) dissimilar, where, all that the fire can do, is to divide the body into very minute parts, which are of the same nature as one another, and with their totum, as their reduction by condensation evinces. And even when the fire seems most *congregare homogenea, et segregare heterogenea*, it produces that effect but by accident; for the fire does but dissolve the cement, or rather shatter the frame, or structure, that kept the heterogeneous parts of bodies together, under one common form... (*S.C.* 488, *E.* 54)

As a result the particles, being set at liberty,

> do take those places, which their several degrees of gravity and levity, fixedness or volatility (either natural, or adventitious from the impression of the fire) assign them. (*S.C.* 488, *E.* 55)

The grain of truth in Aristotle's principle is that fire will not separate substances of equal volatility even though these substances differ in many other properties. It follows that the products of treatment by fire may be by no means homogeneous.

The fact that the principles may apparently be produced *de novo* may be explained by saying that compound bodies differ from one another merely in 'the various textures resulting from the bigness, shape, motion and continuance of their small parts' so it is reasonable to suppose that one and the same portion of matter may be altered so as to be sometimes sulphurous, sometimes earthy and sometimes igneous (*S.C.* 494, *E.* 64).

Any 'principles' separated by fire may be composed of clusters of corpuscles and so not be elementary; their differences may be explained in terms of differences between the corpuscles themselves or their arrangements in the clusters. Such clusters may be of many more kinds than three or five and it need not be that each 'principle' has every kind of corpuscles as a constituent. Corpuscles may form such stable clusters that the constituents may not be separable by fire; so the failure of fire in practice to disintegrate salt, sulphur or mercury does not show that they are elementary. Moreover, the fact that gold and other substances are not disintegrated by fire need cause no alarm to those who do not want to regard them as elementary (*S.C.* 511–12, *E.* 96–7). Thus the products of analysis that seem homogeneous may or may not be so, for all the experiments show; they may be composed of combinations of either like or unlike corpuscles (*S.C.* 525–7, *E.* 121–5). The advantages of this kind of theory, in contrast to those of Boyle's opponents, are great flexibility, generality and explanatory power. Not only may properties such as permanence, solubility and volatility, allegedly explained by the presence of principles, be better explained in terms of corpuscles but so may many other not so easily explained properties (*S.C.* 547, *E.* 160). For example, it is not clear how the properties of the lodestone, the mechanism of the formation of a chick in an egg and the growth of plants from water could be explained in terms of the *tria prima.* Boyle regards it as conceivable that they could be explained on the corpuscular hypothesis (*S.C.* 548–50, *E.* 163–5). Although it must be admitted that this is a mere expression of faith, it at least involves a possibility, in contrast to 'explanations' already given and regarded as satisfactory but which Boyle regards as worthless.

The culmination of *The Sceptical Chymist* in the final main section, the fifth, leaves behind the matter of experimental confirmation and turns to some general considerations about the logic of explanation. The preceding sections support the idea that Boyle's opponents are so unclear about the explananda they use that it is impossible to see how they relate to any explanans. Their language often suggests that, whatever show of precision they make, they are, in fact, relying upon occult entities and qualities. Their alleged explanations fail to be explanatory because they are empty, circular or unintelligible. This is the view he develops in later works, to which I shall turn in the next chapter.

2

BOYLE AND THE PERIPATETICS

I

Boyle's most comprehensive and developed account of the corpuscular hypothesis is contained in *The Origin of Forms and Qualities*.[1] The first part is called 'The Theorical Part' (*sic*) and it begins with criticism of those he is opposing aimed rather more directly at the peripatetics than the spagyrists. His view is that his hypothesis avoids logical weaknesses to be found in both their views and it has for him the overwhelming advantage that all its concepts have empirical content; so it allows at least the theoretical possibility of empirical confirmation. He also believes that it avoids weaknesses to be found in earlier atomistic accounts of natural phenomena although he spends much less time on this because the problems of those accounts are fewer and less precise. The second main part of the book is called 'The Historical Part' and largely contains accounts of experiments intended to give empirical support to the hypothesis.

In order to see what Boyle was trying to achieve and avoid it is necessary to consider briefly how he saw the scholastic views, allegedly derived from Aristotle, which he was attacking. As is well known it is not easy to be sure what Aristotle's own views on various relevant matters were and this is the subject of continuing controversy among scholars. There is even more uncertainty about the views of medieval and later philosophers who were roughly classified as 'Aristotelians'. Boyle frequently says that it is Aristotle's later followers, particularly the contemporary schoolmen, that he is attacking and fortunately it is fairly clear how Boyle interpreted what he found in their recent influential writings. This is what matters for an understanding of his corpuscular hypothesis so if scholars of Aristotle, later scholasticism and alchemy find in my account views that they believe are not to be attributed to any of these then my reply is that, for my purpose, it is enough if this is how Boyle saw them.

In the Fifth Part of *The Sceptical Chymist* Boyle glances at the kind of theoretical problems about which he goes into much more detail in *The Origin of Forms and Qualities*. He is quite clear that many scholastic 'explanations' of natural phenomena depend upon occult qualities and he sees it as a worthy aim of some chemists to avoid this dependence in basing their explanations on the *tria prima*. However, he questions the adequacy of even these explanations and doubts that they really do, in spite of first appearances, avoid reference to occult qualities. A central aim of the corpuscular philosophy is to succeed in this where they failed.

The most important discussion in *The Sceptical Chymist* concerns the following passage from Daniel Sennert.[2]

Wherever the same affections and qualities exist in many things, it must be that they exist in them through some common principle; as for example, all heavy things are heavy on account of earth, all hot things hot on account of fire. But colours, smells, tastes, inflammability, and other such, are present in minerals, metals, gems, stones, plants and animals. Therefore they are in them through some common and underlying principle. But such a principle is not the elements; for they have no power to produce such qualities. Therefore other principles, from which they may flow, must be looked for.

Carneades says that Sennert's view is based upon 'a precarious supposition, that seems to me neither demonstrable nor true' and asks 'how does it appear that where the same quality is to be met with in many bodies, it must belong to them upon the account of some one body whereof they all partake?' (*S.C.* 551, *E.* 167). Has anything been done, for example, to establish that the heaviness of bodies is due to the presence in them of the element earth? As we shall see, Boyle does not object to the idea that a quality common to different bodies depends upon a common *structure* of corpuscles; what he does object to is the idea that it depends upon a common *substance*, such as elementary sulphur or earth, and the suggestion that the qualities of those substances are responsible for qualities similar to them, but observable, in the bodies in question. What is in doubt is whether this would *explain* those qualities because the explanation would contain the very term to be explained.

Carneades goes on to say that some of his opponents' explanations are empty, some are inconsistent and some just fail to explain because of ineradicable *lacunae* in them. Inconsistency requires no elucidation but some comment about the other two points is in order. An explanation is empty if the explanandum is no better understood than the explanans; it has a *lacuna* if no connection can be shown between the explanandum and the explanans. Sennert is a useful example for Carneades because he accepted the three alchemical principles but held that they were composed of the four elements. Thus his view may be used to bring out a similar unsatisfactoriness of explanations using each set of concepts.

Carneades cites common quicksilver as an example. It is said by the peripatetics to be fluid because it 'participates much in the nature of water'; the chemists attribute its fluidity to the presence of elementary mercury. There is one important aspect of this example on which Boyle does not comment here; in view of things he says elsewhere it is plausible to suppose that he is glancing at the point that such explanations are empty since they claim to explain the fluidity of one substance in terms of the fluidity of another and thus fail to give an explanation of fluidity as such.

His main point here, however, appears to concern the need for contradictory explanations on both the theories of peripatetics' elements and chemists' principles. The fluidity of quicksilver is explained by the predominance of water in it; but it is also heavy and this is explained by the fact that 'earth abounds in it'. However, quicksilver is heavier than an equal volume of earth. How can it be heavier than that one of its ingredients responsible for heaviness. Moreoever, if this involves saying that both earth and water *predominate* in it, how can that be possible? A similar objection can be made to the chemists' explanations in terms of their principles.

There are various properties that the chemists simply fail to explain. Which chemical principles can explain fluidity, motion, light and sound? (*S.C.* 551–2, *E.* 168–9). Fluidity, for example, is alleged by some to be explained by the presence of mercury. I believe that Boyle has two points in mind. The more explicit is that there are some qualities of substances that the chemists do not even attempt to explain. The less explicit, mentioned elsewhere by him, is that even when they offer explanations they fail; if *A* is said to explain *B* we cannot understand this as an explanation unless a clear connection is made between *A* and *B*. We cannot see as an explanation the bare statement that a substance is fluid because it contains mercury unless we are shown a mechanism, that is, shown precisely *how* the presence of mercury causes fluidity. Even if we could establish a close correlation between fluidity and the presence of mercury, which Boyle thinks has not been established, we should still need something more to see that mercury is the cause of fluidity.

I may illustrate this by a very different example but one which is related to Boyle's mechanistic idea. Even if I have established that every time a coin is put into a certain machine a cup of coffee is delivered I have not, by this alone, understood how this happens. I have not explained the emergence of the cup of coffee. The less we know about machines in general, the more reasonable is this demand. It is analogous to one that neither chemical nor peripatetic explanations satisfy.

Later, Carneades says

what is it to me to know, that such a quality resides in such a principle or element, while I remain altogether ignorant of the cause of that quality, and the manner of its production and operation? How little do I know more than any ordinary man of gravity, if I know but that the heaviness of mixt bodies proceeds from that of the earth they are composed of, if I know not the reason why, the earth is heavy? (*S.C.* 557, *E.* 178)

We learn nothing if we are told that a body is heavy because it has heavy ingredients and it does not require a scientist to tell us this. It must not, by the way, be thought that Boyle is taking the untenable position that all attributes of all bodies or substances must be explained; he is well aware that in order to get any explanation going we must take some attributes as primitive. What I think he is objecting to is taking as primitive, attributes too complex to be thought incapable of explanation. The vulgar chemists ascribe colours to mercury, Paracelsus ascribes them to salt and Sennert ascribes them to sulphur; but none of them explains how they arise from their chosen principle and so none gives a reason for choosing one rather than another (*S.C.* 556, *E.* 176). Carneades refers to Boyle's alternative hypothesis, which does propose a mechanism. Experiments show that

> bodies exhibit colours, not upon the account of the predominancy of this or that principle in them, but upon that of their texture, and especially the disposition of their superficial parts; whereby the light rebounding thence to the eyes is so modified, as by differing impressions variously to affect the organs of sight. (*S.C.* 556, *E.* 176)

More detailed accounts, for example in Boyle's *Experiments and Considerations Touching Colours*, fill this account out to assert that the texture of light is so related to the texture of reflecting bodies, that some particles of light are absorbed by the 'pores' in the texture of a body, others are reflected. Those particles reflected are 'fit to produce' sensations of particular colours in us by operating on the textures of our sense-organs.

The chemists are forced to relate more than one quality to each principle in order to explain the many observable qualities of substances (*S.C.* 552, *E.* 169) but, even worse, they must relate some qualities to all their principles since, for example, all substances and all principles are more or less heavy. Which principle does sound depend upon, since oil falling on oil, salt on salt, water on water, and earth on earth all make a noise? The general point being made here is the one mentioned earlier about the need for flexibility in our theory if it is to allow explanations of all the wealth of phenomena and the qualities involved in them. What underlies this is the belief, to be dealt with later, that our fundamental bodies or substances must have as few qualities as possible and that as many other qualities as possible be analysable into them so that as much as possible may be explained in terms of as little as possible. The aim is to achieve great generality in our fundamental theories.

This has various aspects. If we attempt to base our explanations on principles or elements having many properties of observable substances we shall not be able to explain those properties. It will be difficult to specify the principles, or elements, and to give criteria for identifying them, just because of their complexity. We are likely to have difficulty in proposing mechanisms by which various combinations of them produce the observable properties of things.

Boyle frequently uses the old analogy, explained by Lucretius, of the letters of the alphabet which are few and simple but may be combined to form enormous numbers of complex words and sentences.[3] Similar features will produce the most fruitful kind of scientific theory. Some theory based on simple fundamental entities will be best fitted to explain enormous numbers of complex properties and phenomena since the simplicity of these entities will make them neutral between large numbers of observable properties. These entities should, of course, be clearly conceivable and definable and their properties should be, in some sense, familiar to us so that they do not have to be invented purely *ad hoc*. The elements need not be observable, indeed, had probably better not be, but their properties will be most clearly conceived if they are somehow like some of the simplest observable qualities. This will also help to ensure that we can understand causal mechanisms involving them. We are concerned primarily with material phenomena partly because their observability makes them the most fruitful starting point for an account of the world that is intelligible to all.

It is clear that many observable material phenomena depend upon motion and collision. A moving ball may be seen to change its direction when, and because, it hits a bump or is hit by another object. Phenomena involve change and many changes are of this sort. Thus

in a clock, the hand is moved upon the dial, the bell is struck, and the other actions belonging to the engine are performed, not because the wheels are of brass or iron, or part of one metal and part of another, or because the weights are of lead; but by virtue of the size, shape, bigness and coaptation of the several parts; which would perform the same things, though the wheels were of silver, or lead, or wood, and the weights of stone or clay; provided the fabric or contrivance of the engine were the same... (*S.C.* 559, *E.* 182–3)

Boyle perhaps underplays the importance of using particular materials for making clocks but his main point is sound. Within certain limits what is important in explaining the working of the clock is not different ingredient substances but rather the shapes, sizes, motions and organization of those ingredients, whatever they are. He is stressing structures rather than materials. And so, he says

I am apt to think, that men will never be able to explain the phænomena of nature, while they endeavour to deduce them only from the presence and proportion of such or such material ingredients, and consider such ingredients or elements as bodies in a state of rest; whereas indeed the greatest part of the affections of matter, and consequently of the phænomena of nature, seem to depend upon the motion and the contrivance of the small parts of bodies. For 'tis by motion, that one part of matter acts upon another; and it is for the most part, the texture of the body, upon which the moving parts strike, that modifies the motion or impression, and concur with it to the production of those effects, which make up the chief part of the naturalist's theme. (*S.C.* 557, *E.* 178–9)

The relative unimportance to explanation of different ingredient substances with many different properties suggests explanation in terms of roughly similar entities differing only in degrees of a few basic properties.

In the Sixth Part of *The Sceptical Chymist* Boyle outlines the corpuscular hypothesis though not exactly in the form in which he put it later. I shall not dwell on that here but move on to his more detailed logical criticisms of the peripatetics.

II

Whereas in *The Sceptical Chymist*, except for the Fifth Part just discussed, Boyle criticizes the peripatetics mainly on empirical grounds and for their accounts of phenomena in terms of the four elements, in *The Origin of Forms and Qualities* he has moved on to consider the logical character of their explanations and especially to consider their use of Aristotelian ideas of forms and qualities. The theory of *real* qualities and *substantial* forms that he is attacking is, though couched in Aristotelian terms, largely that of the later peripatetics rather than Aristotle.

A major change introduced by Aristotle's later followers into his view concerns the doctrine of form and matter. Aristotle himself appears to have used the four elements, earth, water, air and fire, when he was doing natural philosophy whereas the distinction between form and matter arose in his metaphysical philosophy when he was studying the concept of change itself and perhaps the nature of predication.[4] One thing that, I believe, happened during the Middle Ages is that this distinction between metaphysical and natural philosophy was blurred so that a conception of forms very far from that of Aristotle came to have great importance in the explanation of natural phenomena. Here there is a clear connection, as we shall see, between at least some chemists and the later peripatetics.

To help make clear what was at issue I shall briefly sketch a kind of explanation which was current in Boyle's day and which he attacked. Both peripatetics and chemists were inclined to give explanations of this form although their terminology differed.

Consider a kind of substance, say gold. What makes it what it is? What properties must it have in order to be gold? Well, it is yellow, fusible, malleable, chemically stable (for example, only *aqua regia* among acids will dissolve it), and so on. These are observable, discoverable by the unaided senses and, of course, no very effective aids to the senses were available in the medieval period. They were called *qualities* and some combination of them, taken together, was regarded as determining the nature of gold and enabling us to identify it. However, taken separately, other substances than gold have them; sulphur is yellow and lead is malleable, for instance. Moreover, these qualities were not further analysable by means of the senses or by any other means; given a uniform patch of yellow, colour discriminations could not be made within it and no other observable feature of yellow things could be regularly correlated with this colour. So if we want to know what in a substance makes it yellow, what explains yellowness in all the various yellow things, the only available answer seemed to be some such thing as 'the cause of yellowness', 'a yellow-making feature' or 'the real quality of yellowness'. If we tried to conceive of this as some inner structure of all yellow things underlying, causing and so explaining the seen colour we would face the difficulty that the unaided senses cannot discover such structure and there are neither any other empirical means available for obtaining evidence, nor *a priori* arguments, for one structure rather than another. So it would seem to be indescribable except as 'what makes a thing yellow' or some equivalent. Questions such as 'What makes this substance yellow?' were likely to be regarded as the most fundamental kind of question about natural phenomena and as representing the point at which questioning had to stop because we are at the limits of observation.[5]

The idea of the transmutation of metals was related to this kind of view; if you work on a metal in the right way you may transform it into a more useful or valuable metal. A favourite pursuit of the alchemists was to try to turn lead, or some other 'base' metal into gold. Lead has some but not all of the qualities of gold. For a metal it is soft and fusible, like gold but, unlike gold, it is not yellow, for example. So one part of the task of transmutation is to get the yellow colour needed for gold into the lead, and so on for the other qualities that gold has and lead lacks. Bacon, in opposing this view and arguing in favour of searching for mechanisms by which substances may be transformed mentions two alleged ways of transforming substances. The first way

regards the body as an aggregate or combination of simple natures. Thus in gold are united the following circumstances: it is yellow, heavy, of a certain weight, malleable and ductile to a certain extent; it is not volatile, loses part of its substance by fire, melts in a particular manner, is separated and dissolved by particular methods, and so of the other natures observed in gold. An axiom, therefore, of this kind deduces the subject from the forms of simple natures; for he who has acquired the forms and methods of superinducing yellowness, weight, ductility, stability, deliquescence, solution, and the like, and their degree and modes, will consider and contrive how to unite them in any body, so as to transform it into gold.[6]

Bacon does not here use 'form' in the way that Aristotle does, as we shall see, but the character of the process he is attacking is similar to that attacked by Boyle. The task as a whole is seen thus: one starts with a lump of matter having some of the desired qualities and tries to replace unwanted qualities by wanted ones, that is, according to Bacon, to replace the underlying forms of some qualities by those of others. This was sometimes attempted by trying to get the original lump of matter to react with some substance that has one or more of the desired qualities in such a way that they will be transferred to the original substance. For example, one might start with lead and try to get a reaction with sulphur which would transfer the form of yellow from the sulphur to the lead. If that succeeds then part of the aim of transmuting lead into gold is achieved.[7]

This seems strange to us for two main reasons. In the first place, it confuses properties or qualities with ingredients; sulphur may be a constituent of something without making it yellow and you can transfer a constituent from one substance to another but not a quality, such as colour. In the second place, the alleged explanation underlying all this is really no explanation. To say that a substance is yellow because of the presence in it of the form or quality of yellowness gives us no further information, just as it gives us no further information to say that a certain chemical cures boils because it has the power of curing boils. This is like the *virtus dormitiva* of opium given as an explanation of its inducing sleep and parodied by Molière in *Le Malade Imaginaire* (1673) in the Finale of Act III.

Boyle gives examples of such explanations put forward by schoolmen. He says

As if (for instance) it be demanded how snow comes to dazzle the eyes, they will answer, that it is by a quality of whiteness that is in it which makes all very white bodies produce the same effect; and if you ask what this whiteness is, they will tell you no more in substance, than that it is a real entity which denominates the parcel of matter to which it is joined, white; and if you further inquire what this real entity, which they call a quality, is, you will find, as we shall see anon, that they either speak of it much after the same rate that they do of their substantial forms ... or at least they will not explicate it more intelligibly. (*O.F.Q.* 12–13, *S.* 16)

And again

for if it be demanded why jet attracts straws, rhubarb purges choler, snow dazzles the eyes rather than grass, etc. to say, that these and the like effects are performed by the substantial forms of the respective bodies, is at best but to tell me what is the agent, not how the effect is wrought... (*O.F.Q.* 47, *S.* 68).

Marie Boas says of the later schoolmen that 'Almost every individual property, physical or chemical, from gravity to colour, had become a distinct and real form, almost concrete; to the late mediaeval Peripatetic even light had become a form...'[8] Boyle, however, usually puts it rather differently; he regards the schoolmen as distinguishing between real qualities and substantial forms, saying that colours, for instance, were regarded as real qualities but then, he adds, they treat them as they treat substantial forms as if they could exist independently of matter (*O.F.Q.* 12–13 and 16–17, *S.* 16 and 2). This not only blurs the distinction but renders the idea of real qualities unintelligible and useless.

How did such explanations seem plausible to intelligent and learned men? It may very well have come about through a misunderstanding and consequent misapplication of Aristotle's analysis of change; certainly both Boyle and Leibniz appear to have thought that Aristotle's doctrine had been distorted. It will be as well to consider Aristotle's view briefly to bring out this point.

One of the vexed questions of Aristotelian scholarship is whether Aristotle believed in 'prime matter', an absolutely featureless and indeterminate somewhat embedded in all things, the receptacle of all qualities or the subject of all predicates.[9] If he did so it may only have been in connection with the transmutation of his elements since he did not need it at any other level of explanation. I am not competent to pronounce on this controversy, although it seems to me that Aristotle was postulating prime matter in *de Generatione et Corruptione*, I.4. It is clear that Boyle and Locke regarded those they were attacking as accepting prime matter in a way that went well beyond Aristotle.

Aristotle was concerned to understand change, generation and corruption, not only particular observable changes so described, but the concept of change itself of whatever sort, whether exemplified by the growth of living organisms or changes in inanimate matter. He considered in an abstract philosophical manner what 'change' means, what must be involved in any change, what are the necessary and sufficient conditions of something's being a change rather than, say, a substitution.

Consider some examples. In the hands of a sculptor a rough lump of clay becomes a statue of a woman; in cold weather a cupful of water freezes; in the course of time a man grows old, wrinkled and grey. These are indisputably changes. We can separate in thought the causes of the changes from the changes themselves and ask what constitutes the changes and what they all have in common. They take place *in* the material in question: the statue is still a lump of clay but with some properties, especially of shape, different from the original; ice is still water but now solid; the old man we know is still the same human being as the child we knew long ago. In common parlance we often say such things as that the sculptor in making the statue has given the clay a different form, or imposed a new form on the existing material. It is not unnatural to say that when water freezes it acquires a new form, namely, a crystalline form. Ignoring, for the moment, all we know about changes in the cells of the human body, we may say that the same matter, the person's flesh, bones, etc. has a different form from what it had fifty years ago, at least in certain respects. Here matter is being thought of as that out of which something is made. Clay, water, flesh and bones are the matter which takes on different forms.

This is the basis of Aristotle's analysis of change. We may say that in each of these examples some material has gained one form and lost another while remaining the same matter. That is, in anything we count as a change we can distinguish

1. a form lost;
2. a form acquired;
3. a subject that loses and acquires the forms.[10]

These are features that must be present in any change whatsoever; this is a logical analysis of the concept of change. So far there is nothing very mysterious about the terminology I have used. Form involves such items as shape, size, colour, ordinary observable qualities. Matter is the sort of thing we would normally call a material or a substance, clay, water or flesh. We talk of sticky coloured substances and brittle, shiny materials.

Since in these changes we can say that the material or thing remains the same material or thing we can call them 'accidental changes' to indicate that nothing *essential* to being clay or water or a human being has disappeared. The forms gained or lost are accidental rather than essential; a piece of clay remains a piece of clay whether it is in the shape of a rough ball or a woman, and so on. Also, in these changes, indeed, in all accidental changes the matter that endures through them is observable. When we say that the same matter now has this form or quality and now that, the matter is always accessible to observation. We can discover by observation that we still have clay or water or human flesh. However, we are familiar with changes in which this does not appear to be so and there is at least the logical possibility of changes in which all observable features are changed and so whatever endures through change is absolutely inaccessible to observation. Such changes have been called 'essential' or 'substantial' changes.[11]

Consider, first, some familiar examples. What was an egg three weeks ago is now a chick; we cannot say it was a chick or, *a fortiori*, the same chick, three weeks ago or that it is an egg today. Eggs are not chicks. The man we knew is now a small heap of dry ashes; we cannot say that the man was a heap of dry ashes a week ago or that the heap of ashes is a man today. Men are not heaps of dry ashes. Nevertheless the man has changed into the heap of ashes and the inwards of the egg have changed into a chick. This is to say something other, clearly stronger, than we should say if we thought that someone had *substituted* a heap of dry ashes for a man, or a chick for an egg; substitution would allow that the man or the egg still existed but elsewhere and is quite different from one thing *changing into* another. Now if these are really changes and nothing observable endures through them then they are examples of substantial or essential changes; the form lost was essential to being a man or an egg. But it is of the nature of change of whatever sort that something must endure through it or we could not say '*It* was an egg and is now a chick' or '*It* was a man and is now a heap of ashes.' These are changes 'in the category of substance' as Aristotle would say.

This argument depends upon our beginning with accidental changes and regarding our analysis of them as being an analysis of change as such. Can we conceive of change without the three features pointed to by Aristotle? It seems clear that to say that something has changed is to say that *its* description now is different from what it was and that may be expressed by saying that a form has been lost and another gained. What about the subject that loses and gains the forms? We distinguish between change and substitution or replacement: the change of object *A* into object *B* is not the same as the substitution of object *A* by object *B*. Of the first we can say that *it* or *this*, changed into *that*, of the second we cannot. If *A* is changed into *B* then *A* no longer exists as *A*; if *A* is replaced by *B* then *A* may still exist but elsewhere. Card-sharpers practice sleight of hand, not magic, substitution, not change. It is the presence through the change of some subject which loses and gains forms that makes the distinction between change and replacement of objects. We must, of course, be careful how we fill in *A* and *B*. If *A* is a green leaf and *B* is a brown leaf then if *A* changed into *B* it is because although it is still that leaf it is now brown instead of green; 'that leaf' does not mean the same as either 'that green leaf' or 'that brown leaf'; that green leaf no longer exists. But *A* is replaced by *B*, on a table say, if *A* is removed to somewhere else and *B* is placed on the table; that green leaf (*A*) now exists somewhere else and that brown leaf (*B*), which previously existed somewhere else, now exists on the table. Replacement requires the existence, for the duration of the transaction at least, of two leaves; change of one into another requires the existence of only one at any given time. It is important, also, that when one *object* changes into another, some *quality*, say green, is replaced by another quality although when we are talking of qualities it won't do to say that the same (token) quality still exists elsewhere when it has been replaced. The analysis of change that I have been outlining concerns changes of objects specifiable in one way into objects specifiable in another.

Since this is simply a logical analysis of the concept of change it is quite general and must apply to both essential or substantial change and accidental change. Since accidental change is distinguished from substantial change just because in accidental change the enduring subject observably remains the same in some respect while in substantial change it does not, the subject in substantial change must be in principle unobservable and so the change, that it *is* a change rather than a substitution, is also in principle unobservable.[12] The subject of such a change cannot be observable because it is what loses and gains the most basic of forms and without these nothing would be observable; of itself the subject does not have the basic determinations upon which observability depends. Thus the subject in substantial change must be matter of a more basic kind than the observable materials out of which things are made; it was therefore called 'prime' or 'primary' matter in contrast to secondary matter such as clay, water or flesh.

The examples I have used may appear, at least to the modern mind to be open to objections. We can, in fact, find some observable matter which endures through the change of a man into ashes or of an egg into a chick even if this involves analysing the objects down as far as ingredient chemicals. However, this does not make any difference in principle; it follows merely that these are not, after all, examples of substantial change but not that there are no such changes. Suppose that we consider the most basic elements of things, on the Aristotelian view, earth, water, air and fire. They have no observable qualities in common and are not themselves observable; the qualities hot, cold, moist and dry attributed to them are not observable but are the unobservable causes of corresponding qualities in observable things. Now if, say, earth may be changed into water, or *vice versa*, then these would be true substantial changes and prime matter would be the only possible enduring subject. It has indeed been held that Aristotle needed the conception of prime matter only because he held that the transmutation of one element into another is possible.[13]

In his doctrine of categories Aristotle classified terms or predicates which we use for the description of the world or, as W. D. Ross puts it, gave 'a list of the widest predicates which are predicable essentially of the various nameable entities, i.e. which tell us what kinds of entity at bottom they are'. The list, in its most complete form, is Substance, Quantity, Quality, Relation, Place, Date, Posture, Possession, Action and Passivity. These are answers to questions of the form 'What is so-and-so?' For example, 'What is Aristotle?' has the answer 'A substance'; 'What is redness?' has the answer 'A quality'; and 'What is running?' has the answer 'An action'. Individual predicates in all the categories except the first may be predicated of individuals (substances) in the first category. For example, 'This pineapple is one, green, larger than this apple, here, now and standing upright.' 'The primary category is substance which is the substratum presupposed by all the others'[14] but Aristotle distinguishes two sorts of substances, individuals such as particular men or particular stones, which are called 'primary' or 'first' substances, and the genera or species to which these belong, such as *men* or *inanimate objects*, which are called 'secondary' substances.

Prime matter cannot be described by any of the usual predicates from the list of categories, or by their negations. But matter can never exist without form so some predication must be possible. The only things that can logically be predicated of it, and that in a special way, are forms and such forms were later called, by Aquinas and others, *substantial forms.*[15] If you say 'Socrates is a man' you are predicating a universal (man) of a singular (Socrates), who is in fact a man. As Owens says

> Subject and predicate are really the same when a universal substance is predicated of a particular substance.

However, when one predicates something of prime matter, as in 'Matter is a man', a different kind of predication is involved. For, Owen says,

> Matter does not coincide in reality with a man in the way Socrates does. A really distinct principle, the form of man, is added. From this viewpoint the predication resembles rather the assertion of an accidental form in regard to substance, as when one says that a man is pale or fat. The accidental form is really distinct from the substance, as the substantial form is really distinct from its matter.[16]

So the doctrine is, according to Owens, that 'as accidental forms are predicated of substances ... so the substantial form is what is predicated of matter within the category of substance'.

How is this to be understood? A leaf may be green or brown according to whether the informed matter essential to a leaf has received in addition the form of greenness or the form of brownness. Prime matter, itself featureless and indeterminate, becomes a man or gold by having impressed upon it, receiving, the substantial form of man or gold. The matter itself is not anything determinate without some substantial form. It must, however, be remembered that the verb here must be interpreted with care. Prime matter cannot exist without some form any more than a leaf can exist without some colour; we cannot take a piece of unformed prime matter and impress upon it the required form although we may be able instantaneously to replace one substantial form qualifying prime matter by another. If we are thinking in metaphysical rather than scientific terms this may seem to give us the basis of the most fundamental kind of explanation. What is it to be a man or gold? It is to be prime matter of which the substantial form of man or gold is properly predicated. A substantial form is conceived as the one characteristic that, at a given level, makes a thing the thing it is; for example, the form of man is rationality at the level at which one is thinking of men among the animals.

In attacking the doctrine of substantial forms Boyle often mentions, in the same breath, 'real qualities' and, indeed, in *The Origin of Forms and Qualities* his attack appears to be mainly against the doctrine of real qualities. However, he treats these doctrines as clearly connected, either regarding them as rival hypotheses of the schoolmen or regarding the second as derived from the first. He refers to 'those grand disputes [among the schoolmen], whether the four elements are endowed with distinct substantial forms, or have only their proper [*sc.* essential] qualities instead of them' (*O.F.Q.*6, *S.* 6) and claims 'that almost all sorts of qualities, most of which have been by the schools either left unexplicated, or generally referred to I know not what incomprehensible substantial forms, may be produced mechanically' (*O.F.Q.* 13, *S.* 17). He also refers to 'real qualities and other real accidents, which not only are no moods of matter, but are real entities distinct from it, and, according to the doctrine of many modern schoolmen, may exist separate from all matter whatsoever' (*O.F.Q.* 16, *S.* 21). And again he says

> Now when any body is referred to any particular species (as of a metal, a stone, or the like) because men have for their convenience agreed to signify all the essentials requisite to constitute such a body by one name, most of the writers of physicks have been apt to think, that besides the common matter of all bodies, there is but one thing that discriminates it from other kinds, and makes it what it is, and this, for brevity's sake, they call a form: which because all the qualities and other accidents of the body must depend on it, they also imagine to be a very substance, and indeed a kind of soul... (*O.F.Q.* 27, *S.* 37–8)

He also refers to a substantial form as that 'from which all its properties and qualities immediately flow' (*O.F.Q.* 46, *S.* 67).

I believe that Boyle is here suggesting, as he was in the passage about whiteness quoted at the beginning of this chapter, that many 'modern schoolmen' have both extended and misunderstood Aristotle's doctrine of form.

Aristotle usually spoke of the form as some one feature that made a thing what it was and distinguished it from everything else. For example, the form of a man is the soul. The modern schoolmen, however, work also with the conception of 'real qualities' and Boyle contends that they treat real qualities as if they were substantial forms in relation to particular observable qualities. Boyle shows signs of thinking that their motives, at least, were praiseworthy since this represents an attempt, however misguided, to bring more detail and precision into what they regarded as an Aristotelian pattern of explanation. It was misguided because it resulted in empty explanations owing to the schoolmen's inability to characterize real qualities except in terms of the observable qualities they were intended to explain. The feature of real qualities that Boyle sees as assimilating them to substantial forms is that they are treated as if they were entities or ingredients which could be transferred from one parcel of matter to another; this amounted to regarding them as capable of existing independently of matter. There was a double mistake here since this was an illegitimate way of treating substantial forms in the first place.

This conception of real qualities probably arose in the following way. According to Aristotle, being a man is being matter informed by the form of man. However, the natural philosopher is interested in the explanation of the more detailed characteristics and behaviour of human beings and seizes on the point that we recognize human beings by various particular qualities that we take to be essential to being human such as certain shapes, colours, sounds and movements. These are observable and there is pressure to explain them in terms of real qualities actually possessed by the pieces of matter that we call human beings and independent of our perceptions. If we accept the doctrine of forms and misconceive it in a certain way, it is a short step to supposing that the substantial form of man can be characterized by, or analysed into, a collection of real qualities necessary and sufficient to being a man so we have a basis for a more detailed study of the nature of a man; to talk about the form of man seems to be to give some information. Now if we consider the simplest qualities among those necessary to being human, we can say that human beings *and other things* resemble one another in respect of some of these. All objects can be analysed into various simple shapes, sizes, colours, and so on, which, as far as unaided sense-experience reveals, are not further analysable. So it may seem that we can explain the fact that a human being and a flower are the same shade of pink by saying that they are both matter having the real quality of pinkness, where this real quality is regarded as an unobservable cause of the appearance of pinkness. This may seem to be an explanation of the same kind, and to give information in the same way, as the explanation of the nature of a man in terms of a substantial form and to be acceptable if no other way of explaining pinkness or humanity has been conceived.

Similarly, if the substantial form of gold is thought of as being analysable into real qualities underlying and explaining the observable qualities such as yellowness, fusibility and fixity by means of which we recognize gold it may seem that we can explain the nature of gold in terms of these real qualities. Moreover, it may seem reasonable to suppose that we can change lead into gold by replacing, all together or one at a time, the real qualities responsible for the nature of lead by those responsible for the nature of gold, in the matter provided by the lead. This way of thinking is facilitated if we think of real qualities, as Boyle has it, as 'real entities distinct from' matter and, as capable of existing 'separate from all matter whatsoever'. This would seem to allow the transferability of qualities from one portion of matter to another and so, among other things, the transmutation of metals.

However, as Boyle points out, such explanations are empty and unintelligible if the substantial forms and real qualities are not characterizable except in terms of the kind of thing ('form of *man*') or the observable quality ('real quality of *whiteness*') they are meant to explain. This difficulty would be removed if we could describe, say, inner structures of collections of corpuscles composing objects and then describe mechanisms by which they produce perceptions of colours, shapes, and so on. Our explanations would then at least be intelligible and informative. One aim of *The Origin of Forms and Qualities* is to show how, in principle, such structures and mechanisms could be proposed and would provide explanations.

So much, for the moment, for the way in which Boyle sees the modern schoolmen as having extended the Aristotelian doctrine of forms. I now turn to the way in which he sees them as having fundamentally misunderstood it.

Boyle says in several places (e.g. *O.F.Q.* 4–5, *S.* 3) that the peripatetics put too much weight on logical and metaphysical niceties and not enough on reasoning from physical observations.[17] He also says that he respects Aristotle's writings on natural philosophy more than his writings on physics, which is a main source for his account of forms and their part in change, generation and corruption (*O.F.Q.* 6–7 and 9, *S.* 6–7 and 9–10). He says

> divers of the ancient, especially Greek commentators of Aristotle, seem to have understood their master's doctrine of forms much otherwise, and less incongruously, than his Latin followers, the schoolmen and others, have since done. (*O.F.Q.* 37, *S.* 53)

He also says that Aristotle makes 'very little use, if any, of substantial forms to explain the phenomena of nature' (*O.F.Q.* 38, *S.* 53) and that the arguments used by the schoolmen for substantial forms in explaining natural phenomena are 'rather metaphysical or logical, than grounded upon the principles and phenomena of nature, and respect rather words than things' (*O.F.Q.* 40, *S.* 57–8). He appears to criticize Aristotle's physics 'or rather ... the speculative part of them' for not being the sort of thing that he, Boyle, recognizes as physics, that is, the study of particular natural phenomena (*O.F.Q.* 9, *S.* 10).

The trouble is that the analogy between forms, as used by Aristotle in the analysis of change, and real qualities, as used by the later schoolmen in explaining natural phenomena, is not a good one. The natural philosopher is concerned with *causal* explanation and the concept of real qualities is regarded by the schoolmen as a causal concept but Aristotle's use of forms in analysing change is not causal. Thus I believe that Boyle's fundamental criticism is that what for Aristotle was merely a logical analysis of change leading to the metaphysical assertion of prime matter and 'substantial' forms was mistakenly interpreted as providing the pattern for the explanation of particular natural phenomena. It was therefore looked upon also as providing a recipe for producing changes, especially substantial changes, such as that from lead to gold, in the context of natural philosophy. The assimilation of real qualities to substantial forms, of which Boyle also accuses the later schoolmen, just compounded the error.

There is an essential difference in kind between a logical analysis of change, the attempt to delineate the necessary and sufficient conditions for anything's being a change, and the explanation within natural science of particular changes. As far as logic goes it is perhaps not objectionable to say that if there are predicates then there must be subjects and even that if there are properties then there must be substances to have them. If we have a change, as distinct from a substitution, then there must be something that endures through it, even if we cannot give a description of that something. But if the subject or substance is inaccessible to scientific investigation and description then merely to say that it must nevertheless exist is of no help in the scientific explanation of particular changes. Their explanation requires that the enduring thing be described somehow and that its features be connected with the features of the things involved in the particular change in question. Boyle was well aware of this (see, for example, *O.F.Q.* 38, *S.* 54–5).

Such criticisms of scholastic arguments were very much in the air during the seventeenth century. For example, a clear version of the general interpretation I have been sketching is to be found in Leibniz's *Discourse on Metaphysics.*[18] In Section IX Leibniz explains the basis of his conception of monads and then begins Section X by likening his conclusions to those of the ancients who believed in substantial forms 'which to-day are so decried'. This is 'something, knowledge of which is so necessary in metaphysics' where we are concerned about ultimate principles. He is preparing to mount a limited defence of substantial forms.

However, he also refers to the 'failure and misuse' of forms by the schoolmen and the physicians (or doctors) who followed their example. Consideration of substantial forms is, he says

of no service in the detail of natural philosophy (*physique*), and must not be used for explaining phenomena in particular. (p. 15)

The schoolmen were at fault for

believing that they could account for the properties of bodies by mentioning forms and qualities without going to the pains of examining the manner of operation; as if one were willing to content oneself with saying that a clock has the horodictic quality resulting from its form, without considering in what this consists. (p.15)

This is very close to one of Boyle's objections to explanation in terms of substantial forms.

Leibniz goes on to say that just as the geometer does not have to trouble himself with problems of the continuum and the moralist does not have to worry about the reconciling of free will with the providence of God, so the physicist

can account for his experiments using now simpler experiments already made, now geometrical and mechanical demonstrations, without needing general considerations which belong to another sphere (p.16)

such as considerations of a soul or animating force.

Souls are, Leibniz says elsewhere, comparable to substantial forms. The essence of a body cannot consist in extension but must be conceived in terms of a substantial form which 'corresponds in some way to the soul'. He says

Still, approve as I may of the Schoolmen in this general and, if I may so put it, metaphysical explanation of theirs of the principal bodies, I still subscribe fully to the corpuscular theory in the explanation of particular phenomena; in this sphere it is of no value to speak of forms or qualities.[19]

Of course, Boyle's views are different from Leibniz's in radical ways and Boyle was hardly interested in metaphysical explanation. Nevertheless their views are very similar in respect of their understanding of scholastic views about substantial forms. They both accuse the scholastics of having misinterpreted and misapplied Aristotle's views and the grounds of the accusation are similar: Leibniz holds that they mistook a metaphysical account and Boyle that they mistook a logical account for a physical, scientific one. They saw the unfortunate consequences for scientific explanation in a similar way.

It is interesting also that Leibniz's chief desire, here, is to urge that it is sensible to talk of the soul as the form of the human being at least in metaphysical contexts, as Aristotle did, and that Boyle says that this is the one intelligible use of substantial forms by the schoolmen.

There is one aspect of Boyle's attack on the idea of 'real qualities' that I have not so far mentioned. If we rely upon the unaided senses for information about the world we are provided with no reason for not regarding the simplest observable qualities of things as really qualities of them. As far as the senses go it is difficult, perhaps impossible, to distinguish between the colour and the shape of a lump of gold in such a way as to justify our regarding its shape as a quality *of it* and its colour as not a quality *of it* but as an effect upon us of its qualities. All the observed 'qualities' of things appear to our senses to be equally *in* those things, equally to belong to them. Those whom Boyle was attacking are clearly taken by him to be saying that the shape, size, colour, odour, malleability, and so on, which under favourable conditions we observe a body to exhibit, are really qualities of the body or, more precisely, are explained by corresponding *real* qualities of bodies that resemble the observed qualities. There are two important points to be made about this.

In the first place, Boyle argues not only that qualities of bodies are not real qualities, in the scholastic sense, that is, *entities* separable altogether from matter, but also that some observed 'qualities' apparently belonging to bodies do not belong to them in the way that others do; they are not, just as they appear to us, qualities at all, let alone real qualities. The theoretically based distinction between primary and secondary qualities is partly intended to allow this.

In the second place, if a real quality is regarded as a constituent or ingredient of a body our senses do not give us any warrant for going further in saying anything about that quality. They give no basis for asserting that the real quality yellow is itself either yellow or not yellow or that it is of a nature fit to cause the body to look yellow. At least, that last clause gives no information. The tautological or empty explanation 'this body is yellow because it is informed by the real quality of yellowness' signals the unavoidable end to our chain of questions and answers or, putting it differently, it indicates what we are taking as our basic categories for classification, explanation and description.

As we shall see, Boyle argues that we can go further and give a different meaning to 'form' and 'quality' which allows us to provide non-tautological explanations. For him, what the peripatetics took as basic categories are not basic enough and stop far short of what is scientifically possible. His approach involves conceiving expressions such as 'the real quality of yellowness' as names for internal structures of material particles which may be describable in detail in intelligible terms even if those structures are not directly observable. It may still be possible to collect empirical evidence for the nature of these structures. For his hypothesis about their general character Boyle builds on, and modifies, ancient and contemporary atomic theories.

III

Boyle begins *The Origin of Forms and Qualities* by arguing against the peripatetics' doctrines of 'forms and qualities, and generation, and corruption and alteration' which, he says, 'are wont to be treated ... in so obscure, so perplexed, and so unsatisfactory a way'. Their discourses in these subjects consist

> so much more of logical and metaphysical notions and niceties than of physical observations and reasonings, that it is difficult for any reader of but an ordinary capacity to understand what they mean... (*O.F.Q.* 4–5, *S.* 3)

The peripatetics are, besides, much taken up with rival theories about the generation of forms; they have good arguments against one another's accounts but unsatisfactory arguments for their own. 'They all confute well', he says, 'and none does well establish' (*O.F.Q.* 38, *S.* 54). This he finds unsurprising because, in spite of their disagreements, they all treat forms and real qualities as independent of matter and not to be understood in material terms. In consequence they preclude the possibility of saying anything either precise or having empirical content about them, and so of giving any explanation of material phenomena on their basis.

He admits that a human being is made up of an immaterial form (the soul) and a body but holds that nothing else in nature need be regarded as composed of matter and a substance distinct from matter (*O.F.Q.* 40, *S.* 57). His aim in this work, and elsewhere, is to show that his corpuscular philosophy can provide the basis for the analysis and explanation of everything else in nature. The advantages of this will be that explananda can be made intelligible, clear connections will be able to be made between explanandum and explanans and there will be at least the theoretical possibility of empirical confirmation of postulated entities and explanations.

He attacks the peripatetics on four main grounds: the vagueness of their talk about qualities, real qualities and substantial forms; the occultness of real qualities and substantial forms; the alleged necessity of these for explanation; and the possibility of giving explanations in terms of them.

He now considers and rejects three abstract arguments and three physical arguments in favour of substantial forms. I shall outline these in turn. The three abstract arguments and Boyle's replies to them are as follows.

1. Believers in substantial forms argue that 'every substantial composite requires matter and substantial form of which it is composed; every natural body is a substantial composite; *therefore*, every natural body requires matter and substantial form of which it is composed' (*O.F.Q.* 40, *S.* 58). Some deny the consequence of this syllogism but Boyle prefers to deny the minor premiss: every natural body is a substantial composite. His reason is that he does not know of any body in nature that is composed of matter and a substance distinct from matter, which is what 'substantial composite' means, except man. The combination of matter and substantial form is smuggled in by the 'uncouth term' '*compositum substantiale*', which does more than name something; it implies an analysis of it which is at issue.

2. The next argument is that without substantial forms all bodies would be 'accidental entities' or 'entities by accident' (*entia per accidens*) and that is absurd. Learned men, Boyle says, conceive an *ens per accidens* as consisting of things 'not ordered in a unity' (*non ordinantur ad unum*) in contrast to things 'ordered in themselves and intrinsically' (*ordinantur per se et intrinsicè*) to form a natural body. He replies that we need not admit natural bodies as *entia per accidens* because their 'constituents', such as matter, shape, situation and motion '*ordinantur per se et intrinsicè* to constitute one natural body' (*O.F.Q.* 41, *S.* 58). (In the first edition 'intrinsicè' appears as 'intrinsecè'.)

This passage is a little difficult to interpret because Boyle refuses to attempt to translate these 'uncouth' terms, which are scholastic technical terms. However, it seems clear that things *quae ordinantur per se et intrinsicè* are of a sort that Leibniz calls 'real unities' and things *quae non ordinantur ad unum* are not real unities. We have seen that Leibniz regards substantial forms as providing real unity in a metaphysical context and that Boyle agrees with this for the special case of human beings. But in a scientific context Leibniz agrees with Boyle that talk of substantial forms is unnecessary; although a pile of sand is not a real unity a grain of sand is, purely because of its physical structure, a real unity.

Boyle concludes his consideration of this argument by pointing to the dangers that lie in the incautious naming of things. He says

> that which I am solicitous about is, that what nature hath made things to be in themselves, not what, [a] logician or metaphysician will call them in the terms of his art; it being much fitter in my judgment to alter words, that they may better fit the nature of things, than to affix a wrong nature to things that they may be accommodated to forms or words that were probably devised, when the things themselves were not known or well understood, if at all thought on. (*O.F.Q.* 41, *S.* 58)

3. The third abstract argument is this: without substantial forms there could be no substantial definitions but that is absurd. Substantial definitions would seem to be definitions of substances in terms of their forms and Boyle rightly argues that definitions need not be of that sort. In contrast, for natural things it may suffice to have 'essential definitions' by which he means

> such as are taken from the essential differences of things, which constitute them in such a sort of natural bodies, and discriminate them from all those of any other sort. (*O.F.Q.* 41, *S.* 59)

We need, for example, only such definitions of gold and silver as will suffice to tell us what properties things must have in order to be properly called 'gold' and 'silver', respectively, and to enable us to distinguish between samples of the two. These must clearly rest on the discoverable properties of things.

This last condition cannot be met by substantial definitions. To define natural things in terms of substantial forms gives no enlightenment because as even some peripatetics admit, most substantial forms are 'things that we know not'. Boyle quotes Scaliger[20] as supporting this by saying that the wisest peripatetics do not even pretend to give a substantial definition of any natural composite except man.

Some modern peripatetics have also put forward some physical arguments for substantial forms, fearing that these abstract arguments, depending as they do upon words rather than things and upon precarious suppositions, may not appeal to 'naturalists'. I mention four of these.

1. Such phenomena as the spontaneous return of heated water to coldness are explained by the presence of a substantial form, the function of which is to preserve the body in its natural state. Boyle objects that the idea of a natural state is not clear and, by implication at least, that there is no evidence that Nature attempts to preserve bodies in their natural states. The 'natural state' of water would seem to differ between the tropics and the Arctic. Changes of the sort under consideration can be

explained as the effects of external conditions and so need not be explained in terms of inner principles. A body continues in a given state unless it is affected by something external to it; no explanation is needed for its not changing but only for its changing and even then substantial forms are unnecessary.

It is interesting to note here a connection with the developing theory of motion at the hands of Galileo and Newton. A special case of the need to explain change rather than rest or uniform motion is Newton's principle of inertia, which implies a rejection of the Aristotelian idea of natural states to which things naturally return.

2. Matter is 'indifferent to' all accidents (or qualities) so substantial forms are needed to explain why the set of accidents belonging to a body stay together in that body so as to preserve it in its natural state. In other words, substantial forms are needed to explain the stability of bodies and the fact that the world is not just a chaos of freely floating accidents (*O.F.Q.* 42, *S.* 61).

Boyle once again questions the conception of natural states. Everything that happens in nature is natural and it is a feature of nature that things are constantly changing. If we seize on one state of a thing as natural we may find that most of the time it is in an unnatural state. The most we can mean by a natural state is the state a thing is usually in and that is a useful idea only for stable bodies. If we account for the stability of a piece of brass by reference to the substantial form of brass then it is mysterious how it can be melted and yet remain brass. Boyle claims that anything that is allegedly explained by substantial forms can be better explained without them in terms of corpuscular structures.

3. Next, it is alleged that there can be no reason why whiteness is 'separable' from a wall and not from snow or milk without recourse to substantial forms. The point is that essential qualities are allegedly accounted for by substantial forms; remove whiteness from snow or milk and we no longer have snow or milk. On the other hand, a wall may be painted blue and remain a wall (*O.F.Q.* 44, *S.* 64).

Boyle refrains from such obvious questions as 'If a painted wall remains a wall why doesn't dyed milk remain milk?' but instead replies with a different view of essential properties and thus precludes a search for better examples. The reason why a quality is essential to a substance, for example, milk, is that men have agreed not to call anything milk if it does not have that quality. As we shall see, this is closely related to Locke's view of nominal essences. We regard whiteness as an essential quality of snow, if we do, because its inner structure is such that heat will remove its whiteness but at the same time makes it liquid so we no longer call it snow. If we found some means of making snow yellow without making it liquid then we should no doubt stop regarding whiteness as an essential quality of it.

4. Boyle appears to regard this argument as a more serious argument requiring more careful consideration. It is

> that there seems to be a necessity of admitting substantial forms in bodies, that from thence we may derive all the various changes to which they are subject, and the differing effects they produce, the preservation and restitution of the state requisite to each particular body, as also the keeping of its several parts united into one *totum*. (*O.F.Q.* 45, *S.* 65)

I presume that this argument has a greater appeal than the others for Boyle because it is more general and does not mention such suspect ideas as natural states and essential qualities or the existence of substantial forms independently of matter. Boyle, and natural philosophers generally, aim to find fundamental properties of bodies or substances which will facilitate explanations of their observable properties and their interactions with one another. This argument suggests a similar aim and does not mention the mysterious nature of the usual conception of substantial forms. Boyle could accept this conception: substantial forms could be taken to be identical with the fundamental properties sought and to be in some way accessible, at least indirectly, to scientific investigation and inference since this would mean that it might serve the function suggested in the argument. He merely says here that much of the rest of his book will be devoted to discussing this matter and that, in the meantime, he has three points to make.

First, he claims that changes in bodies proceed from 'their peculiar texture' together with the action of other bodies on them and that this kind of explanation will work even if there are no substantial forms in nature. It seems to me that these textures have the advantage that they are, at least in principle, capable of being described and known while substantial forms, as usually conceived, are not and that this influenced Boyle.

Second, the operations of a substance may be deduced from 'the peculiar texture of the whole, and the mechanical affections of the particular corpuscles or other parts that compose it'. Substantial forms give little promise of *mechanical* explanations. Moreover, it can be shown that artificially produced vitriol has the same properties as vitriol 'made by nature in the bowels of the earth'; it is reasonable to suppose that this is due to their having similar inner structures. This avoids such arguments as raged among the schoolmen about whether man-made substances had substantial forms.

Third, the unity and stability of a body may be explained in terms of 'the contrivance of conveniently figured parts' and without reference to substantial forms. Once again, a mechanical account of cohesion may be obtainable and sufficient. A plum grafted on to an apricot thrives and produces good fruit and is a unity in spite of the fact that the schoolmen would not allow that it has only one substantial form but would insist that it has two. Mechanical cohesion will suffice to explain this also, so substantial forms are irrelevant.

Boyle then enunciates a general principle of explanation which reads

> to explicate a phænomenon being to deduce it from something else in nature more known to us than the thing to be explained by it... (*O.F.Q.* 46, *S.* 67)

The trouble with substantial forms, as conceived by the schoolmen, is that they are less known to us than any of the phenomena they purport to explain as is admitted even by many of their champions.

The conclusion from all this is that we should either reject altogether the terminology of substantial forms and real qualities, replacing it by a terminology of corpuscular structure, or keep the terminology of substantial forms and real qualities and re-interpret it in such a way that these terms refer to nothing but corpuscular structures. One of the ways in which Boyle describes his project, and explains his title *The Origin of Forms and Qualities*, suggests the second possibility. What he aims to do is to give an analysis of substantial forms and real qualities in corpuscular terms which will be acceptable to natural philosophers and give them some hope of discovering and describing their characteristics. For example, he talks of 'expounding ... what, according to the corpuscularian notions, may be thought of the nature and origins of qualities and forms' (*O.F.Q.* 4, *S.* 2). On the whole, however, Boyle appears to be in favour of the first course, dropping the terminology altogether and replacing it by his own. How he sets about doing this is the subject of the next chapter.

3

BOYLE'S CORPUSCULAR PHILOSOPHY

I

Believing that he has exposed serious defects in some current methods of explanation, Boyle, especially in *The Origin of Forms and Qualities,* puts forward a positive view. This redefines 'form' and 'quality', seen as the concepts to be used in explanation, by relating them to the mechanical qualities of material particles and the fine structures of complexes of such particles.

At the end of *The Sceptical Chymist* he side-steps charges of irreligion often directed against atomists and, in so doing, indicates some doubts about the possibility of finding mechanical explanations for *all* natural phenomena whatever. He denies that he is an Epicurean and asserts that he has never read the whole of Lucretius (*S.C.* 569, *E.* 199). The point of this, I believe, is to reject the idea that the whole of the natural world, including living things, could have arisen from a 'fortuitous concourse of atoms' by some cosmic accident. If the universe be but matter in motion an intelligible account of it requires reference to the guidance and determination of 'the most wise Author of things' and perhaps even to final causes. He says

> For I confess I cannot well conceive, how from matter, barely put into motion, and then left to itself, there could emerge such curious fabricks as the bodies of men and perfect animals, and such yet more admirably contrived parcels of matter, as the seeds of living creatures. (*S.C.* 569, *E.* 200)

For all other natural phenomena the hypothesis is that mechanical explanation, in terms of matter, motion and rest, is possible. Qualities such as colours, odours, tastes, fluidity, hardness and softness that 'diversifie and denominate' bodies, may be deduced from these. Matter, however, implies the more fundamental qualities magnitude (size), figure (shape) and weight. The smallest particles of bodies, the corpuscles, must be material and have motion or rest as necessary properties. Observable bodies are organized collections of these corpuscles and their observable qualities depend upon the fundamental properties of the corpuscles and their various patterns of organization in groups. Boyle refers to these patterns as structures or *textures*, a word that he uses as a technical term, as we shall see (*S.C.* 570, *E.* 201). Changes in the observable qualities of bodies are brought about by rearrangements of their constituent corpuscles, that is, by changes of texture, which are describable in terms of the shape, size, motion, situation and connection of the corpuscles (*S.C.* 579, *E.* 217).

In *The Origin of Forms and Qualities*, published some five years after *The Sceptical Chymist*, Boyle describes his corpuscular theory in much greater detail and in somewhat different terms. For example, he pays little attention to weight in the later work and the theory becomes more purely geometrical in pursuit of a view that was current among natural philosophers at the time. One finds the view, for example, in both Descartes and Leibniz.

He begins by making it clear that his favoured explanations are to be in terms of concepts derived, even if indirectly, from experience and the observation of natural phenomena, and to be explicit in showing in detail how things actually happen. In direct opposition to the peripatetics he aims to describe the mechanisms underlying observable events and, at least in this part of his work, to avoid final causes and teleological explanations.[1] In the Preface to *The Origin of Forms and Qualities* he says

> the knowledge we have of the bodies without us, being for the most part fetched from the informations the mind receives by the senses, we scarce know anything else in bodies, upon whose account they can work upon the senses, save their qualities... And as it is by their qualities that bodies act immediately upon our senses, so it is by virtue of those attributes likewise that they act upon other bodies, and by that action produce in them, and oftentimes in themselves those changes that sometimes we call alterations and sometimes generation or corruption.

He is anxious to say

> what changes happen in the objects themselves, to make them cause in us perceptions sometimes of one quality and sometimes of another. (*O.F.Q.* 11, *S.* 13)

This is not an aim that Boyle pursues in this book. He will later attribute to the unobservable corpuscles only qualities that we discover in observable bodies. There are to be no 'occult' qualities, but only such as we know from observation, attributed to the unobservable corpuscles. Although it does not emerge from this quotation, there is room for controversy about what Boyle means by 'qualities' and, indeed, later he refuses to define the word but leaves us to discover for ourselves how he uses it. The above passages make it clear, I think, that he here uses it to refer to whatever it is in bodies that enables them to act upon one another as well as upon our senses. He later concludes that not all the qualities we apparently find in observable bodies need to be attributed to their unobservable constituents because some of them can be explained in terms of others.

II

The general form of Boyle's project is this. He takes it for granted without argument that there are external material objects which act upon one another independently of us and which act also upon us to produce perceptions. Such knowledge of the natural world as we have comes through sense-experience. This is the most plausible assumption from which to develop an intelligible account of natural phenomena. The aim of the natural philosopher is, given this assumption, to investigate and describe those external material things and their actions upon one another and upon us in as much detail as is possible and necessary.

The clearest explanations of material phenomena with which we are familiar are mechanical explanations of middle-sized phenomena in which the behaviour of complexes are explained in terms of the properties and interactions of their component parts. The favourite example is that of a watch; we come to understand the working of a watch by discovering how the cogs, springs, levers and other parts operate upon one another and finally produce the appropriate motion of the hands. We can describe all this in terms of the shapes, sizes, motions, hardness, flexibility and organization of those various parts. If we could do the same for natural phenomena we should have the clearest, most intelligible explanations of their nature and occurrence. The workings of the watch can be examined in greater and greater detail so that eventually we may be able to explain some of the qualities of the metallic parts, such as their hardness or flexibility, in terms merely of the shapes, sizes and motions of even smaller, unobservable parts. Moreover, the qualities of different metals, such as bronze or iron, may be explainable in terms of different arrangements or structures of relatively similar parts. In other words, the ideal of explanation to which we are thus led is an atomistic one, although Boyle prefers not to use that word because he sees problems about the alleged indivisibility of atoms and because certain unwelcome associations cling to it. The grand hypothesis, then, is that all material bodies and substances are composed of enormous numbers of unobservable and minute particles each having a very limited range of properties, namely, those regarded as essential to matter. The aim is to explain as many of the phenomena as possible in such terms and the value of, and support for, the grand hypothesis will depend upon the success and comprehensiveness of these explanations. It is not assumed from the beginning that every observable phenomenon will be mechanically explained although it is expected that at least some part of their explanation will be mechanical; the project is to take mechanical explanation as far as possible.[2]

In his short work *The Excellency and Grounds of the Corpuscular or Mechanical Philosophy* (1674)[3] Boyle summarizes his theory and its aims and advantages. He attaches great importance not only to 'the intelligibleness or clearness of mechanical principles and explications' (*E.G.* 69, *S.* 139) but also to their economy. He says 'there cannot be fewer principles than the two grand ones of mechanical philosophy, matter and motion' (*E.G.* 70, *S.* 141). These principles are also primary, simple and comprehensive. The aim is to explain as many phenomena as possible in terms of as few factors as possible and these factors are to be as simple, fundamental and closely related to experience as possible.

Boyle now returns to the analogy of the alphabet. Just as the twenty-six letters of the alphabet will suffice to build all the works of the poets in English, Latin and many other languages, so a few basic properties of matter will suffice for the explanation of an enormous variety of material phenomena. It has often been argued that mechanical principles will suffice for the explanation of some of the phenomena of nature, such as the planetary motions, but cannot be applied to all phenomena. Boyle says

> I am apt ... to look upon those, otherwise learned, men as I would do upon him, that should affirm, that by putting together the letters of the alphabet, one may indeed make up all the words to be found in one book, as in *Euclid*, or *Virgil*; or in one language, as in Latin or English; but that they can by no means suffice to supply words to all the books of a great library, much less to all the languages in the world. (*E.G.* 70–1, *S.* 142)

Such 'otherwise learned' men would be ignoring the obvious.

Boyle has stressed the importance of observation in discovering what the world is like and it would normally be said, perhaps somewhat misleadingly, that he set out to give an empiricist account of scientific investigation and knowledge in accordance with the ideals of the newly formed Royal Society. How then, it might be asked, is he entitled to postulate unobservable particles and attribute properties to them? This is not a problem of which he was unaware and he has two ways of dealing with it. The first, already briefly mentioned, is that the hypothesis of unobservable particles will be empirically tested by means of observation, experiment and prediction on the basis of explanations of phenomena that can be deduced from it. However, that is programmatic and for the future. The second, and more immediate, answer is that it makes sense to postulate unobservables with specific properties as long as we are careful to attribute to them the only *kinds* of properties with which we are familiar from everyday observation; we can understand what we are saying when we talk of, say, unobservable *spherical* particles because we have observed spheres of various sizes and we can conceive of spheres that are just very much smaller, in particular, too small to be seen. Since we cannot conceive of a sphere having no particular size so we at the same time conceive of the invisible spheres as having some dimensions.

This is supported by many of Boyle's statements. He says 'as we plainly see in the universe or general mass of matter, there is really a great quantity of motion, and that variously determined, and that yet divers portions of matter are at rest' and 'that there is local motion *in many parts* of matter is manifest to sense' (*O.F.Q.* 15, my italics, *S.* 18) and then he goes on

> since experience shews us (especially that which is afforded us by chymical operations, in many of which matter is divided into parts too small to be singly sensible) that this division of matter is frequently made into insensible corpuscles or particles, we may conclude, that the minutest fragments, as well as the biggest masses of the universal matter, are likewise endowed each with its peculiar bulk and shape. (*O.F.Q.* 16, *S.* 20)

In *The Sceptical Chymist* he says

> there may be several sorts of bodies which are not immediate objects of any one of our senses; since we see, that not only those little corpuscles, that issue out of the loadstone, and perform the wonders, for which it is justly admired; but the effluviums of amber, jet, and other electrical concretes, though by their effects upon the particular bodies disposed to receive their action, they seem to fall under the cognizance of our sight, yet do they not as electrical immediately affect any of our senses, as do the bodies, whether minute or greater, that we see, feel, taste, etc. (*S.C.* 516, *E.* 104–5)

This is an exemplification of the view about what properties may be legitimately attributed to unobservable particles. When we observe electrical phenomena we do not observe them *as* electrical; what we do observe is, for example, small pieces of paper moving towards pieces of amber, that is, portions of matter in motion. Here is Boyle refusing to attribute electrical properties to the unobservable particles; it may be that we can explain the observed motion of pieces of paper in terms of the shape, size and motion of their particles and those of the electrified body. Hence the talk of 'effluviums' which are streams of tiny material particles from the 'electrical concretes'. Electrical and magnetic qualities may be explained as 'emerging' when particles, themselves having no electrical properties, are organized in particular ways in large groups.

It is important, in the interests of economy and generality in explanation, that not all the qualities we observe in bodies be attributed to their constituent particles. We do not observe individual particles so we need not attribute to them colours, tastes and smells, *on condition that* we can find explanations for the colours, tastes and smells of the bodies we *do* observe, or, at least for our perceptions of these, on the basis of other properties we attribute to the particles and to our organs of sense. Hence arises the distinction between primary and secondary qualities. Boyle depends heavily upon this distinction although he does not use this terminology consistently. Primary qualities are those qualities attributed to every individual particle in terms of which every other quality is analysed and our perceptions explained. I shall not now go into secondary qualities except to say that they are analysable into the primary qualities of the constituent particular bodies and to add the warning that, in my view, many commentators on Boyle and Locke have misconstrued the distinction.

It would perhaps be as well, at this point, to explain some of Boyle's terminology in order that I may keep as close as possible to what he says in expounding his view. He uses the term 'body' both for material objects, such as stones or trees, and for what we normally call 'substances', such as gold or common salt. In order to avoid confusion with a more technical sense of 'substance' I shall usually follow him in this. The word 'property' had a technical use for Aristotle and his followers, and many other seventeenth-century philosophers, in which it meant, roughly, 'essential property'. Boyle usually uses instead the words 'quality' and 'accident' and also, in some contexts, 'attribute'. Unfortunately, his use of 'quality' appears to vary: sometimes he appears to use 'quality' or 'sensible quality' for the observed properties of bodies other than shape, size, and motion or rest, *i.e.* for colours, tastes and odours; at other times he uses it for all properties of bodies *including* shape, size and motion or rest. For example, he refers to 'shape and other qualities' of bodies (*O.F.Q.* 16, *S.* 21). He finds that current uses of the word are various and confused but he thinks that his use of it will be understood in the light of his theory (*O.F.Q.* 26, *S.* 37). I shall return to this at the end of the chapter. He also uses the word 'accidents' for qualities of bodies. He points out that accidents are sometimes distinguished from the essential properties of bodies and sometimes from the substances of which they are properties, that is, to distinguish *all* properties from that which has them. Boyle usually uses 'accident' in the second sense, that is, to refer to all properties or qualities, whether or not they are essential to the thing in question. I shall also return to this.

In writing of the distinction between primary and secondary qualities I shall use this terminology, following Locke, but it must be noted that Boyle talks of primary qualities also as, variously, 'moods or primary affections of body', 'inseparable accidents' and 'essential properties' (*O.F.Q.* 16, *S.* 20–1). He refers to colours, tastes and odours as 'less simple qualities' and 'secondary qualities' (*O.F.Q.* 24, *S.* 32). This last reference may later appear to present some problems for my interpretation; here I merely point out that he is here referring to colours, tastes and odours as 'sensible qualities' by which he frequently appears to mean 'qualities as sensed' or, in Locke's terminology, 'ideas'.

Perhaps Boyle's most important technical term, as I shall argue later, is 'texture' which, in my view, has been much misunderstood. Its technical sense must not be confused with the sense in which we feel the texture of a surface when we touch a piece of sandpaper or velvet; it is rather a *structure* of unobservable particles and so is itself not directly observable.

Boyle calls his minute particles 'corpuscles' rather than 'atoms' because, as I have mentioned, he regards his corpuscular hypothesis as different in some important respects from the traditional atomism developed from Democritus and Leucippus through Epicurus and Lucretius (*O.F.Q.* 7, *S.* 7). However, he uses the word 'corpuscle' in a wide sense, applying it to both the smallest particles that actually exist and stable complexes of them below the level of observation. For example, he refers to the smallest undivided particles as '*minima* or *prima naturalia*' and then says that there are also multitudes of corpuscles 'which are made up of the coalition of several of the former *minima* or *prima naturalia*' (*O.F.Q.* 29–30, *S.* 44). This suggests that he is using 'corpuscles' only to refer to small stable collections, such as a minute particle of gold, of the smallest particles. But elsewhere, when he is clearly talking of the smallest parts of matter he refers to them as 'corpuscles or particles' (*O.F.Q.* 16, *S.* 20) and he talks about 'intire and undivided corpuscles' in the passage in which he asks us to conceive one of them alone in the universe and says that each would have three essential properties (*O.F.Q.* 22, *S.* 30). When he talks of 'undivided' particles he always seems to mean the smallest actual particles and it is these that he here calls 'corpuscles'. He also says

> Now, since a single particle of matter, by virtue of two only of the mechanical affections, that belong to it, be diversifiable so many ways; how vast a number of variations may we suppose capable of being produced by the compositions and decompositions [i.e. *further* compositions] of myriads of single invisible corpuscles that may be contained and contexed in one small body ... (*E.G.* 70, *S.* 142)

This looseness of terminology runs right through his work and it is probably possible for the argument to go either way. However, for my purposes this will be a minor terminological point which will not affect my interpretation. I shall take it that Boyle uses 'corpuscles' for both the *minima naturalia* and for small stable aggregates of them and when I use the word, unqualified, in the rest of this account I shall mean it in the former sense. Besides it seems likely to me that Boyle's name for his hypothesis, the 'corpuscular hypothesis', is most likely to have been invented to characterize the most fundamental entities of the hypothesis.

III

Boyle sets out his corpuscular hypothesis most systematically in *The Origin of Forms and Qualities* in eight 'particulars' (*O.F.Q.* 15–35, *S.* 18–50) having stressed that it is merely a hypothesis for consideration and, it is hoped, empirical confirmation.

The implicit assumption is one usually made by natural philosophers that they are attempting to discover the features of a material world which is independent of our perceptions but which acts upon us causally to produce those perceptions. It is thus a realist assumption. The method of discovery is to rest heavily on observation and experiment. It is clear, however, from the very nature of the corpuscular hypothesis that Boyle also attaches considerable importance to theorizing, which is acceptable as long as it is firmly based in 'natural history', the description of nature arrived at by empirical methods.

All bodies have in common 'one catholick or universal matter' which Boyle says is 'a substance extended, divisible and impenetrable' (*O.F.Q.* 15, *S.* 18) but clearly our world is not just a solid block of undifferentiated stuff. To explain the fact that we find in it enormous numbers of distinguishable, separable and changeable bodies we must refer to something other than the common matter of which they all consist. Changes in individual bodies suggest that their parts are not permanently at rest in relation to one another so the differentiation of the common matter into relatively independent bodies may be regarded as depending upon motion. We in fact see the observable bodies around us sometimes in motion and sometimes at rest.

We may say that the minute parts of stable bodies are in motion even when the bodies themselves are at rest, and it may be necessary to say this in order to explain such things as chemical interactions between stable bodies. All that is necessary to make this intelligible is that various parts of a stable body at rest are moving but in different directions with various speeds and within certain limits; the various motions of the parts cancel one another out at the level of ordinary observations so the resultant is apparent stability and rest of the observable body as a whole.

We do not regard a stone that starts rolling down a hillside as thereby having its nature changed or, even more, as becoming more or less *material*. There is no *logical* impossibility in the conception of a universe which is just a solid block of undifferentiated and unchanging matter but it is difficult to see our universe in that way.

Boyle says that he is following Descartes in regarding motion in matter as having been initiated by God but he is prepared to go even farther than that. The universe is 'beautiful and orderly' to the extent that we cannot conceive of its having occurred by accident as a result of God's having merely given matter an initial push; God must have designed it. He thinks

> that the wise Author of things did, by establishing the laws of motion among bodies, and by guiding the first motions of the small parts of matter, bring them to convene after the manner requisite to compose the world, and especially did contrive those curious and elaborate engines, the bodies of living creatures, endowing most of them with a power of propagating their species. (*O.F.Q.* 15, *S.* 19)

Boyle says that this is strictly not relevant to his present purpose; nature can be investigated along the lines suggested by his theory of matter whether or not we accept his views about the creation of the universe. It is not his aim to give here a complete account of the principles of natural philosophy but only to investigate the origin of forms and qualities. However, it is worth noting that the idea of a designed universe has not been utterly irrelevant to the scientific project; it is one of the beliefs that can give a scientist the faith that there are laws, waiting to be discovered, that govern all phenomena. This is strengthened when it is coupled, as I think it is by Boyle, with the belief that God, being benevolent, must have given us the faculties and abilities to discover these laws, at least for the most part. One does not need to share these beliefs to see that they have played an important and fruitful part in the development of science.

Local motion is regarded by Boyle as 'the principal among second causes, and the grand agent of all that happens in nature' (*O.F.Q.* 15, *S.* 19); the first cause is God, who set up the whole universe in the first place. The importance assigned to motion is to be expected in an account that aims to be mechanical. Whatever intrinsic and fundamental properties the minute parts have, without motion these are merely necessary conditions of phenomena; without motion there would be no interaction between the parts of bodies and between bodies, so they would not be involved in events or phenomena. There must be something with characteristics that allow motion but there must be motion in order that anything should happen. It is only when things move and collide that their fundamental properties become effective. We might say, in Aristotelian terms, that for Boyle the intrinsic properties of the minutest particles are *material* causes while their motions are *efficient* causes. He believes that most material phenomena may be explained in terms of just these two sorts of causes once we have accepted that God set up the whole system. In this connection he uses one of his favourite illustrations: if a key is too big or too small for a lock then no amount or kind of motion will enable it to open the lock but, however well it fits the lock, without motion it won't open it either.

In his third 'particular', which is of central importance because it elucidates the fundamental details of the corpuscular hypothesis, Boyle calls matter and motion the 'two grand and most catholick [i.e. *universal*] principles of bodies' (*O.F.Q.* 16, *S.* 20). Matter must be actually divided into parts and each 'primitive fragment', like each distinct observable body, must have two attributes: 'its own magnitude, or rather size, and its own figure or shape'. The point here is that we cannot conceive of a part's being material without its having some shape and size, however small it may be. Boyle believes that many of his chemical experiments show that matter is frequently divided into 'parts too small to be singly sensible', that is, invisible parts. So, since matter is frequently divided into 'insensible corpuscles or particles' we may conclude that

> the minutest fragments, as well as the biggest masses of the universal matter, are likewise endowed each with its peculiar bulk and shape. (*O.F.Q* 16, *S.* 20)

In experience we cannot divide a material object with a given shape and size into observable parts without some shape and size; we cannot *conceive* of a material object's being divided into unobservable parts without shape and size because we cannot conceive its being divided into immaterial parts.

Now he says that

> we have found out, and must admit three essential properties of each intire or undivided, though insensible part of matter; namely, magnitude (by which I mean not quantity in general, but a determined quantity, which we in *English* oftentimes call the size of a body) shape, and either motion or rest (for betwixt them two there is no mean): the two first of which may be called inseparable accidents of each distinct part of matter ... (*O.F.Q.* 16, *S.* 20)

Boyle calls shape and size 'inseparable accidents'; because any portion of matter must have *some* shape and size they are inseparable, and because no specific shape and size is essential to matter they are accidents. Different particles of matter may have different shapes and sizes.

He also calls shape and size, along with motion or rest, 'essential properties' of any portion of matter. This may, at first sight, seem to conflict with his earlier statement that 'local motion ... is not included in the nature of matter, which is as much matter when it rests as when it moves' (*O.F.Q.* 15, *S.* 19). However, there is a crucial difference between motion and motion-or-rest. It is essential that any particular body be in some state of motion-or-rest but not that it be in motion, where motion excludes rest. I believe that Boyle's thought can be captured by saying that what is essential to a body is that it has *mobility*. This means that matter, *as such*, has the characteristic that any portion of it must be either in motion or at rest.

Why does Boyle say here that only two of the essential properties, magnitude and shape, are inseparable accidents, seeming to withhold this description from 'motion or rest'? He has just said that there are *three* essential properties. The answer may be that neither motion nor rest, taken separately, is essential or an inseparable accident. What is essential is mobility or 'motion-or-rest'. In the rest of the work Boyle treats mobility in the same way, in this respect, as magnitude and shape. So although he is, quite correctly, not prepared to treat motion, or rest, as an inseparable accident he is prepared to treat mobility thus, although he does not explicitly say he is.

Boyle mentions two ways of thinking of inseparable accidents: they may be thought of as 'the moods or primary affections of bodies, to distinguish them from those less simple qualities (as colours, tastes, and odours) that belong to bodies on their account' or they may be thought of, as by the Epicureans, as 'the conjuncts of the smallest parts of matter'. This presumably means that we may think of them as in any bodies large or small, observable or not, but contrasted with other qualities of those bodies or we may think of them as just the qualities possessed by the smallest parts of matter (atoms or corpuscles) rather than by aggregates of them. He proposes not to settle this here but only wishes to make it clear that he rejects a third way of thinking of them, namely, that of the modern schoolmen who regard them as 'real qualities' which are 'no moods of matter, but are real entities distinct from it' which 'may exist separate from all matter whatsoever' (*O.F.Q.* 16, *S.* 21).

It is clear from what follows that Boyle thinks of inseparable accidents as 'moods or primary affections of bodies', whether the smallest particles or aggregates of them, but that a criterion for deciding what they are depends on considering 'the smallest parts of matter' so his view is a sort of compromise between the two views he mentions. This is the basis of his version of the distinction between primary and secondary qualities. The inseparable accidents or primary qualities are those into which more complex qualities of bodies are analysable or by means of which they are explained. Observable bodies all consist of minute particles or corpuscles, which are the smallest actual portions of matter; the primary qualities which are the basis for explanations of the properties of bodies, simple or complex, observable or not, must belong to these minute corpuscles. The primary qualities are in fact the defining properties of matter which Boyle took to be shape, size and mobility; the secondary qualities are other qualities of bodies that depend upon these. With good fortune and hard work the natural philosopher will be able to show us the details of this dependence. We must note that the more complex qualities are said by Boyle to *belong to bodies*; there is no suggestion that secondary qualities are, unlike primary qualities, somehow illusory, subjective or 'in the mind'. I shall have much to say about this later.

Boyle spends some time here arguing against the schoolmen and their real qualities, so convinced is he of the necessity of avoiding their mistakes. He uses his celebrated, but sometimes misinterpreted,[4] analogy of a key and a lock. This analogy throws considerable light on his positive theory so I think it is worth discussing in some detail. He refers to Tubal-Cain[5] or whoever made the first lock and key. Boyle says that when he made his first lock

> that was only a piece of iron contrived into such a shape; and when afterwards he made a key to that lock, that also in itself considered was nothing but a piece of iron of such a determinate figure: but in regard that these two pieces of iron might now be applied to one another after a certain manner, and that there was a congruity betwixt the wards of the lock and those of the key, the lock and the key did each of them now obtain a new capacity; and it became a main part of the *notion and description* of a lock, that it was capable of being made to lock or unlock by that piece of iron we call a key, and it was *looked upon* as a *peculiar faculty and power in the key*, that it was fitted to open and shut the lock; and yet by these new attributes there was not added any real or physical entity either to the lock or to the key, each of them remaining indeed nothing but the same piece of iron, just so shaped, as it was before... To carry this comparison a little farther, let me add, that though one that would have defined the first lock and the first key would have given them distinct definitions with reference to each other; and yet... these definitions being given but upon the score of certain respects, which the defined bodies had one to another, would not infer that these two iron instruments did physically differ otherwise than in the figure, size, or contrivement of the iron whereof each of them consisted. And proportionably hereunto, I do not see why we may not conceive, that as to those qualities (for instance) which we call sensible, though by virtue of a certain congruity or incongruity in point of figure (or texture or other mechanical attributes) to our sensories, the portions of matter they modify are enabled to produce various effects, upon whose account we make bodies to be endowed with qualities; yet they are not in the bodies that are endowed with them, any real or distinct entities, or differing from the matter itself, furnished with such a determinate bigness, shape, or other mechanical modifications. (*O.F.Q.* 18 my italics, *S.* 23–4)

I believe that this passage is to be interpreted in the following way. We can describe the lock's physical characteristics completely in terms of the shapes and sizes of the pieces of metal composing it and their relations to one another. We can similarly describe the key in terms of its shape and size. The lock and key are nothing beyond what can be so described. Now if the key is applied to the lock in a certain way we find that it will open and close it; we can add to the description of the lock and the key that one will open and close the other and this is a capacity or power not mentioned in our earlier descriptions. This then, becomes a 'main part of the *notion and description*' of the lock and key. However, at this point some people make the mistake of thinking that this part of the description indicates a peculiar faculty and power somehow omitted from the original description; this is how it is 'looked upon'. It is a mistake because these 'new attributes' do not 'add any real or physical entity' to the lock or the key. Attributes are things we attribute to something; there was nothing physical omitted from the original description. We do not necessarily 'add anything' to a body or point to something in it by attributing to it something that we did not formerly attribute to it.

To discover a power, in this situation, is not to discover any new part of a body but merely to discover what the parts previously known will do, that is, to discover that the description we formerly gave *implied* that power. The key and the lock, after we have discovered that one will open the other, are each of them 'nothing but the same piece of iron, just so shaped as it was before'; the key's having a certain shape in relation to the shape of the lock just is, among other things, its having a certain ability and its fitting the lock is nothing over and above that. Indeed, if we had formerly examined the complete description of lock and key we would have seen, without any trial, that the key had the power to open and close the lock. More than that, a locksmith in possession of the full description of the lock would discover, with certainty and without trial, what shaped key it required. The thing to which we newly attribute some power when we already know all its qualities has still only the qualities it previously had. I believe that Boyle is here attacking an idea of powers, held by some modern schoolmen and chemists, that makes powers occult. At the end of his previous section he has discussed the doctrine of real qualities (*O.F.Q.* 16–17, *S.* 20–3) which involves such an idea of powers.

The example of the lock and key is used by Boyle as an analogy for the more fundamental explanations of natural phenomena on the basis of the corpuscular philosophy and he immediately goes on to apply it in that way to explanations of colours, the solubility of gold and the power of poisons. When a body appears to us to be coloured this may be explained by the operation of its constituent corpuscles, by virtue of their shape, texture or other mechanical qualities, upon the corpuscles of light falling on it and then by the operation of this stream of light corpuscles, modified accordingly, upon the corpuscles composing our sense-organs, nerves and brain. Because of their effects upon our senses we take bodies to be really endowed with colours and other sensible qualities; 'yet they [*sc.* these sensible qualities] are not, in the bodies that are endowed with them, any real or distinct entities, or differing from the matter itself, furnished with such a determinate bigness, shape, or other mechanical modifications' (*O.F.Q.* 18, *S.* 23). As I shall show later, Boyle is here suggesting that there is no need to regard the corpuscles as having colours, in the sense in which they have shapes, since the appearance of colours to us can be explained by means of the interactions of uncoloured particles.

Goldsmiths contemporary with Boyle regard solubility in *aqua regia* and insolubility in *aqua fortis* as among the 'most distinguishing qualities of true gold', yet these attributes are not, in the gold, '*anything distinct from its peculiar texture*' (my italics). Pliny did not know about these qualities of gold but our concept of gold was not changed by our discovery of them. If we do not accept this we shall have to admit 'that a body may have an almost infinite number of new real entities accruing to it without the intervention of any physical change in the body itself' as each previously unknown quality is discovered (*O.F.Q.* 18–19, *S.* 23–4). If Pliny had known all the details of the 'peculiar texture' of gold and of *aqua regia* he could have known from that alone that gold would dissolve in *aqua regia*.

The final example concerns the lethal qualities of powdered glass. The power of 'poisons' is looked upon as a real quality and a particularly abstruse one. Yet, as Boyle says, 'this deleterious faculty, which is supposed to be a peculiar and superadded entity in the beaten glass, is nothing really distinct from the glass itself'. Its effect can be explained entirely in terms of the texture of the glass, together with the texture of human tissue.

It is important for my purpose to note here that there is a passage in Locke's *Essay* that bears a striking resemblance to this passage. He mentions in connected passages, just as Boyle does, the examples of the lock and the key, the appearance of objects as coloured, the solubility of gold and the effects of poisons (IV.iii.25).

There are two fundamentally important points involved in these passages from Boyle. The first is that they contain the basis of both his and Locke's belief in necessary connections in nature; full knowledge of a corpuscular structure would enable us to *deduce*, without trial, particular powers of interaction. The word 'deduce' was frequently used in the seventeenth century merely to mean 'infer', whether inductively or deductively, but when Boyle writes that various sensible phenomena are

> but the effects of the often-mentioned catholick affections of matter, and deducible from the size, shape, motion (or rest), posture, order, and the resulting texture of the insensible parts of bodies (*Q.F.Q.* 26, *S.* 36)

I think he is using 'deducible' in the narrower sense of 'deductively inferrable'. I believe that Locke also used it in this way. How else could they be using it when they say that a complete knowledge of corpuscular structure would enable us to *know*, without observation of effects, what those effects would be? I suspect that it is such views against which Hume is arguing when he denies that there are entailments in nature.

The second point is that these corpuscular structures, or *textures*, are *identified* in the example concerning colour, with qualities in bodies producing appearances of colours to us and, in the example concerning gold, with the ability of gold to dissolve in *aqua regia*. That is textures are identified with powers and, as I shall argue in more detail later, with secondary qualities and Locke's 'third sort' of qualities.

Boyle is here outlining a kind of explanation depending upon the distinction between primary and other qualities and upon the conception of texture. We can distinguish between qualities that are essential to matter and qualities that are not. Corpuscular structures, *i.e.* textures, are to be analysed entirely into qualities essential to matter and are to be the basis of the explanation of all qualities that are either inessential or merely apparent. The rest of Boyle's 'particulars' help to make this clear.

The fourth of these 'particulars' makes more explicit a criterion for deciding what are the 'inseparable accidents' of bodies. Bodies are composed of unobservable corpuscles the smallest of which are physically indivisible at least by us. If we imagine a body divided progressively into smaller and smaller parts we will reach a point at which physical division is no longer possible because there are no spaces between material parts into which a dividing instrument could penetrate; we have reached a single, simple corpuscle. There is no logical impossibility in conceiving all the corpuscles in the universe to be annihilated except one of these simplest ones so there is nothing unintelligible in supposing a universe containing just one corpuscle in otherwise empty space. This would still be a material body so it would have the defining properties of matter but need not have any property which can be exhaustively explained as arising from a complex collection of such corpuscles. The explanation of the colour of bodies already sketched is of this kind so we do not need to attribute colours to the corpuscles. The same is true of many other qualities that appear in observation to be possessed by bodies. In fact Boyle regards the essential properties, or primary qualities, as just those that a single corpuscle, alone in the universe, would have, and must have if it is to be material. Thus he says

> if we should conceive that all the rest of the universe were annihilated, except any [sc. any *one*] of these intire and undivided corpuscles ... it is hard to say what

could be attributed to it, besides matter, motion (or rest) bulk, and shape. Whence by the way you may take notice that bulk, though usually taken in a comparative sense, is in our sense an absolute thing, since a body would have it, though there were no other in the world. (*O.F.Q.* 22, *S.* 30)

It must be pointed out that unless one believes in absolute space as providing a frame of reference there would be no sense in saying that a single corpuscle, alone in the universe, was either at rest or in motion. There is little evidence that Boyle believed in absolute space in this sense and, indeed, the rest of this passage strongly suggests that he did not. This supports the view suggested earlier that by 'motion (or rest)' he meant 'mobility'; the lonely corpuscle without any change occurring in it would be found to be either at rest or in motion as soon as other corpuscles were introduced into the universe. There are problems about whether it makes sense to attribute even the primary qualities to a lone corpuscle but these will be discussed later. Boyle appears to have no qualms. It may be wondered why solidity is not mentioned here as one of the primary qualities; that will also, I hope, become clear as we proceed.

Boyle now turns to spatial relations. The universe in fact contains enormous numbers of corpuscles and so

there arise in any distinct portion of matter, which a number of them make up, two new accidents or events: the one doth more relate to each particular corpuscle in reference to the (really or supposedly) stable bodies about it, namely its posture (whether erected, inclined, or horizontal); and when two or more of such bodies are placed one by another, the manner of their being so placed, as one beside another, or one behind another, may be called their order ...(*O.F.Q.* 22, *S.* 30)

Posture and order, he says, seem reducible to situation.

He is surely taking it, in the first part of this passage, that it would not make sense to attribute situation or any spatial relation to a single corpuscle alone in the universe; they 'arise' only when it is related to other corpuscles. Situation is thus not an essential property of a body and Boyle is clearly rejecting the idea of absolute space as providing a frame of reference.

At the end of this 'particular' Boyle leads up to the next by saying

And when many corpuscles do so convene together as to compose any distinct body, as a stone or a metal ... there doth emerge a certain disposition or contrivance of parts in the whole, which we may call the texture of it. (*O.F.Q.* 22, *S.* 30)

There are three things in this passage to which I should like to draw attention. First, 'disposition' here simply means 'arrangement' and must not be thought of as referring to dispositional properties; Boyle has not yet reached that stage. Second, the use of the word 'emerge' is of historical interest; given that the individual corpuscles have only the three primary qualities, shape, size and mobility, he is here moving on to his account of the 'less simple qualities' to which he has already referred, that is, the non-primary qualities which belong to bodies 'on account of' the primary qualities, and these, he is now saying, 'emerge' in bodies more complex than the simplest corpuscles. The *Oxford English Dictionary* under 'emergent quality' suggests an origin in the nineteenth century but here there is the conception, if not the precise expression, in the seventeenth century. Third, we have here perhaps the clearest use so far of the word 'texture' as a technical term. Complex bodies are 'conventions' of corpuscles in specific arrangements and these arrangements are called 'textures'.

Boyle's account of primary qualities is now more or less complete so I shall recapitulate some of the central points, exposing some features previously unexposed, and reinforce my view of the importance of textures. The primary qualities are just those that any corpuscle, the smallest *actual* body, must have even if it is alone in the universe. Such a corpuscle is, of course, material and composed of whole matter with no empty spaces within it; it is, as he says, 'extended, divisible, and impenetrable'. This needs some explanation since 'divisible' and 'impenetrable' might be thought to conflict. They do not conflict, I believe, because 'divisible' means 'divisible, in thought or in principle' or 'mathematically divisible' while 'impenetrable' means 'impenetrable in practice' or 'physically impenetrable'. Anything with dimensions, that is, anything material must be mathematically divisible because any given length can be conceived of as halved, and so on. But, according to Boyle and most atomists, anything with dimensions is physically divisible only if there are empty spaces within it which can be penetrated by a dividing instrument. Anything without such spaces is physically indivisible or impenetrable. One reason why Boyle rejects the term 'atom' is that it had been taken to mean something logically or in principle indivisible as well as physically indivisible; for him, the fundamental particles, or corpuscles, are simply the smallest portions of matter which naturally exist or could be obtained by us by any physical means. God might be able to divide even these.

This interpretation is supported by what Boyle says elsewhere. For example, later in *The Origin of Forms and Qualities* he says of his smallest particle

> though it be mentally, and by divine Omnipotence divisible, yet by reason of its smallness and solidity, nature doth scarce ever actually divide it; and these may in this sense be called *minima* or *prima naturalia*. (*O.F.Q.* 29, *S.* 41)

I take solidity to be a central notion here. Each corpuscle, being nothing but matter, is *absolutely* solid. No two portions of whole matter can occupy the same portion of space at the same time. Solidity, in this sense, is absolute impenetrability and fullness. Bodies that we meet in experience are in practice divisible and penetrable but we call some of them solid. This, however, is *relative* solidity, perhaps better called hardness; such bodies are *more or less* solid and penetrable, harder or softer, because although they consist of absolutely solid corpuscles there are empty spaces between these constituents. They have 'pores' as Boyle puts it. They are in practice divisible and penetrable because the corpuscles of a finer and harder cutting instrument can work their way into the body's empty spaces and even push its corpuscles aside. In *The Sceptical Chymist* Boyle says

> there are few bodies, whose minute parts stick so close together, to what cause soever their combination be ascribed, but that it is possible to meet with some other body, whose small parts may get between them and so disjoin them... (*S.C.* 506, *E.* 87)

Now, what qualities must the smallest corpuscle have intrinsically, without reference to other corpuscles? Boyle argues that it must have shape, size and mobility. He actually says, in a passage quoted earlier (*O.F.Q.* 16, *S.* 20), that it must have just *three* essential properties. This he arrives at by conceiving of one such corpuscle alone in the universe. They are his primary qualities. It may be wondered, since he says that matter is not only extended and divisible but also solid, why he does not include solidity among the primary qualities or accidents. The reason is to be found, I believe, in the meaning he attaches to 'moods and accidents' or 'qualities'. Qualities or accidents are what differentiate, and sometimes allow us to distinguish, one body from another. Thus Boyle says

> bodies having but one common matter can be differenced but by accidents... (*O.F.Q.* 35, *S.* 49)

and

> all bodies thus agreeing in the same common matter, their distinction is to be taken from those accidents that do diversify it. (*O.F.Q.* 35, *S.* 50)

Since the matter of all bodies is the same, two corpuscles cannot be differentiated by differences in solidity; their matter is equally solid because it is absolutely solid. So although solidity is essential to matter it is not a quality or accident of it. What will differentiate the simplest bodies, different portions of matter, is differences in shape, size and motion-or-rest. Since these are the only things that could differentiate single corpuscles from one another, they are the primary qualities. They are, moreover, intrinsic to any body whatsoever, whether simple or complex, considered in isolation from other bodies.

In our universe there are many complex bodies made up of 'conventions' of simplest corpuscles. The spatial relations between these corpuscles are part of the structure or arrangement of the corpuscles which, themselves, have only the three primary qualities. This arrangement in a particular convention or body is what Boyle calls the 'texture' of a body; it can be described entirely in terms of the shapes, sizes and motions (including rest) of the corpuscles and the spatial relations between them. It is to be noted that spatial relations *between* corpuscles are, as bare extension, exactly like spatial relations between points in corpuscles so can be considered as primary qualities in the description of conventions. Strictly a single corpuscle has no texture since 'texture' is defined as a structure of a collection of corpuscles. This accords with the idea that textures are responsible for the observable interactions of bodies with us and with other bodies. Single corpuscles have no directly observable effects; their only effects arise from collisions with one another, and penetrations of groups of them.

The felt texture of body, its feeling smooth or rough, granular or oily, may give little direct guidance to its texture in the technical sense; that is partly because felt texture depends upon the texture of the part of the body doing the feeling as well as upon that of the body felt, whereas texture in the fundamental sense does not, and partly because texture, in that sense, is too minute to be felt, as it were, directly. Textures in the fundamental sense are structures of objects that cause, and correspond to, sensations of colours, sounds, odours, tastes and 'feels' in us and that cause changes in inanimate objects by interacting purely mechanically with the textures of those objects. Textures are intrinsic to complex bodies but they are not primary qualities because they are not intrinsic to the simplest bodies, individual corpuscles. Textures can be analysed into, or described in terms of, primary qualities.

The following quotations support this interpretation.

from these more catholic and fruitful accidents [*sc.* the bulk and figure of the smallest parts] of the elementary matter may spring a great variety of textures, upon whose account a multitude of compound bodies may very much differ from one another (*S.C.* 476, *E.* 33)

however we look upon them [*sc.* colours, etc.] as distinct qualities, [they] are consequently but the effects of the often mentioned catholick affections of matter, and deducible from the size, shape, motion (or rest) posture, order, and the *resulting* texture of the insensible parts of bodies (*O.F.Q.* 26 my italics, *S.* 36–7) the aggregate of these corpuscles may be further diversified by the texture resulting from their convention into a body, which, as so made up, has its own bigness, and shape, and pores ... (*E.G.* 70, *S.* 142)

there can be no ingredient assigned, that has a real existence in nature, that may not be derived either immediately, or by a row of decompositions, from the universal matter, modified by its mechanical affections (*E.G.* 75, *S.* 149–50)[6]

It is important to note that Boyle refers to the 'resulting' texture. This is not a casual notion. He is saying that the texture is a *resultant* of an arrangement of particles in something like the sense in which we talk of a resultant of two opposed forces; the resultant can be resolved into (or analysed into) two or more components that are jointly equivalent to it. It is important that a description of a texture would also involve descriptions of relative *motions* of corpuscles.

After this digression I return to Boyle's 'particulars' outlining the corpuscular philosophy. In his fifth he moves on to other qualities of complex bodies or qualities they may appear to have, especially in relation to other bodies and to our sense-organs. If the universe contained just one complex body, such as a metal or a stone, it would be 'hard to shew that there is physically any thing more in it than matter and the accidents we have already named' (*O.F.Q.* 22, *S.* 30). But our world contains, besides such bodies, men, that is, rational beings with sense-organs which can be acted upon by the textures of bodies outside them. As a result we see objects as coloured, we smell them, and so on, and we call 'these operations of the objects on the sensories' colours, sounds, odours, etc. Because sense-experience occurs even before reasoning we have from infancy been 'apt to imagine that these sensible qualities [*sc.* green, blue, sweet, bitter, etc.] are real beings in the objects they denominate' (*O.F.Q.* 23, *S.* 31). However

there is in the body, to which these sensible qualities are attributed, nothing of real and physical but the size, shape, and motion or rest, of its component particles, together with that texture of the whole, which results from their being so contrived as they are; nor is it necessary that they should have in them any thing more, like to the ideas they occasion in us, those ideas either being the effect of our prejudices or inconsiderateness, or else to be fetched from the relation that happens to be betwixt those primary accidents of the sensible object and the peculiar texture of the organ it affects: as when a pin being run into my finger causeth pain, there is no distinct quality in the pin answerable to what I am apt to fancy pain to be ... (*O.F.Q.* 23, *S.* 31)

Here Boyle probably uses 'sensible qualities' to mean, not qualities in objects but their effects on us; that is, to mean what Locke means by 'ideas' when he uses the word in such contexts. Earlier in the passage

Boyle talks of 'such things, as, *for the relating to our senses*, we call sensible qualities' (*O.F.Q.* 23, my italics, *S.* 31) and he also talks of sensible qualities as being 'attributed' to bodies. Later he refers to them as 'ideas' and compares them to the pain felt when a pin pricks a finger. Locke, of course, uses the same analogy.

Another important connection with Locke is that Boyle is arguing that the colour we see is an apparent or sensible quality of the object not resembling any quality in the object that is independent of our perceiving. The force of the expression that it is not *necessary* to suppose a quality in the object resembling the colour we see is that the colour we see can be accounted for in terms of the corpuscular structure of the objects, as Boyle goes on to explain.

Light is regarded also as corpuscular and the effect of a body on a ray of light is to reflect and absorb different light corpuscles depending upon both the texture of the light and the surface texture of the body. The texture of the reflected light acts upon the texture of our eyes, nerves and brain in such a way that the object looks blue or red or some other colour. (See *E.C.T.C.* 674, *F.* 21.) The actual colour it looks is just a sensible quality or idea.

We can see how similar explanations might plausibly be thought possible for our perception of sounds, tastes, smells and 'feels'. If one object strikes another and we hear a sound of a particular pitch this is because the textures of the bodies, including the motions of their corpuscles, are modified and these modifications are transmitted by means of corresponding modifications of the texture of the air surrounding the bodies to the ear. Modifications in the texture of the sensitive parts of the ear are transmitted in turn by the nervous system to the brain and then we hear a particular note. We taste a particular substance because its texture modifies the texture of our tongues, nervous systems and brains and at the end of this process we experience a salty or a bitter taste. It was proposed by some atomists that a sour taste might be caused by sharply pointed particles in the substance but that allows far less flexibility and variety than an account that involves a texture, depending upon the shape, size and motion of a collection of particles rather than just one of these. Such an explanation would no doubt depend in part upon the fit or failure of fit between the texture of the substance and the texture of the tongue. The smell of a flower could be accounted for by its giving off minute collections of corpuscles in a specific pattern which modify the texture of the sensitive parts of the nose until, the pattern having reached the brain, we smell the scent of a rose or a violet.

Felt textures are, as I have said, not the same as the corpuscular patterns referred to by the technical use of 'textures'. The smooth, soft 'feel' of velvet we experience when we say we feel its texture depends immediately upon the many small flexible filaments that make up the pile; but it depends ultimately upon the pattern of corpuscles in each filament, which patterns we cannot distinguish by touch. These corpuscular textures modify the textures of our fingers, or other tactile organs and this sets up a train of modifications that eventually lead to our feeling softness and smoothness.

Boyle next anticipates objections and in the course of it supports the interpretation I am suggesting. It will be said that it is evident that colours 'have an Absolute being irrelative to us, since snow would be white even if there were no living creatures, and, moreover, bodies work upon other bodies by such qualities as heat, as when wax is melted by a fire'. He says

> I do not deny but that bodies may be said in a very favourable sense to have those qualities we call sensible, though there were no animals in the world: for a body in that case may differ from those bodies which now are quite devoid of quality, in its having such a disposition of its constituent corpuscles, that in case it were duly applied to the sensory of an animal, it would produce such a sensible quality which a body of another texture would not: as though if there were no animals there would be no such thing as pain, yet a pin may, upon the account of its figure, be fitted to cause pain, in case it were moved against a man's finger ... (*O.F.Q.* 24, *S.* 33)

That is, colours are no more really in bodies than pain is in a pin. The difference between a coloured and a colourless body is simply that the coloured body has a certain texture differing from that of the colourless body in such a way that the first can cause colours to appear to us while the second cannot. We can properly call an object, say, red but we should realize that what this must mean is that it has a certain texture suitable, in conjunction with light of a given texture, to produce a sensation of red in us. It is to be noted that the word 'disposition' in this passage just means 'arrangement'.

Boyle says

> if there were no sensitive beings those bodies that are now the objects of our senses would be but dispositively [*sc.* dispositionally], if I may so speak, endowed with colours, tastes, and the like; and actually but only with those more catholick affections of bodies, figure, motion, texture, etc. (*O.F.Q.* 25, *S.* 34)

Moreover

> when one inanimate body works upon another, there is nothing really produced by the agent in the patient, save some local motion of its parts or some change of texture consequent upon that motion: and so, if the patient come to have any sensible quality that it had not before, it acquires it upon the same account upon which other bodies have it, and it is but a consequent to this mechanical change of

texture, that, by means of its effects upon our organs of sense, we are induced to attribute this or that sensible quality to it. (*O.F.Q.* 25, *S.* 35)

Thus there would be no colours, as we perceive them, without human beings and other animals but if all animals were annihilated this would not change anything in bodies we see as coloured; there would remain in them the powers to produce sensations of colours in animals, if there were any, these powers residing in, or being, the textures of the bodies. To say that a body melts wax, is not to say that its heat, as we sense it, is in the body; it is to say that its corpuscles are in rapid motion, which motion is communicated to the corpuscles of the wax so that 'their agitation surmounts their cohesion' and the wax becomes fluid (*O.F.Q.* 26, *S.* 36).

I shall argue later that the doctrine being put forward here, and in Locke's *Essay*, may be summarized as follows, in spite of some waywardness of expression on the part of both Boyle and Locke. Bodies have three sorts of qualities, primary, secondary and a third sort. All these qualities are really qualities of bodies independent of our perceptions; they produce in us sensations or ideas or the so-called 'sensible qualities'. What we see as colours are, in bodies, textures. The qualities by means of which inanimate bodies act upon one another, the third sort, are also textures. Thus bodies have 'powers' to affect us and other bodies and these powers are textures or secondary and the third sort of qualities. They are complex textures of conventions of corpuscles, analysable into the primary qualities of those conventions. At this stage there may appear to be both scholarly and philosophical problems attached to this interpretation but I hope eventually to remove most of them.

Boyle now moves on to put some arguments against substantial forms which I have already mentioned. However men may use substantial forms in their theorizing, he says, in practice they rely upon observable collections of accidents or qualities for distinguishing between bodies and for classifying them in genera and species. Although qualities are 'but accidents', in the sense I have explained, they may be essential to a body. An accident, without being essential to matter, may be essential to this or that particular body; sphericity is not essential to brass but it is essential to a brass sphere. A convention of observable qualities may be enough to define a particular kind of body but those qualities 'proceed from those more primary and catholick affections of matter: bulk, shape, motion, or rest, and the texture thence resulting'. The form of a body, being made up of that particular convention of qualities, may likewise be analysed into those primary affections. This indicates how Boyle proposes to use, for brevity, the word 'form' in future (*O.F.Q.* 28–9, *S.* 40). We shall see Locke arguing in a similar way to Boyle about classifying things into genera and species.

Natural things, Boyle says, 'for the most part operate by their qualities'. He has now outlined a general account of forms and qualities which, he believes, does not leave them occult and mysterious. Forms are merely conventions of qualities and qualities are, or are analysable into, the primary qualities of the constituent corpuscles of a body and the body's resultant texture. He thinks that closer observation and better controlled experiments will eventually enable us to discover and describe those textures. Already, he believes, many of his observations, such as the behaviour of shot silk or, as he calls it 'changeable taffity', under various lighting conditions, and many of his experiments, especially chemical ones involving precipitation and sublimation, give strong support to the general theory of the corpuscular structure of bodies.

Boyle's final 'particular' deals with generation, corruption and alteration and introduces the notion of relatively stable clusters of corpuscles, roughly comparable to the more modern idea of molecules. A handleable sample of a substance such as common salt or mercury is, he thinks, an aggregation of relatively stable clusters which are unobservable and, unlike the smallest corpuscles of which each is composed, the smallest possible bits of salt or mercury. These Boyle also calls 'corpuscles'. When individual smallest corpuscles join such a cluster the texture of the cluster will almost certainly be altered and so will its capacity to act upon other bodies. In this change 'it' will 'acquire a congruity to the pores of some bodies (and perhaps some of our sensories) and become incongruous to those of others' (*O.F.Q.* 30, *S.* 42). This idea of the congruity or incongruity of the particles of a body with the 'pores' of another is the basis of Boyle's theory of chemical interaction and of perception.

If a body belonging to a particular species is 'generated or produced *de novo*' it is

> not that there is really any thing of substantial produced, but that those parts of matter that did indeed before pre-exist, but were either scattered and shared among other bodies, or at least otherwise disposed of, are now brought together and disposed of after the manner requisite to entitle the body that results from them to a new denomination, and make it appertain to such a determinate species of natural bodies, so that no new substance is in generation produced, but only that which was pre-existent obtains a new modification or manner of existence. (*O.F.Q.* 32, *S.* 45)

That is, no new matter is created in generation. When sand and ashes are melted together and cooled glass is formed. This process is simply a rearrangement of the corpuscles of the sand and ashes, that is, it is a change of texture.

Since no natural body is able to annihilate matter, corruption is simply the reverse process of reorganization, a scattering of corpuscles formerly clustered together. Putrefaction is merely a kind of slow corruption in which the body's texture is changed by the air 'or some other ambient fluid' entering the pores of the body and, by its agitation, loosening the less cohesive of its corpuscles. Although this is often looked upon as an 'impairing' alteration that is because it is often accompanied by a foul smell which is unpleasant and useless to us, but some putrefactions produce substances which are useful and not unpleasant; musk, obtained from certain decaying animals, is used as a perfume and blue cheese is regarded as a delicacy. These are matters of taste; all such changes are of the same general nature (*O.F.Q.* 33–4, *S.* 47–8).

This account admits the possibility of transmutation and, indeed, 'almost . . . any thing, may at length be made any thing'. Since all bodies have matter in common and differences between them depend upon their textures, an appropriate series of rearrangements of the fundamental corpuscles of, say, lead may eventually produce gold (*O.F.Q.* 35, *S.* 50). Boyle introduces the reservation '*almost* any thing' because corpuscles differ in shape and size and two bodies made up of differently shaped corpuscles, for instance, could not be transmuted into one another simply by rearrangement. One implication of this account is that it is a mistake to think of transmutation, in the way in which it was frequently thought of, as the direct transferring of real qualities, such as that of yellowness, or of forms, such as that of gold, from one body to another as if they were substances or ingredients. It is more intelligible if it is thought of as a rearrangement of fundamental corpuscles whose qualities remain unchanged.

An important question that arises from all this concerns the status of secondary qualities. What precisely are they? Boyle often talks of colours, sounds, odours and tastes as 'sensible qualities' and this is often taken to mean that they are secondary qualities. I believe this to be a mistake. I believe that he intends 'sensible qualities' to refer to the effects of bodies on us, in the form of sensations. Corresponding to and causing such effects, however, there are qualities really in the bodies and independent of our sensations. Our common-sense mistake is to suppose that the quality in a body corresponding to a given sensation is the same as, or like, its effect on us, that is, for example, that colours, *as we see them*, are really qualities of bodies. In his major work on colour Boyle says something very like what he says in the passage quoted earlier in this chapter from *O.F.Q.* 24 and 25, *S.* 33–5. Thus

after all I have said of colour, as it is modified light, and immediately affects the sensory, I shall now remind you, that I did not deny, but that colour might in some sense be considered as a quality residing in the body that is said to be coloured, and indeed the greatest part of the following experiments refer to colour principally under that notion, for there is in the bodies we call coloured, and chiefly in their superficial parts, a certain disposition, whereby they do trouble the light that comes from them to our eye, as that it there makes the distinct impression, upon whose account we say, that the seen body is either white or black, or red or yellow, or any one determinate colour ... we shall (God permitting) ... shew, that the changes and consequently in divers places the production and the appearance of colours depends upon the continuing or altered texture of the object. (*E.C.T.C.* 674, *F.* 21)

In so far as qualities are in bodies, as Boyle clearly says they are, they must all, whether primary, secondary or of the third sort, be modifications of matter, or the textures formed by the primary qualities, and arrangements, of the constituent corpuscles. Non-primary qualities, therefore, are textures, as I have said. Secondary qualities cause sensations in us; colours, sounds, odours and tastes are sensations and therefore correctly called *not* secondary qualities but the effects of secondary qualities on us. So 'red', 'bitter' and 'pungent' do not name secondary qualities but sensations caused by them. The word 'sensible' in Boyle's 'sensible qualities' means, I suggest, not 'able to be sensed' but 'as sensed'.

The fundamental corpuscles, according to this account, are individually unobservable, in principle and not merely in practice, because they do not have textures. Only groups of them have textures. An observable body must have a texture both because it must be large enough to be observed and because observability depends upon secondary qualities. The corpuscles of a body may have empty spaces between them or they may touch one another and cling together (*O.F.Q.* 34, *S.* 49). In terms of this the empirical hardness or softness of bodies and their physical and chemical stability or instability are to be explained. The farther apart the constituent corpuscles of a body are the softer and less stable it is.

As I have said, texture is fundamental for the explanation of chemical interaction. The more modern idea of *chemical* entities with distinctive chemical properties was still in the course of development and there was nothing like the modern idea of chemical interaction. The solubility of gold in *aqua regia* was to be explained in terms of the corpuscles of *aqua regia* being of the right size and shape to fit regularly between the corpuscles of gold, that is, into the *pores* of gold. Exchange reactions such as that between silver nitrate and hydrochloric acid to give a precipitate of silver chloride, i.e.

$$AgNO_3 + HCl \rightarrow AgCl\downarrow + HNO_3,$$

could be explained by saying, for example, that the particles of silver were a better fit in the pores between the corpuscles of chlorine than they were between those of the nitrate radical and the fit was so good that no interaction was possible between the nitric acid and silver chloride formed, so the silver chloride was precipitated. This explanation is, of course, anachronistic in its details but it is of the form envisaged by Boyle. Similar forms of explanation, it was hoped, would serve for all chemical reactions.

We would regard such explanations nowadays as physical rather than chemical. The theory was intended to be as far as possible geometrical, or at most kinematical, rather than dynamical, an idea that had wide currency during the seventeenth century. Boyle says in his 'New Experiments Physico-Mechanical Touching the Spring of Air' (1660) that he has realized 'the usefulness of speculative geometry to natural philosophy'.[7] The advantage of explanations of this kind is that they avoid talk of affinities and antipathies, of attraction and repulsion between bodies; Boyle regarded such talk as occult. His theory allows a conception of interaction that can be visualized and even modelled. The relation of this to the example of the lock and the key should be obvious. For this reason, Boyle retained the old idea of 'effluvia', streams of minute particles issuing from magnets and electrified bodies to account for their action on iron filings and fragments of paper; they act by the mechanical collision of particles rather than by mysterious forces and fields propagated somehow over a distance through empty space. Boyle was by no means clear how such explanations would work in detail but they fit his idea of explaining what is not understood by deducing it from something else in nature better understood (*O.F.Q.* 46, *S.* 67).

It seems strange to us nowadays to have so general a theory involving the interaction of minute particles which does not anywhere rely upon forces between them. This all depends upon Boyle's anxiety not to rely upon occult entities, qualities or influences. Forces and their operation could not be observed but only, if they were considered at all, their effects. Moreover, the idea that material bodies could act upon one another at a distance was widely regarded as unintelligible because no mechanism could be given or even conceived. It is salutary to remember that even Newton had such doubts when he introduced forces into his system of mechanics[8] and Locke appears never to have been fully convinced of their intelligibility.

Boyle perhaps makes his attitude to explanation clearest when he talks about corrosiveness. He says

> The attributes, that seem the most proper to qualify a liquor to be corrosive, are all of them mechanical, being such as are these, that follow: First, that the menstruum consist of, or abound with corpuscles not too big to get in at the pores or

commissures of the body to be dissolved; nor yet be so very minute, as to pass through them, as the beams of light do through glass; or to be unable, by reason of their great slenderness and flexibility, to disjoin the parts they invade.
Secondly, that these corpuscles be of a shape fitting them to insinuate themselves, more or less, into the pores or commissures above-mentioned, in order to the dissociating of the solid parts.
Thirdly, that they have a competent degree of solidity...
Fourthly, that the corpuscles of the menstruum be agile and advantaged for motion...[9]

It is plausible to suppose that Boyle was relating chemistry to physics in an attempt to make chemistry a respectable part of natural philosophy in a way that the conceptions of the peripatetics and alchemists could not. The development of chemistry during the next century suggests that he succeeded in this aim.

I have chosen to restrict myself, in this first part, to those of Boyle's ideas that I take to have had a direct or indirect influence on central ideas in Locke's *Essay*; I have not dealt with all the ideas that Boyle and Locke held in common. It may nevertheless appear that I have not used explicitly in Part II some of the material in Part I; I hope the reader will be able to make more connections than I have explicitly made. My aim in Part I has been to give as coherent and comprehensive an account of Boyle's views as I am able in the compass of this book in order to make clear the character of his epistemology and methodology and of his theory of the natural world.[10] The influences of Boyle upon Locke were, I claim, both general and particular.

When I speak of 'general' influences I have in mind such matters as Boyle's anti-Aristotelianism and his related views about scientific methods of investigation and acceptable forms of explanation, which I believe affected the general tenor of Locke's *Essay*. As I have indicated, Boyle's anti-Aristotelianism is complex because he is attacking features he regards as common to scholasticism and alchemy and because he accepts, as does Locke, certain ideas derived from Aristotle while attacking Aristotle's followers. I have tried to give a clear exposition of Aristotelian views of forms and to account for implausible extensions of them in the hands of the scholastics because Boyle and Locke were rejecting central features of these views that they found currently in favour. I have tried to explain how such views, upon which Boyle poured such scorn, could have had so wide a currency among people he thought worthy of attack. He was not attacking straw men nor unintelligent men and the formulation of a critique of the faults found in their thought by Boyle and Locke did much to shape their own accounts of the structure of the world and our knowledge of it.

The rejection by Boyle and Locke of 'explanation' in terms of occult entities was largely what commended a corpuscularian theory to them and was partly responsible for Locke's worries about substance-in-general and for his preference for an account of human experience and knowledge in terms of '*ideas*' having an empirical basis. This epistemological atomism is a fitting accompaniment to a corpuscularian view of the structure of nature; both seek explanations that can be given in terms of simple and intelligible units and both seek to provide an account of what 'intelligible' can mean. Boyle's consideration of particular observations and experiments and of methods of scientific investigation in *The Sceptical Chymist* are largely directed to showing what constitutes intelligible explanation and how empirical confirmability bears on it.

I have occasionally gone into details about Boyle's accounts of experiments partly because it appears that Locke drew heavily on these for his examples and partly in order to expose Boyle's tendency to argue that observational and experimental results may be interpreted, or explained, in radically different ways from those that seemed obviously acceptable to many of his contemporaries. Locke, in his philosophical arguments, has a tendency to argue in a similar way against 'vulgar' or everyday interpretations of natural phenomena and our experience of them. This is clearly seen in some of his arguments about ideas, qualities, substance and language.

When I speak of 'particular' influences of Boyle upon Locke I have in mind mainly matters that arise from Boyle's corpuscular philosophy and the possibility of explaining at least all physical phenomena in purely mechanical terms. For example, Boyle's conception of a common matter in all bodies, his technical sense of 'texture', treatment of qualities and the distinction between kinds of qualities were all, I believe, used directly by Locke. It is with these that I mainly deal in Part II because of the important implications he saw them to have for his account of experience and knowledge. They constitute a connecting thread running through the whole of the *Essay* but it must never be forgotten that all this was developed in opposition to very different conceptions of scientific investigation and explanation. It is those conceptions and the reasons for rejecting them that I have tried to make clear in Part I.

PART II

4

IDEAS

I

The word 'idea' figures very largely in Locke's philosophy and his various uses of it have seemed an easy target to commentators. From Stillingfleet onwards critics have regarded him as radically confused, careless and inconsistent.[1] Stillingfleet pours scorn on the 'way of ideas' and Locke replies that Stillingfleet has neither read him properly nor understood what he has read. John Yolton has made it clear that a number of Locke's contemporaries regarded his use of 'idea' as unfamiliar, technical or obscure.[2] I am inclined to think, and wish to argue, that the central use of the word has been little understood either by his or our contemporaries because it is indeed a technical use which Locke did not sufficiently explain. Yet the materials for an explanation are to be found in the *Essay*.

Locke opens Book I by saying that he proposes to examine the understanding since it is this that 'sets Man above the rest of sensible Beings' (I.i.1). His purpose is 'to inquire into the Original, Certainty and Extent of humane Knowledge; together with the Grounds and Degrees of Belief, Opinion, and Assent...' (I.i.2). He continues

> I shall not at present meddle with the Physical Consideration of the Mind; or trouble my self to examine, wherein its Essence consists, or by what Motions of our Spirits, or Alterations of our Bodies, we come to have any Sensation by our Organs, or any *Ideas* in our Understandings; and whether those *Ideas* do in their Formation, any, or all of them, depend on Matter, or no. (I.i.2)

That is, he is leaving to the natural philosophers the task of examining the physical mechanisms related to the workings of the mind. What he hopes to do is, by examining 'the discerning Faculties of a Man', to discover how far knowledge is possible, what are the grounds of belief and the necessary conditions of knowledge and why men hold so tenaciously to different and contradictory opinions. There are distinct similarities with the way in which Descartes saw his own task and distinct differences between their approaches to it.

Despite Locke's disclaimer, however, it seems clear that underlying his investigations there are firmly held assumptions drawn from contemporary natural philosophers, especially corpuscularians, concerning general types of mechanism involved in perception and the formation of ideas.

He begins by outlining what is in fact an important part of the project of the whole *Essay*. This concerns the distinction between knowledge and opinion and the grounds of rational assent. He says

First, I shall enquire into the *Original* of those *Ideas*, Notions, or whatever else you please to call them, which a Man observes, and is conscious to himself he has in his Mind; and the ways whereby the Understanding comes to be furnished with them.
Secondly, I shall endeavour to shew, what *Knowledge* the Understanding hath by those *Ideas*; and the Certainty, Evidence, and Extent of it.
Thirdly, I shall make some Enquiry into the Nature and Grounds of *Faith*, or *Opinion*: whereby I mean that Assent, which we give to any Proposition as true, of whose Truth yet we have no certain Knowledge: And here we shall have Occasion to examine the Reasons and Degrees of *Assent*. (I.i.3)

The first two Books of the *Essay* are largely concerned with the first item in this project. I shall deal hardly at all with Book I since this consists mainly of arguments against the view that we have innate ideas and Locke himself attaches less importance to these arguments than to the rest of the *Essay* which, he believes, shows that the postulation of innate ideas is unnecessary for an account of our knowledge.

At the end of his first chapter Locke gives his first explanation of his use of the word '*Idea*' thus

It being that Term, which, I think, serves best to stand for whatsoever is the Object of the Understanding when a Man thinks, I have used it to express whatever is meant by *Phantasm, Notion, Species*, or whatever it is, which the Mind can be employ'd about in thinking... (I.i.8)

'Phantasm', 'Notion', and 'Species' mean three different things although commentators have seldom made much of this. I believe that 'Phantasm' for Locke means 'mental image or percept without present external cause' and that 'Notion' means 'concept'.[3] 'Species' is more complex and I shall have more to say about it shortly but here it will perhaps suffice to say that it primarily means 'appearance of an external object or quality to the mind' and that its most central referent for Locke is what he calls 'simple idea of sensation'.[4] As we shall see, Locke quite consciously and intentionally uses 'idea' for at least these three things. When I see a yellow book I have, according to him, a simple idea of yellow and a complex idea of a book and both are ideas of sensation. When I talk of my idea of democracy or jealousy I am using the word for concepts. A vivid memory of the yellow of the book or a vivid imagination of a strange animal that I have never seen may, at least in some people, be a mental image or picture, although it need not be.

An idea is 'the Object of the Understanding when a Man thinks', so it is presumably anything that a man can operate with in thinking. There is a possible ambiguity in Locke's statement: I can think about St Paul's Cathedral but St Paul's Cathedral is not, even for Locke, an idea. However, what I operate with or manipulate when I do so is not the building but, according to Locke, my idea of it. I can think about St Paul's Cathedral as I see it or as I remember it and in both cases I am having and operating with ideas produced by it. I am doing this when I compare St Paul's Cathedral with an office building across the road while looking at them or while remembering them later. While looking at it I am operating with 'species'; while remembering it I am operating with a 'phantasm' if I am having a mental picture of it; and with a 'notion' if I am considering it merely as 'a domed stone building with turrets and columns' without picturing it. I shall return to this but I should perhaps say here that the central and most fundamental use of 'idea' for Locke is to mean, 'whatever we are justified in meaning by *species*'.

At the beginning of Book II Locke gives a list of examples of words standing for ideas: '*Whiteness, Hardness, Sweetness, Thinking, Motion, Man, Elephant, Army, Drunkenness*...' (II.i.1) and then says that all our ideas come from experience. He says of experience

In that, all our Knowledge is founded; and from that it ultimately derives it self. Our Observation employ'd either about *external, sensible Objects; or about the internal Operations of our Minds, perceived and reflected on by our selves, is that, which supplies our Understandings with all the materials of thinking.* (II.i.2)

There are two points in this passage to which I wish to draw attention. The first is that the expressions 'is founded', 'ultimately derives it self' and 'the materials of thinking' leave open, as I believe Locke intended, the possibility that we may justifiably think about things that we do not or cannot observe, as long as the ideas we operate with are at least indirectly derived from, or related somehow to, ideas we have in observation. Locke is allowing that we may think and talk sensibly about, for example, unobservable material corpuscles.

The second, and perhaps even more important, point is that our observation is sometimes 'employ'd about' external sensible objects. He engages in no argument, here or elsewhere, to justify his reference to external sensible objects and he immediately goes on to distinguish between ideas of sensation and ideas of reflection on the grounds that the former are caused by things outside us and the latter by things inside us. Thus he says

First, *Our Senses*, conversant about particular sensible Objects, do *convey into the Mind*, several distinct *Perceptions* of things, according to those various ways, wherein those Objects do affect them: And thus we come by those *Ideas*, we have of *Yellow, White, Heat, Cold, Soft, Hard, Bitter, Sweet,* and all those which we call sensible qualities, which when I say the senses convey into the mind, I mean, they from external Objects convey into the mind what produces there those *Perceptions.* This great source of most of the *Ideas* we have, depending wholly upon our Senses, and derived by them to the Understanding, I call *SENSATION.* (II.i.3)

Later in the same chapter he talks of the impressions made on our senses 'by outward Objects' (II.i.23) that are 'extrinsical to the Mind' (II.i.24) and says that with respect to the resulting ideas the understanding is passive (II.i.25).

It is clear that from the beginning Locke accepts the assumptions of the natural philosophers he favours that there are external objects, independent of minds, which natural philosophy investigates and which cause ideas of sensation in us. It may very well be that this is one of the 'principles' to which he refers when he says, at the end of Book I,

I warn the Reader not to expect undeniable cogent demonstrations, unless I may be allow'd the Privilege, not seldom assumed by others, to take my Principles for granted; and then, I doubt not, but I can demonstrate too. (I.iv.25)

If I am right about this then although it may be justifiable to criticize his assumption of external objects it cannot be justifiable, given that this *is* an assumption, to criticize him, as is so often done, as if he were trying to *infer* external objects from ideas. I believe that he sees the only possible justification of the assumption as lying in the comprehensiveness and adequacy to our knowledge and experience of the world of the overall account he bases upon it. The assumptions are among the premisses of a hypothetico-deductive system which may be justified as a whole by its explanatory success.

The last sentence of Book II, Chapter i reads

As the Bodies that surround us, do diversly affect our Organs, the mind is forced to receive the Impressions; and cannot avoid the Perception of those *Ideas* that are annexed to them. (II.i.25)

He makes it clear in several places that the mind can neither create nor destroy simple ideas. This may be intended as a supporting argument: where the mind has no control over its ideas it cannot be the cause of them; their cause must be something independent of it.

One feature of that passage that is worthy of note at this point is that Locke is distinguishing between *impressions* and the perception of ideas for he says that the ideas are 'annexed to' the impressions. This points to a feature of the early part of the *Essay* that critics have largely ignored. He makes certain distinctions between various terms relating to sense-experience which critics regard him as having conflated. It is true that he added to the confusion later by not sticking clearly to the distinctions he makes here but the later passages can often be understood as involving the distinctions although this is obscured by carelessness of expression.

This makes it at best incautious to say, as D. M. Armstrong does,[5] that for Locke 'idea' covers, at least, (a) sense-perceptions (sense-impressions); (b) bodily sensations (such things as pains and tickles); (c) mental images; (d) thoughts and concepts, and that this leads Locke into many errors.

For one thing, Locke sometimes distinguishes 'impressions', on the one hand, from 'perceptions' and 'ideas', on the other. He says that 'Sensation', that is, the faculty, is such an Impression or Motion, made in some part of the Body, as produces some Perception in the Understanding (II.i.23). He talks about '*Ideas* got by *Sensation*' (II.i.24) and then in the next section, quoted above, talks of the perception of ideas *annexed* to impressions.

It is true that problems about the mind/body relation, as Locke later points out, infect these statements but at least the official doctrine appears to be that external objects affect the sense-organs by making physical impressions on them, as a seal does on wax, stamping them with certain patterns. These patterns, impressions, are conveyed *via* the nerves to the brain, upon which ideas or perceptions mysteriously occur in the mind. Thus impressions are bodily; ideas or perceptions are mental. The sentence 'The Mind is forced to receive the Impressions' (II.i.25), which seems to conflict with this, may be interpreted as meaning that the mind cannot avoid being affected by the impressions in the brain, that is, cannot resist the formation of the correlated idea. When he used these words in ways that appear to conflict with this account it is instructive to attempt to read him as if he were consistently adhering to it. What he says often makes better sense when we do this.

Locke certainly uses 'idea' for such things as pains, tickles, and other bodily sensations but he does not *confuse* ideas like those of square or red with ideas which are pains. Indeed, he suggests that because of the character of these two groups of ideas we are misled into thinking that, for example, colours are less like pains than they really are. Their similarity is not phenomenological but epistemological and causal.

There is little evidence that Locke uses 'idea' frequently to mean 'mental image'. He sometimes does when he is talking of memory. It is particularly misleading to suppose, as Berkeley and many other critics appear to, that he means this when he talks of concepts or ideas of sensation. What is 'in the mind' need not be, usually is not, a mental image. He seldom uses the word 'image'. In two places he compares ideas with mirror images, once to point out that mirror images disappear, just as some ideas do, when their objects are removed (II.i.15) and once to point out that minds can't choose the simple ideas they have, any more than mirrors can choose the images they have (II.i.25). It does not follow that ideas are like 'pictures in the head' as mirror images are like pictures in a mirror. It may, however, be important, as we shall see, that mirror images can be pointed at, just as their objects can. (If a mirror could point it would point away from itself, whether it was pointing at the object or the image.)

He once talks of images as 'resemblances' but this is in the course of *denying* that ideas are all resemblances or images (II.viii.7) and once of images as 'representations' of what exists but, here again, he is denying that ideas are all representations, which must mean, in this context, 'resemblances' (II.xxx.2). The main context in which he talks of ideas as images, meaning pictures, is when he is talking of memory (e.g. II.x.5) and it is plausible to suppose that he regards remembered ideas of sensation as, at least sometimes, mental images.

Locke undoubtedly uses 'idea' for 'concept' but it is usually clear when he does so. On the other hand, he distinguishes between thoughts and ideas. Ideas are what the mind is employed about in thinking, thoughts are 'made up of' ideas (III.ii.1) and ideas are 'whatsoever is the Object of the Understanding when a man thinks' (I.i.8). However, when we think, we use propositions and, as we shall see, not every element of a proposition is, or stands for, an idea. We cannot think without ideas which are the basis of, and are combined in, thoughts, but thoughts are not composed only of ideas; rather, they relate ideas.

I think, incidentally, that there is no evidence at all that he takes ideas *as concepts* to be mental images or that he supposes that we do our abstract thinking in pictures. If I think about my experience of a red book later, in the absence of the book, Locke thinks that I move from having simple ideas of sensation to operating with ideas in the sense of concepts of, for example, oblong shape and red colour. Sometimes I may have images but this is not necessary for thought. Ideas regarded as concepts are, I suspect, what he has in mind in referring to ideas as 'secret' or 'invisible' (III.ii.1). When he comes to talk about communication he has to find a special account of how we may convey such ideas to one another in view of their particular privacy.

Critics have delighted in moving rapidly from the not very remarkable discovery that Locke uses 'idea' in various senses to the conclusion that he confuses these senses. This does not, of course, follow. It is true that he might have distinguished the senses more clearly and that he might have adhered more explicitly to the distinctions he did make but nevertheless I believe that it is usually possible to see in what sense he is using the word in particular passages by careful consideration of the context.

The most central and important way in which Locke uses 'idea' is to mean 'percept'. When his usage allows us to say, for example, that when I am seeing a yellow flower I am having an idea of yellow he is using it in a way that is unfamiliar to us and that was unfamiliar to many of his contemporaries. I call this his 'technical' sense and I shall have more to say about it because I think it is more strange than we tend to think it is and that is why it has been so widely misunderstood. I wish at least to put forward a hypothesis about what Locke had in mind when he used it in this way.

II

Before I do this I should like to deal with another question which has received little or no attention, perhaps because it appears to be trivial. I think that it is not trivial because it helps to throw light on the way in which Locke thought of ideas. The question is 'Why did Locke always italicize the word "*Idea*" in the *Essay*?'

The answer I propose is that he italicized it because it is a foreign word, namely the Greek word ἰδέα, simply transliterated. Locke attached considerable importance to the derivation of words for establishing their meanings (see chapter on Substance-in-General, below) so it may well be that he consciously intended to stress the original Greek meaning, 'form' or 'pattern'. It may also be important that this is the sense translated into Latin as '*species*' and into the English of the period as 'species' which can mean the appearance or shape of things.

Kemp Smith drew attention to this in connection with Descartes. He says that, for Descartes in the *Regulae*,

> The mind, *quâ* cognitive, is taken as being sheerly receptive, purely contemplative, of the objects, whatever they be, mental or physical, which present themselves to it. He is here, in terms of his new physical teaching, substituting brain-patterns for the *species sensibiles* which in the Scholastic theory of sense-perception were supposed to mediate the sensing of external bodies; and if we bear in mind that the Latin translation of the Greek ἰδέα and εἶδος was *species*, the look or appearance, the figure or shape of things, we shall be less surprised to find him speaking of these brain-patterns as *ideae vel figurae*, and also as *ideae corporeae*. Giving as he does a mechanical account of the processes intermediate between the external bodies and the brain-patterns – citing the analogy of the die and the waxen tablet – he rejects the Scholastic view of the species as issuing from external bodies and as being transmitted, *quâ* species, in some unexplained fashion to the mind ... [the patterns] 'inform' the mind ... as being the 'objects' [Lat. *objecta*] which the awareness is disclosing to us.[6]

It is well known that Locke was influenced by Descartes in a number of ways and that they, together with Boyle and other natural philosophers, were rejecting many scholastic doctrines and replacing unanalysed scholastic concepts by concepts analysed in mechanical terms. Descartes' mechanical account of perception, as reported by Kemp Smith, has much in common with those of Boyle and Locke. One aspect of Kemp Smith's interpretation to which I wish to draw attention is that brain-patterns were supposed to have features in common with the patterns in nerves, sense-organs and the external objects initiating the causal chain. This view was espoused by Locke.

A doctrine held by numbers of scholastics up to the sixteenth and seventeenth centuries which descended, in one form or another, from Democritus, Epicurus and Aristotle is that of *species* or *sensible species*. The Aristotelian view was that in perception the sense-organ receives, and passes on, the *form* of a perceived object without its matter[7] or, as D. W. Hamlyn puts it, 'the sense-organ receives a quality of the object without the material in which the quality inheres'.[8] Democritus and Epicurus had held that fine images or 'films' were continuously thrown off from the surfaces of bodies and that they impinged on the sense-organs to produce perceptions of the bodies.[9] These images were called εἴδωλα and, in Latin, *species* or *effluvia* . Although Aristotle does not appear to have held a doctrine of effluvia many of his medieval followers did.

According to Sir William Hamilton[10] the opinion that 'generally prevailed among the Peripatetic philosophers of the Middle Ages' was that in explaining knowledge a medium was required which was different from both the mind and the (material) external object of thought. The medium consisted of objects called *Intentional Species*

> *Species (formae, similitudines, simulacra, idola)*, because they represented the object to the mind – *intentional*, to express the relative and accidental nature of their manifestation. (p. 952)

These species were necessary for both the intellect and the senses and were distinguished into *species impressae* and *species expressae*. A *species impressa* was 'the vicarious existence itself, as emitted by the object, as impressed on the particular faculty'; a *species expressa* was 'the operation itself elicited by the faculty and the impressed species together . . .'.

Finally

> a species fitted to affect the sense, was called a *sensible species* . . .: it proceeded immediately from the object . . . and, if not altogether immaterial, was of an intermediate nature between matter and spirit. (p. 953)[11]

There was considerable controversy among the scholastics about what form of this doctrine was acceptable but it is clear that it was held until the seventeenth century, by some at least, in roughly the form here outlined. That is, species were said to be given off by perceived objects and they conveyed the forms or patterns of the objects to the sense-organs. It is worth noting that among the names mentioned by Boyle are Francisco Suárez (1548–1617) and Jacopo Zabarella (1538–89) both of whom held some form of this view, as did Pierre Gassendi (1592–1655) who is generally regarded as having influenced Boyle.

Leibniz, in the *Monadology*, says 'accidents cannot separate themselves from substances nor go about outside of them, as the "sensible species" of the Scholastics used to do'. Latta comments that Leibniz is thinking of

> a theory (*not* that of Thomas Aquinas), according to which sense-perception means that particles are detached from the body perceived and pass into the percipient, in whom they are reconstructed into images or representations of qualities in the thing perceived

and he mentions Descartes's desire 'to rid people's minds of all these little images, flying through the air, called *intentional species*, which give so much work to the imagination of philosophers'.[12] Descartes said

> there is no need to assume that something material passes from the objects to our eyes to make us see colors and light, nor even that there is anything in these objects which is similar to the ideas or the sensations that we have of them: just as nothing comes out of the bodies that a blind man senses, which must be transmitted along the length of his stick into his hand; and as the resistance or the movement of these bodies, which is the sole cause of the sensations he has of them, is nothing like the ideas he forms of them. And by this means your mind will be delivered from all those small images flitting through the air, called *intentional species*...[13]

The word 'idea' was in use in Locke's time, at least since Descartes, in what is often called its modern sense, to mean 'concept'. Locke used it in this way but he also used it for the other contents of the mind including, most fundamentally, percepts, the products of sense-perception. This is fundamental because percepts are the source, directly or indirectly, of all other kinds of ideas. Locke wished to distinguish ideas from qualities of objects and to reject any notion of sensible species which were a compromise between the mental and the material. The Greek word ἰδέα means form or pattern and it may be that he wished to retain this suggestion since the notion of the transmission of the patterns of objects to our minds was of great importance to him. Sensible species were conceived by some as conveying a form or pattern from objects to us but those words also carried the unwelcome suggestion of a medium that was neither quite mental nor quite material. Locke, following Boyle, wanted to replace that curious medium by material corpuscles which could convey patterns from objects to us without being themselves transferred from the objects to our bodies; the word 'idea' would do very well for what was before the mind as a result. It would retain part of the suggestion of the Greek word but avoid some of the unwanted accretions of the much more frequently used Latin word 'species'. It is possible that Locke intentionally relied on the fact that Aristotelian and Platonic notions of form or pattern clung to the word 'idea' and that he stressed this by italicizing it to stress its origin.

It is perhaps of some importance that Locke makes the connection between his theory of perception and scholastic theories by saying that he uses '*Idea*' to 'express whatever is meant by *Phantasm, Notion, Species*' (I.i.8) and that 'species' was often taken, in contrast to the other two, to refer to something external to minds. So my further suggestion is that '*Idea*' as used by him is intended to be a compromising word involving suggestions not only of mind-dependence and of the transmission of patterns without the transfer of matter but also of externality in some sense. This does not, of course mean that it is sensible to talk of a pattern of corpuscles in a light ray as an idea or as mind-dependent; what is called an idea is dependent upon a pattern that can be transported by material means but also upon that pattern acting upon a mind. It also makes sense to say of an idea that it *has* a pattern which either resembles that material pattern or is 'appropriate' to it. I shall now explain how I think that is possible.

III

(The contents of this section will be more intelligible to those not familiar with the *Essay* after they have read the next two chapters.)

As I have said, Locke, probably influenced by Descartes, uses 'idea' as a technical term when he talks of simple ideas of sensation. I suspect that this use is almost as strange to us, especially outside philosophy, as it was to many of the contemporaries of Descartes and Locke. I don't think that we ever, in the ordinary way, refer to an idea as something we have when we see, for instance, a colour. Even in philosophy I think that it still seems strange to us and that much of the controversy surrounding this use springs from our not having fully grasped Locke's meaning.

There are numerous things in Locke's philosophy that still puzzle us and I should like to try the hypothesis that this is not because Locke was radically confused but, at least sometimes, because of our failure fully to understand the very unusual way in which he uses the word 'idea' in his technical sense. My strategy will be to mention some of those puzzles here and then to look for a sense of 'idea', consistent with what I have already said, which would enable him to deal with them. Even when Locke does not explicitly deal with one of these puzzles it may be that he would have dealt with it in the way I suggest; the fact that he appears not to regard some of them as puzzles may support my hypothesis by suggesting that it was just because he accepted this view that they do not appear to be puzzles to him.

One of the strengths of my view, I think, is that the alleged puzzles, which arise in various parts of the *Essay*, are of very different sorts. I mention five of them.

1. Locke holds that when I see, for example, a red ball I may be correct in supposing that the spherical shape I see is actually a property (quality) of the ball but that I am always mistaken in supposing that the red colour I see is actually a property (quality) of the ball. Why do I make this mistake?
2. Locke says that in perception some of the ideas I have *resemble* properties (qualities) of the object I am perceiving and some do not. Since ideas are mental and perceived objects are material, how is it even logically possible for there to be a resemblance, or a failure of resemblance, between an idea and an object? The defining properties of what is mental are in a different category from the defining properties of what is material.
3. Locke sometimes talks of ideas as if they were in objects, for example, when he talks of finding simple ideas together in substances such as gold; but he also explicitly says that ideas are *in the mind* and *not* in objects. Is this a sheer mistake on his part?
4. In his account of language and communication Locke clearly thinks that we can use words for simple ideas of sensation in the same way as other people do, that is, to stand for the same ideas. But each person has access directly only to his own ideas because ideas are in the mind and are not in the public world. If the ideas for which I make words stand are in *my* mind, are private to me, how is it possible for anyone else to make a word stand for the same idea as I do and know that he is doing so?
5. Locke distinguishes between words for 'common sensible ideas' such as colour-words and words for 'Things that fall not under our Senses' (III.i.5) such as the operations of our own minds like thinking or doubting. He also distinguishes between parts of certain complex ideas, some of which are 'visible' and some not. The most notable example is actions. Some parts of actions are 'visible in the Action it self' (III.ix.7) and some are not. An example is the murdering of someone by shooting: pulling the trigger of a gun is visible in the action whereas the intention to kill, which is a necessary part of the idea of murder, is not visible. *Why* does Locke make this distinction and *how* can he make it, if no idea or part of an idea is supposed to be a part of the external world to be visible or invisible?

On various conventional interpretations of Locke there are serious difficulties in answering these questions. Some of them have been taken as reasons for rejecting his views out of hand, especially those attached to his assertion of the resemblance between ideas and qualities and to his account of language and communication. I believe that if we clear our minds of some of the associations the word 'idea' normally has for us we can find a technical sense of it which removes these difficulties and for which there is some warrant to be found in Locke's text.

At one point, Locke explains this sense of 'idea' and sufficiently indicates what is technical about it, when he talks about the sources of our ideas, thus

First, *Our Senses*, conversant about particular sensible Objects, do *convey into the Mind*, several distinct *Perceptions* of things, according to those various ways, wherein those Objects do affect them: And thus we come by those *Ideas*, we have of *Yellow, White, Heat, Cold, Soft, Hard, Bitter, Sweet*... (II.i.3)

That is, ideas, in this sense, are a certain sort of *perceptions*, namely, perceptions of external objects. Locke does not, as we might wish, go on to define, or even to say much about, perceptions, which would throw more light on this sense.

However, in his later chapter on perception he does say something which I think important and revealing. Thus

What Perception is, every one will know better by reflecting on what he does himself, when he sees, hears, feels, *etc.* or thinks, than by any discourse of mine. Whoever reflects on what passes in his own Mind, cannot miss it: And if he does not reflect, all the Words in the World, cannot make him have any notion of it. (II.ix.2)

It is a familiar point to Locke, as will amply appear later, that in defining a word we do not say everything that can correctly be said about its meaning and reference. To accept a definition of a man as a rational animal is not to preclude our saying that a man has a heart or a liver. Here, in defining 'idea', Locke leaves us free, even encourages us if we attend to the passage about perception, to say further things that we know already about ideas. What we must not do is say anything that contradicts the definition. We are free, in particular, to attribute to ideas, in this sense, features of perceptions with which we are all familiar from experience.

Of course, 'perception' is ambiguous: it means either an activity (my perceiving) or a content (what is perceived in the sense of the resulting experience). In the above-quoted passage about perception Locke is clearly using it in the first sense for what a man 'does himself', but it seems reasonable to say that if we reflect on our perceiving we are aware, usually, of the content or experience resulting, so I believe I am justified in the use I propose to make of the passage.

What features of perceptions are revealed when we reflect upon perceiving? It will be enough to concentrate mainly on visual perception. When I see a red book it looks to me to be roughly oblong in shape, of a certain size, red in parts, black and white in other parts, and so on. Locke calls the simplest products of such an analysis 'simple ideas of sensation'. He holds that I sometimes see the book, in some respects, as it really is and I sometimes do not. According to the corpuscular hypothesis, which Locke accepts at least provisionally, I never see it as it really is in respect of colour because, on that theory, colour is explained as an appearance due to non-coloured constituents of the book. These simple ideas are the *looks* or *appearances* of the book to me, some showing me accurately and some not showing me accurately how the book really is.

When I examine these perceptions, whether colours, or shapes or any other visual features, I find that they all have something in common. It is that they are, or appear to be, all of them, *out there* at a distance of, say, three feet from me; the colours, the shapes and so on, all appear to be *there*, where the book is and where I would say it is if I were asked. That is where I would point if someone asked me to point either at what I see or at the book. If I want to point at the appearances, or looks, of the book I point *outwards*, away from myself; it would never occur to me, in the ordinary way, to point at my head or to say that I can't point at them because they are in my mind and so nowhere or everywhere. The distance from me of the redness I see is part of my perception of it. Now the appearances of individual qualities are simple ideas in Locke's sense. So there is a sense in which I can point at such ideas. They have a sort of minimal publicity in that I can show you at least where they are, where to look in the hope of your having the same ideas as I have. All this, I suggest, is part of what we already understand by 'perceptions' of the sort in terms of which Locke defines his technical use of 'idea'. He suggests that we supply such details for ourselves out of our own experience.

It is interesting now to look back at the quotation from Kemp Smith about Descartes (p. 52). He says 'the Latin translation of the Greek ἰδέα and εἶδος was *species*, the look or appearance, the figure or shape of things.' The look or appearance, the figure or shape, are presented to us as *out there*. The '"objects" ... which the awareness is disclosing to us' (Kemp Smith) are, *as disclosed to us*, out there.

Although Locke does not deal with after-images it is worth noting that similar things can be said even about them. After-images might be regarded, using Locke's terminology, as a special sort of ideas of sensation in the absence of the objects causing them. If I stare at a lighted lamp and then at a wall the after-image appears to be on the wall; that is, it is out there to be pointed at. It looks to be wherever I am looking; if I move my eyes the image moves along the wall. If I close my eyes it appears to be at a distance beyond them in indefinitely extended blackness. In my experience, at least, it does not even seem correct to say that it appears to be on the inside of my eyelids; it appears to be farther away than that.

All this may seem very strange to us because we are used to using 'idea' in a very different way and especially because we think of *ideas*, as we are told by Locke to think of them, as 'in the mind'. However, there have always been problems about what 'in the mind' can mean. It appears to give a location for ideas but it has frequently been argued that it does not make sense to suppose that the mind has any location in space at least and that this is one of the fundamental differences between mind and matter. In consequence, the suggestion that ideas can be pointed at has not been thought to make sense. It is not possible to deny that Locke regarded ideas as 'in the mind' but I have argued that he also thought that they could be pointed at. How is it possible to hold that Locke accepted both these things? What, then, does Locke mean by 'in the mind'? I believe that it is not meant to give a location; Locke could see as well as we can that this would involve trying to make sense of 'somewhere but not in space'. I suggest that it merely means 'mind-dependent', at least for simple ideas of sensation. It makes perfectly good sense to say that something can be mind-dependent and yet spatially located. This is just what appearances of things are. Without minds there would not be any appearances, out there or anywhere else; but, in seeing, a red colour or a spherical shape appears somewhere in space. Such ideas are 'private' not because they are somehow locked up in our heads as in a box (what *could* that mean?) but because they are appearances to, perceptions by, someone and we can have only our own perceptions. I can never be sure that my ideas are qualitatively the same as yours but at least I can show you where they appear to me, so they also have a minimal publicity.

There is one passage in Locke which appears directly to support this interpretation. When he is discussing the mechanism of perception he says

> And since the Extension, Figure, Number, and Motion of Bodies of an observable bigness, may be perceived at a distance *by* the sight, 'tis evident some singly imperceptible Bodies must come from them to the Eyes... (II.viii.12)

Since perceiving is having ideas, of extension etc., what can Locke here be stressing except that these appearances are at a distance from our eyes?

Ideas in the sense of concepts are different; they are 'in the mind' in a different sense and are more radically private. It may be difficult to say precisely what 'in the mind' means for them but at least we can say something about what is not entailed by it. There is no need to try to regard it as having spatial implications because there is no tendency to think of such ideas as spatially located. It may be these ideas to which Locke occasionally applies the words 'secret', 'invisible' and 'hidden'. In talking of language, as we shall see, Locke has most trouble in explaining how these ideas are communicated and how we can discover that two people have the same idea of, say, someone's intentions. He clearly makes a distinction between such ideas and simple ideas of sensation.

I now return to the puzzles I mentioned at the beginning of this section. Perhaps the strongest reason for thinking that Locke thought of ideas of sensation in the way I suggest springs from the first of these puzzles. He surely uses this in connection with the mistake we make in supposing that perceived colours, for instance, show us real properties of objects, unlike pains caused by objects. Why do we unthinkingly take the idea of brown we have on seeing a table to indicate as real a property of the table as, for instance, our idea of its shape? Because, just as much as its shape, it looks to be out there. In this it differs from the pain we have on stubbing our toes against the table. Locke has to use an *argument* to try to convince us that the appearance is misleading and that the seen colour is more like a felt pain than we think it is and than it appears to be. (See *e.g.* II.viii.18, 24, 25.)[14]

The second puzzle is also readily dealt with: resemblance or non-resemblance between ideas and properties (qualities) become intelligible for to say that an idea resembles a quality is just to say such things as that an object has, *e.g.*, shape just as it appears to have and to say that an idea does not resemble a quality is just to say that though an object appears to be, *e.g.*, coloured it is not really coloured, at least in the sense in which it has a shape. Resemblance is just a matter of an object's appearing to us, in some respect, as it really is.

The third puzzle, concerning apparent suggestions by Locke that ideas may be in objects, also becomes less puzzling. There are two sorts of context to consider. Locke warns us at II.viii.8 that he will occasionally, for brevity perhaps, talk as if ideas were in objects when he means to refer not to the ideas but to their causes and in some of the puzzling passages he is clearly engaging in this not strictly accurate practice.[15] However, when in talking of ideas of substances, such as gold, he talks of simple ideas being 'found together' in things he is talking more literally and relying on the minimal publicity of ideas for which I have argued; ideas *are* found occurring together and they *look to be* in things.

These questions will all be dealt with more fully later and the solutions to the last two puzzles can barely be indicated until I come to discuss Locke's view of language and meaning. In the meantime I may just say that they can both be handled with the help of what I have called the 'minimal publicity' of ideas of sensation. The possibility of indicating to another at least the whereabouts of such ideas gives some basis for understanding how communication about them may be possible.

I am aware that it sounds most odd to talk of the 'whereabouts' of ideas but I hope that what I have said so far will have given some plausibility to this way of talking. We are quite familiar, especially from modern physics, with the bewilderment that often arises from the introduction of technical terms. It must be said in Locke's favour that he thinks that his technical use will be more plausible if we examine our experience more closely than we usually do. The advantages of the interpretation I have given will, I believe, become abundantly clear in the sequel.

IV

It is an essential part of my view that, for Locke, simple ideas are usually, perhaps always, the products of the analysis of our experiences; he does not hold that objects or even *all* complex ideas are constructed by us out of simple ideas originally delivered to us in all their simplicity.

The opening chapters of Book II contain a sketch of a causal account of the acquisition of ideas, which all *ultimately* come from experience (II.i.2). Ideas of sensation come from our observation of 'external, sensible Objects' and ideas of reflection from our observation of 'the internal Operations of our Minds'. Thus

> *External Objects furnish the Mind with the* Ideas *of sensible qualities*, which are all those different perceptions they produce in us: And the *Mind furnishes the Understanding with* Ideas *of its own Operations*. (II.i.5)

Locke mentions as ideas we get from sensation '*Yellow, White, Heat, Cold, Soft, Hard, Bitter, Sweet*' (II.i.3) and as ideas we get from reflection '*Perception, Thinking, Doubting, Believing, Reasoning, Knowing, Willing*' (II.i.4).

However, apprehending these ideas as separate and simple, especially those of the operations of the mind, requires attention. Locke says

> For, though he that contemplates the Operations of his Mind, cannot but have plain and clear *Ideas* of them; yet unless he turn his Thoughts that way, and considers them *attentively*, he will no more have clear and distinct *Ideas* of all the *Operations of his Mind*, and all that may be observed therein, than he will have all the particular *Ideas* of any Landscape, or of the Parts and Motions of a Clock, who will not turn his Eyes to it, and with attention heed all the Parts of it. The Picture, or Clock may be so placed, that they may come in his way every day; but yet he will have but a confused *Idea* of all the Parts they are made up of, till he *applies himself with attention*, to consider them each in particular. (II.i.7)

Here I wish to stress particularly the mention of the parts of a picture. The parts of a picture as seen are ultimately shapes and colours. We may, however, be aware of the whole pattern but until we have looked attentively at it we may not have *clear and distinct* ideas of those shapes and colours and the relations between them. Similarly we may be dimly or confusedly aware of our mind's working but unless we attend closely to it we may not have *clear and distinct* ideas of its different operations such as perceiving and thinking. A distinct idea is one that is separate and distinguished from others surrounding it in the experience.

A new-born baby begins to acquire ideas immediately but it takes time and experience to discriminate particular ideas from the presented mass. Locke says

> After some time, it begins to know the Objects, which being most familiar with it, have made lasting Impressions. Thus it comes by degrees, to know the Persons it daily converses with, and distinguish them from Strangers; which are Instances and Effects of its coming to retain and distinguish the *Ideas* the Senses convey to it: And so we may observe, how the Mind, *by degrees*, improves in these, and *advances* to the Exercise of those other Faculties of *Enlarging, Compounding*, and *Abstracting* its *Ideas*, and of reasoning about them, and reflecting upon all these, of which, I shall have occasion to speak more hereafter. (II.i.22)

The mind is passive in its reception of ideas of sensation; it cannot *operate* without something to operate on. The objects of the senses 'obtrude their particular *Ideas* upon our Minds, whether we will or no' (II.i.25) and so give us the materials of thinking and knowledge. The earliest activity of the mind is the sorting out of ideas, discriminating them from one another, and when this has been done other activities come into play. In this passage Locke begins by talking of the discriminating ability and then refers to the mind's 'other Faculties', namely, '*Enlarging, Compounding*, and *Abstracting*'. All these activities are discussed in a later chapter.

In the last section of Chapter i, Locke for the first time mentions simple ideas and he begins Chapter ii with the distinction between simple and complex ideas, making it clear that our apprehending of ideas *as simple* depends upon the discriminating ability of our minds. The ideas we have, he says, are some of them simple and some of them complex. Then he says

> Though the Qualities that affect our Senses, are, in the things themselves, so united and blended, that there is no separation, no distance between them; yet 'tis plain, the *Ideas* they produce in the Mind, enter by the Senses simple and unmixed. (II.ii.1)

This has sometimes been thought to imply that we construct *all* our complex ideas out of ideas which are simple and unmixed in experience rather than that simple ideas are the products of our discrimination or analysis of groups of ideas presented together in experience.

That this cannot be Locke's meaning is evident from many passages in the *Essay*. He immediately continues

> For though the Sight and Touch often take in from the same Object, at the same time, different *Ideas*; as a Man sees at once Motion and Colour; the Hand feels Softness and Warmth in the same piece of Wax: Yet the simple *Ideas* thus united

in the same Subject, are as perfectly distinct, as those that come in by different Senses. (II.ii.1)

Some ideas, such as those of size and shape, can come into the mind by more than one sense while others, such as those of colour and smell, come in only through one sense each. Colour and smell are clearly different and easily recognized as distinguishable because they must come in by different senses. But ideas of colour and motion obtained through the sense of sight differ qualitatively from one another as clearly as do ideas of colour and smell, in spite of their being obtained through one sense. The same is true of felt softness and warmth.

We cannot see an object without at the same time seeing some colour and some shape or extension. These ideas are 'mixed' in the sense that they occur together in that visual experience; they are 'unmixed' because they result from different causal chains and because they can be qualitatively distinguished. The triangular shape we see does not in any way interfere with, or infect, or depend upon the blue colour we see, considered qualitatively. We can see the same colour associated with a different shape or the same shape associated with a different colour.

Locke gives a clue to what he means by 'simple' in the expression 'simple and unmixed' at the end of this section when he says

The coldness and hardness, which a Man feels in a piece of *Ice*, being as distinct *Ideas* in the Mind, as the smell and whiteness of a Lily; or as the taste of Sugar, and smell of a Rose: And there is nothing can be plainer to a Man, than the clear and distinct Perception he has of those simple *Ideas*; which being each in it self uncompounded, contains in it nothing but *one uniform Appearance*, or Conception in the mind, and is not distinguishable into different *Ideas*. (II.ii.1)

An idea is simple if it is 'not distinguishable into different *Ideas*'. If I see a blue triangle I am having (at least) two ideas, a colour-idea and a shape-idea. The colour-idea is simple if I cannot distinguish within it different shades of blue; the shape-idea is simple if I cannot distinguish within it different shapes; but perhaps more importantly, ideas of shape cannot have ideas of non-shapes as parts and ideas of colour cannot have ideas of non-colours as parts, if they are simple ideas.

Thus when Locke says that simple ideas 'enter by the senses simple and unmixed' he does not mean that each occurs alone or that each is immediately seen to be simple and separate. He means that they *are*, however we first apprehend them, distinct and that they are able to be discriminated by the mind as ideas that cannot be further discriminated. Once they are discriminated they are apprehended as different and qualitatively independent parts of a complex experience or idea.

It may be thought that even if Locke's account of simplicity will do for colours, it raises problems in connection with shapes and sizes. I can surely analyse a square into four equal lines enclosing a space or into four equal squares and those squares again into four smaller squares, and so on indefinitely. So can there be a simple idea of a square? However, this is not to the point. We are here considering *visual* ideas; we must concentrate on what I see rather than on what can be done in thought with what I see. If I see a uniformly yellow square patch on a black background I see neither lines enclosing a space nor smaller squares within the square. Nothing I see allows me to discriminate *visually* the original square into smaller ones or into lines enclosing a space. To deny this is to raise some unanswerable questions: how many smaller squares am I seeing? Where do I see their boundaries? Am I also seeing triangles or rectangles as well as squares? Of course, I may divide the original square with ruler and pencil into as many squares as I like and then I would have visual ideas of smaller squares and of lines but then I would be seeing something different. If I continue the process of division until the lines are so close together that I can no longer see yellow spaces between them I have ceased to have simple visual ideas of squares, of yellow and of lines. The last simple idea I have of a square is the idea of the smallest yellow area that I see *as square* and the last simple idea of yellow is that of the smallest area that I see as yellow.

When the mind is furnished with simple ideas by sensation, and apprehends them as simple, it has the power 'to repeat, compare, and unite them' and so to make 'new complex *Ideas*' (II.ii.2). Locke's use of the word 'new' here suggests that there are complex ideas in the mind from the beginning which are distinct from those made by the mind.

We find further material relevant to all this in Locke's important chapter 'Of Discerning, and other Operations of the Mind' (II.xi). It is usual to talk as if Locke mentions only three operations of the mind, that is, comparing, compounding and abstracting. He in fact mentions at least five[16] and insufficient attention has perhaps been given particularly to what he calls 'discerning', and I have been calling 'discriminating', which I take to be one of the most fundamental operations of the mind, according to Locke. He begins the chapter by saying

Another Faculty [*sc.* besides perception and retention, each of which has been given a chapter to itself], we may take notice of in our Minds, is that of *Discerning* and distinguishing between the several *Ideas* it has. It is not enough to have a confused Perception of something in general: Unless the Mind had a distinct Perception of different Objects, and their Qualities, it would be capable of very little Knowledge; though the Bodies that affect us, were as busie about us, as they are now, and the Mind were continually employ'd in thinking. On this Faculty of Distinguishing one thing from another, depends the *evidence and certainty* of several, even very general Propositions, which have passed for innate Truths; because Men over-looking the true cause, why those Propositions find universal assent, impute it wholly to native uniform Impressions; whereas it in truth *depends upon this clear discerning Faculty* of the Mind, whereby it perceives two *Ideas* to be the same, or different. (II.xi.1)

When Locke here mentions 'several, even very general Propositions, which have passed for innate Truths' I think he has in mind such propositions as he mentions in Book IV, for example, 'a man is not a horse' and 'Red is not Blew' (IV.vii.4) or, as he might have added 'a colour is not a sound'. Ideas of man and horse, red and blue, colour and sound are just apprehended by us as different; the truths are the result of the common faculty of discerning in peoples' minds and are not innate or 'native'.

We have, from the beginning, confused perception of objects within which our faculty of discerning enables us to distinguish different, and ultimately simple, ideas. These form the basis of the knowledge we build up with the help of the other faculties of the mind. Discerning is simply distinguishing ideas within sensory experience; apprehending them as different from, and qualitatively independent of, one another. Thus we arrive at clear and distinct ideas by attending closely to them as they are in themselves, whatever their causes (II.xi.3).

The other three faculties of the mind are more properly regarded as intellectual or reasoning faculties. Locke first mentions *comparing*. He begins

The *COMPARING* them one with another, in respect of Extent, Degrees, Time, Place, or any other Circumstances, is another operation of the Mind about its *Ideas*, and is that upon which depends all that large tribe of *Ideas*, comprehended under *Relation*... (II.xi.4)

Whereas discerning is probably a faculty we share with the brutes it is unlikely that they have the faculty of comparing for

though they probably have several *Ideas* distinct enough, yet it seems to me to be the Prerogative of Humane Understanding, when it has sufficiently distinguished any *Ideas*, so as to perceive them to be perfectly different, and so consequently two, to cast about and consider in what circumstances they are capable to be compared. (II.xi.5)

Having *perceived* two ideas as different we may consider *how* they are different and *how* they are similar. This sort of comparing, Locke says, is 'useful only to abstract Reasonings' and concerns general ideas.

The next operation he mentions is *composition*, which includes *enlarging*. In composition the mind puts together simple ideas of sensation and reflection to make complex ideas. When the ideas put together are of the same kind, as when we combine the ideas of twelve units to make a dozen, Locke calls the process 'enlarging'. When he considers whether brutes possess this faculty he makes perhaps his clearest statement of a distinction between *presented* complex ideas and *constructed* complex ideas when he says

For though they [*sc.* brutes] take in, and retain together several Combinations of simple *Ideas*, as possibly the Shape, Smell, and Voice of his Master, make up the complex *Idea* a Dog has of him; or rather are so many distinct Marks whereby he knows him: yet, I *do not* think they do of themselves ever compound them, and *make complex* Ideas. (II.xi.7)

Locke's final operation of the mind is *abstraction* and it is important that in the section before he introduces it he mentions language and then associates abstraction very closely with the use of words (II.xi.8–9). In order that language be possible we must have not only individual ideas and names annexed to them but also general ideas and general names annexed to them. This we do by considering ideas

as they are in the Mind such Appearances, separate from all other Existences, and the circumstances of real Existence, as Time, Place, or any other concomitant *Ideas*. This is called *ABSTRACTION*, whereby *Ideas* taken from particular Beings, become general Representatives of all of the same kind; and their Names general Names, applicable to whatever exists conformable to such abstract *Ideas*. (II.xi.9)

His example is

Thus the same Colour being observed to day in Chalk or Snow, which the Mind yesterday received from Milk, it considers that Appearance alone, makes it a representative of all of that kind; and having given it the name *Whiteness*, it by that sound signifies the same quality wheresoever to be imagin'd or met with; and thus Universals, whether *Ideas* or Terms, are made. (II.xi.9)

The ability to abstract and so to have general ideas is, Locke says, 'that which puts a perfect distinction between Man and Brutes'. We have reason to think that they do not have the faculty of abstraction 'since they have no use of Words, or any other general Signs' (II.xi.10). However, because brutes have senses they do 'some of them in certain Instances reason'; but their reasoning involves only particular ideas 'just as they are receiv'd from their Senses' (II.xi.11). I take it that Locke means here that brutes are capable of *discerning*, which is a rudimentary operation of the mind, but not of abstracting. We can observe that they are capable of discerning. It should be clear that discerning must not be confused with abstracting.[17] In the first place, abstracting depends upon discerning since it involves finding the same idea in different contexts and considering it apart from those contexts. In the second place, abstracting appears to involve giving names to those ideas whereas in explaining the operation of discerning Locke does not make it depend upon language. (See also III.iii. esp. 7–9.)

In his next chapter Locke returns to a detailed discussion of complex ideas and the very first paragraph may seem to conflict with the interpretation I have been putting forward. He says that the mind is passive in receiving simple ideas from sensation and reflection and cannot *make* any simple idea 'nor have any *Idea* which does not wholly consist of them'. Then he continues

> The Acts of the Mind wherein it exerts its Power over its simple *Ideas* are chiefly these three, 1. Combining several simple *Ideas* into one compound one, and thus all Complex *Ideas* are made. 2. The 2d. is bringing two *Ideas*, whether simple or complex, together; and setting them by one another, so as to take a view of them at once, without uniting them into one; by which way it gets all its *Ideas* of Relations. 3. The 3d. is separating them from all other *Ideas* that accompany them in their real existence; this is called *Abstraction*: And thus all its General *Ideas* are made. (II.xii.1)

It is the clause 'and thus all Complex *Ideas* are made' that might be thought to conflict with my view. However, I think that what Locke means is simply that such complex ideas as are *made by the mind* are all made in this way. This does not exclude the possibility that some complex ideas just occur and are not 'made by the mind'.

This view is supported by the rest of this section where Locke says:

> As simple *Ideas* are observed to exist in several Combinations united together; so the Mind has a power to consider several of them united together, as one *Idea*; and that not only as they are united in external Objects, but as it self has join'd them. (II.xii.1)

This clearly implies that we can consider complex ideas that are given in perception or complex ideas that are constructed by the mind, both being collections of simple ideas. We can consider both these kinds of complex ideas as one idea or as a collection of clear and distinct simple ideas. When I see a piece of gold I am having several simple ideas that occur together; I can choose to consider it as one complex idea that I call 'the idea of gold' or to consider separately the simple ideas making up this complex idea. The idea of an army, a complex idea constructed by us by putting together the ideas of individual men, I can also consider at will in either of these ways.

I claim, therefore, that Locke's view is that simple ideas are the products of the analysis of 'naturally occurring' complex ideas and that those complex ideas are not products of our construction out of simple ideas that 'occur naturally' as clear and distinct. However, constructed complex ideas are constructed out of simpler ideas, some of which are simple and the products of our analysis of 'naturally occurring' complex ideas.

I shall argue later that this has important consequences especially for Locke's view of language (see Chapter 12). It also has the advantage over the view I am rejecting that it is more obviously consistent with our experience.

5

QUALITIES

I

Further problems arise in Book II about Locke's use of 'ideas' which require rather different comments from those I have just made. He appears to treat ideas rather more deliberately sometimes as in minds and sometimes as external to minds. The explanation of this lies not in any equivocality about ideas in his technical sense but in the fact that his argument at the beginning of this Book is a developing one leading us from a loose to a more precise way of speaking.

It is of vital importance for what follows that he distinguish qualities from ideas which he does by saying that qualities are features of objects independent of us and ideas are in our minds and caused by those qualities. However, in the very act of making this distinction he has appeared to critics to confuse the two and so contradict himself. For example, at II.viii.7 he talks of distinguishing *ideas*

> as they are *Ideas* or Perceptions in our Minds; and as they are modifications of matter in the Bodies that cause such Perceptions in us...

as if ideas can be either in our minds or in external objects. In an even more famous passage, in the next section, he says

> Thus a Snow-ball having the power to produce in us the *Ideas* of *White*, *Cold*, and *Round*, the Powers to produce those Ideas in us, as they are in the Snow-ball I call *Qualities*; and as they are Sensations or Perceptions, in our Understandings, I call them *Ideas*... (II.viii.8)

Here the problem is that if 'they' in both occurrences refers back to the first occurrence of '*Ideas*' then ideas are both in objects and in the mind whereas if it refers back to 'Powers' then powers are both in objects and in the mind, which seems to conflict with the distinction being made.[1] It is rather unlikely that it refers back on the first occasion to 'Powers' and on the second to '*Ideas*', though perhaps not impossible. If we rule out that possibility then Locke appears to confuse qualities and ideas in the very act of distinguishing them.

It seems to me highly unlikely that he is as confused as this, especially in view of the general heading to these two sections which is 'Ideas *in the Mind, Qualities in Bodies*', and a better explanation is possible. I think that most commentators have failed to notice a development in Locke's argument, especially in this chapter, and to take account of the position of the snowball passage in that argument. Part of what he is doing in these sections is leading us from thinking with the vulgar to thinking with the learned. He is attacking a loose and misleading way of talking about ideas frequently found among philosophers since Descartes.

His detailed discussion of 'the nature of our Ideas' in this chapter begins at section 7, the snowball passage is in section 8, where he distinguishes qualities from ideas, and his celebrated distinction between primary and secondary qualities in sections 9 and 10. He begins the chapter by mentioning simple ideas of sensation and, in section 2, listing as examples 'the *Idea* of Heat and Cold, Light and Darkness, White and Black, Motion and Rest'. Up to section 7 he has not yet introduced into this chapter the term 'quality' although he has distinguished between ideas and their causes

> it being one thing to perceive, and know the *Idea* of White or Black, and quite another to examine what kind of particles they must be, and how ranged in the Superficies, to make any Object appear White or Black. (II.viii.2)

Note that there is a clear reference to the corpuscular hypothesis here. Before section 7, and sometimes after it when it will not cause confusion, he uses '*Idea*' loosely to stand indifferently for both qualities in things and ideas of them in us. This is to speak with the vulgar. It is perhaps important that at I.i.8 he says that he uses '*Idea*' to 'express whatever is meant by *Phantasm*, *Notion*, *Species*' and to remember that 'species' were taken by some to be independent of minds. He is leading us from *this* vulgar way of thinking to the more accurate way of thinking allowed by his later introduction of 'quality'. As I have said, he wishes to retain the idea of pattern and minimal publicity associated with 'species' but to make ideas mind-dependent.

When Locke lists examples of simple ideas in section 2 he is still using '*Ideas*' in the vague, popular and ambiguous sense. He wants to ask such questions as 'What *are* heat and cold?' but he cannot formulate this question or its answer accurately until he has made some distinctions. Having said that the materials of knowledge are simple ideas, but not yet having distinguished qualities from ideas, he formulates his project thus

> To discover the nature of our *Ideas* the better, and to discourse of them intelligibly, it will be convenient to distinguish them, as they are *Ideas* or Perceptions in our Minds; and as they are modifications of matter in the Bodies that cause such Perceptions in us . . . (II.viii.7)

Here, I suggest that '*Ideas*' on its first occurrence is being used in the vulgar sense and on its second occurrence in Locke's more precise sense.

Before the distinction is made we can think of ideas in two ways, *either*, on the one hand, stressing their mentality, their character as experiences or mental occurrences *or*, on the other hand, stressing their content, their character of apparently showing us features of the world, when the words for them can apparently qualify either perceivers or objects. Locke is suggesting that common sense and some of his opponents do not carefully distinguish these. Thus when we think of ourselves feeling a hot object we think of ourselves as having an experience and we think of the content of the experience as somehow duplicating the heat of the object and giving us accurate information about it. We have ideas of heat and cold; objects are hot or cold, *i.e.* heat and cold are in objects; so we can think of those ideas, *as contents*, as being in objects. We use the same words, or at least words from the same bag, to describe the experience and the object experienced. Speaking with and to the vulgar we can use 'idea' as a neutral word to refer to both an experience and its object and ask whether a given idea is the same in objects as 'in the mind'.

However, this equivocal use of 'idea' is a vulgar way of thinking which can lead to confusion and Locke is out to change it. It has connections with a scholastic view, to which both he and Boyle objected, that every externally caused perception resembles a real quality of an object. The beginning of the direct attack on it is the question whether ideas in us and in objects resemble one another. Later Locke says

'Tis true, the Things producing in us these simple *Ideas*, are but few of them denominated by us, as if they were only the causes of them; but as if those *Ideas* were real Beings in them. (II.xxxi.2)

By section 8 the time has come to dispense with the loose way of speaking and to make the word 'idea' more precise. This involves distinguishing formally between ideas and qualities and eliminating ambiguity by restricting the word 'idea' to what is 'in the mind'. So we have

Whatsoever the Mind perceives in it self, or is the immediate object of Perception, Thought, or Understanding, that I call *Idea*; and the Power to produce any *Idea* in our mind, I call *Quality* of the Subject wherein that power is. (II.viii.8)

The passage about the snowball follows immediately. When we consider that it is intended to illustrate the distinction made here quite clearly, it should seem to any well-meaning interpreter that the criticisms of the snowball passage made by some commentators are no more than quibbles.

What have, up to this point, been called 'ideas' actually fall into two classes, those that are in things and those that are 'in minds', and these classes are now separated and given different names: those that are in things are in future to be called 'qualities' and those that are 'in minds' are to be called 'ideas'. If Locke sometimes, inadvertently or for brevity, slips into the old way of talking he warns us to remember that he should be understood in the new way. He says

> which *Ideas*, if I speak of sometimes, as in the things themselves, I would be understood to mean those Qualities in the Objects which produce them in us. (II.viii.8)

All this allows us to understand the passage from II.viii.7 quoted above in a way that is no longer puzzling and, indeed, which makes good sense.

Support for this interpretation is contained in Locke's statement that

> *Flame* is denominated *Hot* and *Light*; *Snow White* and *Cold*; and *Manna White* and *Sweet*, from the *Ideas* they produce in us. Which Qualities are commonly thought to be the same in those Bodies, that those *Ideas* are in us... (II.viii.16)

That is, when we call flame hot, and so on, we are unthinkingly transferring those names from our ideas to their objects and thinking of the objects as having qualities resembling the ideas they produce in us. It is a short and unnoticed step to using the word 'idea' both for what is in the object and for what is in us.

II

As soon as Locke has distinguished between ideas and qualities he proceeds to distinguish between three sorts of qualities, primary, secondary and a third sort of qualities, and to discuss the distinction. This distinction has been widely misunderstood by commentators. The arguments of II.viii.16–21 by which, it is alleged, Locke seeks to support it have been regarded as easy targets for the beginning student of philosophy. The misunderstandings, I believe, stem largely from Berkeley, who failed to see what Locke was doing, and have been perpetuated by a long tradition of discovering what Locke said by reading what Berkeley said about him.

What should first breed suspicions about the correctness of the most usual interpretation of the distinction is the very ease with which the alleged support for it can be knocked down. If this interpretation is correct then Locke must have been both foolish and incompetent. I think that he was neither of these things and that a study of his work in the light of Boyle's work can help us to see this. It is of great importance that Locke be read carefully. In spite of his reputation, I think that he wrote carefully and that every word must be considered important. Abridged versions of the *Essay* are likely to be disastrously misleading just because the decision about what may be safely left out is usually based on the conventional view of his philosophy.

Since, on my view, the corpuscular hypothesis underlies Locke's philosophy and the distinction between primary and secondary qualities is a necessary part of that hypothesis, the distinction is central in Locke's philosophy and a correct interpretation of it has implications for much of the rest of the *Essay*. I shall begin by outlining the distinction roughly and then, with reference to Locke's text, expound it a little more fully and finally, in this chapter, consider the arguments of section II.viii.16–21 which are alleged by many commentators to be the main arguments for making the distinction. In the following three chapters I shall put more detailed arguments in support of my interpretation of the distinction.

The first thing to be said here is that Locke took over and developed the distinction from Boyle, as an essential part of the corpuscular philosophy, and that he was concerned to explain and illustrate it and to consider its implications for language and knowledge.

For Locke, ideas, all of them, are 'in the mind', and qualities, all of them, are in bodies. Primary qualities are those in terms of which body is defined; they are the essential qualities of all bodies. These are qualities both of sensible objects and of the insensible corpuscles of which they are composed. All observable phenomena, it is hoped, will eventually be explained in terms of these corpuscles and their qualities, and secondary qualities can be analysed into the primary qualities of conventions of corpuscles. Colours, sounds, tastes and odours are *ideas* of secondary qualities; they represent and are caused by the secondary qualities, which are in objects. The secondary qualities themselves are 'textures' or arrangements or patterns of the primary qualities of the insensible corpuscles of objects. Colour-words, for example, are names of ideas, not of qualities, the corresponding qualities could only be referred to by complex descriptions of textures or names specially invented for those textures. Our *ideas* of both primary and secondary qualities are caused, respectively, by the primary and secondary qualities of objects. Some of this will not be fully supported until later chapters.

Objects also have powers to act not on our organs of sense but upon other objects. These constitute the third sort of qualities and they too are textures of objects; the interactions between objects are, it is hoped, to be explained in terms of the arrangements of their insensible corpuscles. Objects act upon our sense organs and upon other objects by impulse,[2] by the movements of their insensible corpuscles and the insensible corpuscles of light, producing changes in the state of motion of the corpuscles of our bodies and of other objects. The explanation of such interactions is intended to be entirely mechanical, that is, in terms of the primary qualities, the shape, size and motion-or-rest of the corpuscles. (I am here using Boyle's list of primary qualities. I shall discuss later the propriety of using the same list for Locke.)

So much for the outline. I shall now fill in some of the details by considering Locke's arguments.

Having distinguished between ideas and qualities Locke distinguishes primary, secondary and the third sort of qualities by giving definitions. His first definition is

Qualities thus considered in Bodies are, First such as are utterly inseparable from the Body, in what estate soever it be; such as in all the alterations and changes it suffers, all the force can be used upon it, it constantly keeps; and such as Sense constantly finds in every particle of Matter, which has bulk enough to be perceived, and the Mind finds inseparable from every particle of Matter, though less than to make it self singly be perceived by our Senses ... These I call *original* or *primary Qualities* of Body, which I think we may observe to produce simple *Ideas* in us ... (II.viii.9: I have omitted Locke's example but I shall discuss it later.)

Of the four clauses in this definition, the first, second and fourth appear to go together and to make a conceptual point about matter, or body, as such. The primary qualities are just those that anything considered alone, must have if it is to be counted as a body; however we divide a body, even if we divide it into its constituent corpuscles, each part must have them; even a single corpuscle alone in the universe, if it is to be counted as material, must have them. They are the defining properties of body. That is why they are 'utterly inseparable from the Body', whatever is done to it, and why the *mind* finds them inseparable. We can conceive of a material corpuscle without colour, sound, odour or taste, since we can conceive of these arising only from the interaction of two or more corpuscles, but we cannot conceive of it without, say, extension. Those qualities that we cannot conceive of a material corpuscle as lacking are the primary qualities. So far, this appears to be the same as the view I have attributed to Boyle.

The third clause, 'and such as Sense constantly finds in every particle of Matter, which has bulk enough to be perceived', is more difficult. It is, of course, empirical rather than conceptual and the definition will stand without it. Locke may have included it in order to remind us that the concepts involved are all related to experience and that he was prepared to bring empirical support for some of his conclusions. The qualities attributed to the corpuscles are *of a kind* we meet in experience. Also, we can observe by touch, for example, bodies which are, as far as observation tells us, colourless, odourless, tasteless and silent; we can observe by sight bodies that give us no tactile sensations, such as a cloud of gas, but we cannot observe a body without shape, size and mobility, where 'mobility' may mean just the possibility of its occupying a different portion of space. Bodies have no colour in the dark (II.viii.17 and 19) but we can still observe them and if we can observe them they must have some shape and size. Whenever conditions are such that we can detect a portion of matter we can detect some size, shape and position but we cannot always detect some colour, taste, odour or sound; so these last are not inseparable from our observation of matter, as such. (Given no artificial constraints.) This cannot serve, of course, *as a criterion* of the inseparability of qualities from bodies since, according to the corpuscular hypothesis, there are bodies which are too small to be singly sensible, but Locke does not need it as a criterion and did not, I think, put it forward as one; it concerns only portions of matter with 'bulk enough to be perceived':

In the next section Locke defines secondary qualities. He says

2dly, Such *Qualities*, which in truth are nothing in the Objects themselves, but Powers to produce various Sensations in us by their *primary Qualities, i.e.* by the Bulk, Figure, Texture, and Motion of their insensible parts, as Colours, Sounds, Tasts *etc.* These I call *secondary Qualities*. (II.viii.10)

As we have seen, bodies, according to the corpuscular theory, consist of collections of corpuscles, each having only the primary qualities shape, size and mobility. However, these collections have other, emergent, qualities that can be described entirely in terms of the patterns or arrangement (the 'textures') of their constituent corpuscles. These descriptions would involve only the primary qualities of the corpuscles together with their spatial relations and motions within the collection. Were it not for motion, these descriptions would be merely geometrical, but since motion is to be included time must be brought into them and the descriptions of the textures will be kinematic. They will not be dynamic because the idea of forces between the particles or groups of them does not figure in the theory. Locke does not speak of time as a quality of bodies (see II.xiv) so the only *qualities* used in the descriptions would be of the nature of primary ones; distances between corpuscles are in the relevant respects of the same kind as distances between points on bodies *e.g.* the sizes of the corpuscles.

These collections of corpuscles, if they are large enough and, presumably, have the appropriate patterns, are able to produce sensations in us under certain conditions. Such abilities Locke calls 'Powers to produce various sensations in us'. If we take seriously Locke's statement that qualities are in bodies, as I think we must, then the only things in bodies to cause sensations of 'Colours, Sounds, Tasts, *etc.*' in us are the corpuscles, the patterns in which they are arranged and the resultant powers to act upon us. Notice that, according to this view, there is no warrant for saying that colours, for example, are in bodies in the way in which shapes are but only that bodies have the power to produce sensations of colours in us.

Commentators have frequently held that particular patterns of corpuscles *have* powers to affect us in certain ways and that whereas the patterns are intrinsic properties of collections of corpuscles describable independently of other bodies, the powers they have are dispositional properties and their description involves the mention of other bodies with which they may react.[3] I wish to take the unorthodox view that Locke uses the word 'texture' for the pattern of corpuscles in a group and regards the power of a body to affect our senses or another body as identical with this texture so that powers are, at this level, intrinsic properties of bodies. One piece of support for this is to be found when Locke says

> and in Substances, the most frequent [*Ideas*] are of Powers; *v.g. a Man is White*, signifies, that the thing that has the Essence of a Man, has also in it the Essence of Whiteness, which is nothing but a power to produce the *Idea* of Whiteness in one, whose eyes can discover ordinary objects. (III.viii.1)

Here Locke is *identifying* Essence and Power, and as we shall see, Locke calls textures 'real essences'. Thus I believe that a secondary quality *is* a texture. My discussion of Boyle's theory has prepared the ground for this view. It is also quite usual for commentators to say that words such as 'red' and 'sour' name secondary qualities but, consistently with the view I am putting, I take such words to name *ideas* of secondary qualities and *not* secondary qualities. This interpretation will be more fully supported in subsequent chapters. Here I quote one passage which appears to me to give positive support.

Locke says

> most of the simple *Ideas*, that make up our complex *Ideas* of Substances, when truly considered, are only Powers, however we are apt to take them for positive Qualities; *v.g.* the greatest part of the *Ideas*, that make our complex *Idea* of *Gold*, are Yellowness, great Weight, Ductility, Fusibility, and Solubility, in *Aqua Regia*, *etc.* all united together in an unknown *Substratum*; all which *Ideas* are nothing else, but so many relations to other Substances; and are not really in the Gold, considered barely in it self, though they depend on those real, and primary Qualities of its internal constitution, whereby it has a fitness, differently to operate, and be operated on by several other Substances. (II.xxiii.37)

On my view, this is to be read as saying that yellowness, ductility, and so on, are relational because they depend, *as they are presented to us in sense-experience*, upon other substances. Yellowness depends upon our sense-organs; when we observe gold dissolving we cannot but observe it dissolving in *something*. These, as they are observed by us are not qualities, and so not even secondary qualities, but ideas. The internal constitutions of bodies, describable in terms of the primary qualities of the corpuscles and the relations between them, include secondary qualities which are complex features of these internal constitutions and explain the yellow appearance of the body to us and its dissolving in *aqua regia*. The body's 'fitnesses' to operate differently upon other bodies, and to be operated upon, are its non-primary qualities. This passage is not saying that these qualities are relational; it is about ideas of these, which *are* relational. I shall argue later that just as primary qualities are intrinsic to single corpuscles, so non-primary qualities are intrinsic to collections of corpuscles.

The end of section 10 of Chapter viii deals with the third sort of qualities. Locke says

> To these might be added a third sort which are allowed to be barely Powers though they are as much real Qualities in the Subject, as those which I to comply with the common way of speaking call *Qualities*, but for distinction *secondary Qualities*. For the power in Fire to produce a new Colour, or consistency in Wax or Clay by its primary Qualities, is as much a quality in Fire, as the power it has to produce in me a new *Idea* or Sensation of warmth or burning, which I felt not before, by the same primary Qualities, *viz.* The Bulk, Texture, and Motion of its insensible parts. (II.viii.10)

Non-primary qualities of both sorts are powers of bodies to produce effects. Just as secondary qualities are textures causing ideas of them in me, so those of the third sort are textures causing changes in the textures of other bodies which in turn may cause new ideas in me.

Locke, of course, frequently talks of bodies as coloured, as in the above passage and he would not wish to prevent our talking in this way. However, what he would have us mean is this: when we say a body is coloured we cannot mean that it has in it something which is qualitatively identical with a seen colour; what we do or ought to mean is that the quality it has by virtue of which we see it as red or blue, and call it 'red' or 'blue', is the appropriate texture to cause, in conjunction with light of a certain texture, and the textures of our bodies, this or that idea in us. What the *quality* really is, is not the colour but the corresponding texture.

It is important to notice that Locke does not use the idea of resemblance between ideas and bodies or their qualities in defining, and so distinguishing, primary and secondary qualities. There is still a puzzle about how an idea can resemble a quality but there is no puzzle about how we can make the distinction between primary and secondary qualities on the basis of a forever inaccessible resemblance or non-resemblance. That simply is not the basis of the distinction and is not mentioned in the definition. The basis of the distinction is not to be found within, and by an analysis of, our experience or ideas. It is a theoretical distinction made in the service of a particular form of explanation and does not depend upon qualitative differences between our sensations. It arises as a result of the question: 'Is there any way of separating the qualities that bodies appear to have into two groups, one as small as possible and the other as large as possible, such that the smaller group can plausibly be used to explain the larger?' It is a question addressed to the understanding rather than to the senses and it is answered by a conceptual analysis of the nature of matter.

The tenability of this interpretation of the distinction between primary and secondary qualities depends upon the correctness of the view that Locke accepted Boyle's corpuscular hypothesis and used the word 'texture' as a technical term in the same way as Boyle. The view I am attributing to both can best be put in the form of an example. When two objects look, respectively, red and blue in the same light this is explained by their having different superficial textures and so absorbing and reflecting light corpuscles differently; when the same object looks a different colour under two different lights that is explained by the difference in texture of the light. The descriptions of these textures would have to be in terms of the arrangements of the primary qualities of the constituent corpuscles of the light and of the surfaces of the objects. Textures are what, in objects, correspond to and cause ideas of colours, etc., in us. So secondary qualities *are* textures.

Thus, in III.vi.9 Locke talks of 'that Texture of Parts, that real *Essence* that makes Lead, and Antimony fusible; Wood, and Stones not'. In II.viii.19 he says of porphyry: 'But whiteness or redness are not in it at any time, but such a texture, that hath the power to produce such a sensation in us.' In II.viii.20 he asks 'What real Alteration can the beating of the Pestle make in any Body, but an Alteration of the *Texture* of it?' Again in II.xxiii.11 he says

> Had we Senses acute enough to discern the minute particles of Bodies, and the real Constitution on which their sensible Qualities depend, I doubt not but they would produce quite different *Ideas* in us; and that which is now the yellow Colour of Gold, would then disappear, and instead of it we should see an admirable Texture of parts of a certain Size and Figure.

He also frequently talks of 'patterns' which can probably be regarded as equivalent to 'textures'.

Boyle, as we have seen, says of sensible qualities

> however we look upon them as distinct qualities, [they] are consequently but the effects of the often mentioned catholick affections of matter, and *deducible from* the size, shape, motion (or rest) posture, order, and the *resulting* texture of the insensible parts of bodies. (*O.F.Q.* 26 my italics, *S.* 36–7)

I think we may take it that Locke too regarded textures as *resulting from* the primary qualities of the corpuscles when they convene, and as constituted by their arrangements or patterns. This use of 'resulting from' is not a causal one. When corpuscles convene to form a complex there emerge spatial relations between them and the complex is describable entirely in terms of the primary qualities of the corpuscles together with these spatial relations. The description gives the texture. The texture is a resultant which is analysable entirely into the features mentioned in the description. Just so, a two-inch straight line may be the resultant of the 'convention' of two one-inch straight lines and, as we say, a force of a given magnitude acting in a given direction may be the resultant of two forces of different magnitudes acting at an angle to one another.

I can now turn to Locke's arguments in II.viii.16–21 which are usually taken as intended to give the grounds for making the distinction and as so feeble that any first-year philosophy student can refute them.[4] However, it follows from my view about how the distinction arose and its place in the corpuscular philosophy that the arguments of II.viii.16–21, about manna, porphyry, the pounding of almonds, heat from a fire and the felt temperature of water are not, as they are usually thought to be, intended to *make* and *establish* the distinction, at least in any direct way. The distinction is already made and these sections are intended to give *applications* and *explanations* of the distinction.

Their only place in the establishment of the distinction is that they contribute to showing that all phenomena may eventually be explained, simply and economically, in terms of the corpuscular hypothesis. It is often not noticed that the hypothesis is actually mentioned in each of these sections. Thus, if they are arguments at all, they are simply parts of the grand argument of the whole *Essay* for the corpuscular hypothesis and mechanical explanation. Locke was reporting scientific findings to the layman and showing how we could make sense of some superficially puzzling features of our everyday experience and our everyday descriptions of the world.

It is part of the hypothesis that our ideas of sensation are ultimately caused by complex arrangements of minute corpuscles, having only primary qualities, disposed in an indefinite number of different ways. We cannot observe the causes themselves and Locke later (IV.iii.13) makes it clear that he is doubtful of our ever *knowing*, in his strict sense, the specific texture responsible for a given idea, but we postulate causal relations between corpuscles, and groups of them, on the pattern of observed causal transactions between observed bodies. In terms of Boyle's 'two grand principles', matter and motion, we hope to explain the effects of bodies upon one another and upon our sense organs to produce ideas. This does not involve occult qualities for although the explanations involve things and qualities we do not observe, they do not involve any qualities of a *kind* we do not observe. Locke holds that *analogy* is what allows us to entertain hypotheses that go beyond the senses. He says

> The Probabilities we have hitherto mentioned, are only such as concern matter of fact, and such Things as are capable of Observation and Testimony. There remains that other sort *concerning* which, Men entertain Opinions with variety of Assent, though the *Things* be such, that *falling not under the reach of our Senses, they are not capable of Testimony.* Such are, 1. The Existence, Nature, and Operations of finite immaterial Beings without us; as Spirits, Angels, Devils, *etc.* Or the Existence of material Beings; which either for their smallness in themselves, or remoteness from us, our Senses cannot take notice of, as whether there be any Plants, Animals, and intelligent Inhabitants in the Planets, and other Mansions of the vast Universe. 2. Concerning the manner of Operation in most parts of the Works of Nature: wherein though we see the sensible effects, yet their causes are unknown, and we perceive not the ways and manner how they are produced. We see Animals are generated, nourished, and move; the Load-stone draws Iron; and the parts of a Candle successively melting, turn into flame, and give us both light and heat. These and the like Effects we see and know: but the causes that operate, and the manner they are produced in, we can only guess, and probably conjecture. For these and the like, coming not within the scrutiny of humane Senses, cannot be examined by them, or be attested by any body, and therefore can appear more or less probable, only as they more or less agree to Truths that are established in our Minds, and as they hold proportion to other parts of our Knowledge and Observation. *Analogy* in these matters is the only help we have, and 'tis from that alone we draw all our grounds of Probability. Thus observing that the bare rubbing of two Bodies violently one upon another, produces heat, and very often fire it self, we have reason to think, that what we call Heat and Fire, consists in a violent agitation of the imperceptible minute parts of the burning matter ... a wary Reasoning from Analogy leads us often into the discovery of Truths, and useful Productions, which would otherwise lie concealed. (IV.xvi.12)

I shall now consider in turn the examples in section 16–21.

The first (II.viii.16) concerns the different effects of fire, at different distances, on my hand. If I hold my hand at some little distance from the fire I feel heat but if I move it closer I begin to feel pain. It is common to think of the heat as in the fire and of the pain as not in the fire. This, according to Locke, is a mistake, but he is not, I believe, saying that my sensations show it to be a mistake. He is saying that we have no reason for accepting this view; nothing *in the sensations* shows that heat is in the fire and pain not. Equally, nothing *in the sensations* shows heat and pain to be both in the fire or both not in the fire. However, I do *feel* both; the ideas, the sensations, are both in me. The reason for not distinguishing between heat and pain, in respect of their location, is that both sensations can be explained by the communication of motion to the corpuscles of my hand by corpuscles set in motion by the corpuscles of the fire. So it seems sensible, and adequate, to say that, as ideas, heat and pain are both in me, which is obvious, but that they have correlates and causes in the fire which have quite different characteristics from the ideas. Moreover their causes are of the same general kind since they are both motions of corpuscles differing in speed. The first reference to the corpuscular explanation comes in the final sentence: 'Why is Whiteness and Coldness in Snow, and Pain not, when it produces the one and the other *Idea* in us; and can do neither, but by the Bulk, Figure, Number, and Motion of its solid Parts?' (II.viii.16).

This reference is continued in II.viii.17 where the example of manna is introduced: 'The particular *Bulk, Number, Figure and Motion of the parts of Fire, or Snow, are really in them*, whether any ones Senses perceive them or no: and therefore they may be called *real Qualities*, because they really exist in those Bodies. But *Light, Heat, Whiteness, or Coldness, are no more really in them than Sickness or Pain is in* Manna.'[5] In 18 Locke goes on to say that a piece of manna produces in us the idea of its shape and, by being moved about, the idea of motion. These ideas represent the qualities as they really are in the manna: 'A Circle or Square are the same, whether in *Idea* or Existence; in the Mind, or in the *Manna* ...' However, besides this the manna by the bulk, figure, texture and motion of its parts has the power to produce the sensations of sickness and pain in us. This people readily accept but it is not easy to bring them to accept that sweetness and whiteness are not really in manna and that these are the effects of the operation of the particles of manna on our palate and eye, just as sickness and pain are the effects of their operations on the stomach and gut. There is nothing to distinguish sweetness and whiteness, on the one hand, and pain and sickness, on the other, in respect of location, since both pairs are ideas which can be explained by reference to the corpuscles of the manna without supposing anything in the manna like the ideas.

In II.viii.19 Locke says that under normal conditions porphyry looks red and white but

> Hinder light but from striking on it, and its Colours Vanish; it no longer produces any such *Ideas* in us: Upon the return of Light, it produces these appearances in us again. Can any one think any real alterations are made in the *Porphyre*, by the presence or absence of Light; and that those *Ideas* of whiteness and redness, are really in *Porphyre* in the light, when 'tis plain *it has no colour in the dark?* It has, indeed, such a Configuration of Particles, both Night and Day, as are apt by the Rays of Light rebounding from some parts of that hard Stone, to produce in us the

Idea of redness, and from others the *Idea* of whiteness; But whiteness or redness are not in it at any time, but such a texture, that hath the power to produce such a sensation in us.

Could this be intended to *make* the primary/secondary quality distinction, by pointing to differences in our ideas of them? In the first place, white and red are treated as ideas from the beginning, whereas textures are treated as undoubted qualities of the porphyry. The notion of a configuration of particles clearly does not come directly from experience; what place could this theoretical notion occupy in a mere comparison of different *experiences*? If this passage were intended as it is usually taken is it likely that Locke would not have mentioned the very obvious fact that extension remains when we remove the light, because we can still feel it? He does not do so because, I suggest, what he is supporting is his belief that the appearance to sight of both the colour and the extension in the light, the appearance to touch of extension in the dark and the failure of sight to show us colour in the dark, can *all* be explained in terms of the action upon us of configurations of corpuscles, a theoretical notion postulated for explanatory purposes.

The next example, in II.viii.20, concerns the result of pounding an almond in a mortar. The section reads

Pound an Almond, and the clear white *Colour* will be altered into a dirty one, and the sweet *Taste* into an oily one. What real Alteration can the beating of the Pestle make in any Body, but an Alteration of the *Texture* of it?

Pestles and mortars are size- and shape-changing devices, not colour- or flavour-changing devices. Sometimes their use to change size and shape is accompanied by a change of colour, sometimes not. It is plausible to explain changes of colour, if they occur, in terms of the particular way in which the substance is broken up, the particular change in its texture. Can we conceive of *either* the colour of the pestle and mortar causing a change of texture *or* the primary qualities of the pestle and mortar as operating directly, that is, not *via* textures, on colours to change them? Perhaps we could but the explanation would have to be very much more complex. Locke may be thought of as issuing a challenge: show me how the pounding could produce a change of colour *except through* change of texture.

Note that Locke does not say that the shape and size of the almond remain the same while the colour and flavour change. He is not confusing determinable and determinate properties as O'Connor and Mackie suggest.[6] He is saying that one set of changes can be explained in terms of the other set.

The final example occurs in II.viii.21 and seems to me to show clearly that my interpretation is near the mark. Locke there says

> *Ideas* being thus [*sc.* in this way, *not* by these arguments] distinguished and understood, we may be able to give an Account [*sc.* explain], how the same Water, at the same time, may produce the *Idea* of Cold by one Hand, and of Heat by the other: Whereas it is impossible, that the same Water, if those *Ideas* were really in it, should at the same time be both Hot and Cold. (My glosses.)

This shows that Locke was using the distinction between primary and secondary qualities as central in a manner of explanation. We can explain how the ideas of heat and cold may be set up in us by something that is, properly speaking, neither hot nor cold. The corpuscles of the water are moving at a certain speed which is greater than the speed of the corpuscles of the cold hand and less than the speed of the corpuscles of the warm hand; the consequent increase in speed of the corpuscles of one hand and decrease in speed of the corpuscles of the other hand produce in us respectively the ideas of heat and cold. We cannot suppose that the water is both hot and cold but when the situation is rightly understood there is no need to suppose that; the senses do not mislead us when we see that they give information about the reaction between the water and our hands and not merely about the water.

I believe that my interpretation is supported by II.viii.22 where Locke says that he has just 'engaged in Physical Enquiries a little farther than perhaps, I intended' for the purpose of making clear, not *making*, '*the difference between the Qualities in Bodies, and the* Ideas *produced by them in the Mind*'. He says 'I hope, I shall be pardoned this little Excursion into Natural Philosophy'. He has been explaining and applying a piece of physical science. It was necessary to do this because that is where the basis of the distinction lies; more precisely, it lies in the theoretical requirements of mechanical explanation by means of corpuscles, rather than in any observationally detected differences between the qualities.

Those who are familiar with Locke will have noticed my avoidance of a statement which appears to present difficulties for my interpretation. In II.viii.21 Locke says that by his explanation

> we may understand, how it is possible, that the same Water may at the same time produce the Sensation of Heat in one Hand, and Cold in the other; which yet Figure never does, that never producing the *Idea* of a square by one Hand, which has produced the *Idea* of a Globe by another.

This suggests that primary qualities are never subject to illusions as secondary qualities are, a suggestion that Berkeley was quick to seize upon. However, it does not entail this and Locke cannot have thought it did since he mentions illusions involving primary qualities, and even shape, in various places (*e.g.* II.iv.8 , II.xiv.6 and II.xxi.63). If he were supposing that shape is never subject to this particular kind of illusion this would mean that there were phenomena involving heat which can be explained on the corpuscular hypothesis but no analogous phenomena involving shape. On the other hand, if there were such illusions involving shape they too could be explained on the hypothesis. What we cannot say is that a body is both hot and cold at the same time; this would be contradictory just as it would be to say that a body is cubical and spherical at the same time. The possibility of illusion would be a basis for distinguishing secondary from primary qualities only if primary qualities were never subject to illusion, which is something Locke did not accept.

The upshot of all this is that in II.viii.16–21 Locke was not attempting to give a philosophical justification, based on experience, for the primary/secondary quality distinction and, in particular, he was not arguing for the distinction on the basis of the invariability and freedom from illusion of our perceptions of primary qualities in contrast to our perceptions of secondary qualities. Locke could not have missed either the fact that the shape and size of an object may appear differently from different points of view or the significance of this fact for this argument, if that were what it was. His claim, I suggest, is that our ideas of both primary and secondary qualities, and of variations in them, are all to be explained (if at all), in terms of the corpuscles involved. I think that Locke, if pressed, would have said that having an accurate idea of something does not, and must not, mean having the *same* perception of it under every condition; for a visible thing to be of a given size *just is*, in part at least, for it to 'look smaller' at 50 feet than it does at 10 feet.

If I am right, criticisms of Locke which allege that he has not shown that primary qualities are not subject to illusion, whereas secondary qualities are, are beside the point. Other standard criticisms are also misplaced.

For example, it is said that an empiricist cannot consistently hold that ideas of sensation represent, and are caused by, unperceived external objects such as the corpuscles. *Ex hypothesi* there can be no direct perceptual evidence for them and how could their existence be established *a priori*, by reason alone? Locke's argument is, I believe, such that this criticism does not arise. The existence of external objects, of whatever sort, is a hypothesis; we do not know, through our ideas, that there are such things but we can most adequately and economically explain our ideas on that hypothesis. Thus *knowledge* of these external objects is not claimed, either through sense-experience or reason (*see* IV.ii.); this is probable conjecture. The constitution of external objects, their being collections of corpuscles having certain specified qualities, is a further detailed hypothesis. What makes these hypotheses respectable for an empericist is that the qualities attributed to the corpuscles are all of kinds we encounter in experience and that the hypotheses would be supported indirectly if they enabled us to give an adequate account of the world and our experience.

This, of course, depends upon a certain view of the character of Locke's empiricism. As far as I can see he does not hold the principle that nothing must figure in our account of the world unless it has been, or could be, observed. He wants to show 'whence the Understanding may get all the Ideas it has' (II.i.1); that it is in experience that 'all our knowledge is founded; and from that it ultimately derives itself' (II.i.2); and that sensation and reflection together supply '*our Understandings with all the materials of thinking*' (II.i.2). This implies that the understanding is able to work on these materials and go beyond experience. He nowhere appears to be committed to the slogan 'Nothing in the mind which was not first in the senses or reflection'; at most he is committed to some such more moderate slogan as 'Nothing in the mind that is not somehow connected with the senses'. The only qualities attributed to objects, whether sensible or insensible, are based upon ideas gained through experience; no reference is necessary to qualities *of a kind* of which experiences could never give us accurate ideas. This is the empiricism of the natural philosopher and is perfectly respectable if we recognize, as Locke did, that natural philosophy may not give us knowledge, in the strict sense, but may yet give us something valuable, namely probable 'conjecture' (IV.xvi.12, quoted above).

In his first letter to Stillingfleet Locke said

> I am sure the author of the Essay of Human Understanding never thought, nor in that Essay hath any where said, that the ideas that come into the mind by sensation and reflection are all the ideas that are necessary to reason, or that reason is exercised about... All that he has said about sensation and reflection is, that all our simple ideas are received by them, and that these simple ideas are the foundation of all our knowledge, forasmuch as all our complex, relative, and general ideas are made by the mind, abstracting, enlarging, comparing, compounding and referring, etc. these simple ideas, and their several combinations, one to another. (*L.S.* 11)[7]

6

WHICH QUALITIES ARE PRIMARY?

Locke's critics have been troubled by the fact that he gives various different lists of primary qualities. There is continuing controversy about whether he regarded texture, solidity, number and situation as primary qualities.[1] I now wish to consider whether my contention that Locke accepted from Boyle the corpuscular hypothesis and the distinction between primary and secondary qualities essential to that hypothesis will help us to decide firmly upon a list of these qualities that Locke took to be primary or, at least, upon a list which would have sufficed for his purposes. It should be borne in mind that if I am right Locke's hesitation about the exact list may have been due to his deference to the natural philosophers since the primary qualities would be just those that the natural philosopher finds sufficient for explaining physical phenomena. So the decision about the correct list is for the natural rather than the metaphysical philosopher to make.

Locke is often said to have been just careless and inconsistent about this. My view opens up other possibilities. If it is a hypothesis of natural science that the primary qualities are just those that suffice for explaining physical phenomena then it may be that Locke was simply expounding, and considering the implications of, a currently favoured method of explanation and that he was more concerned with the grounds for the distinction than with the composition of the lists of qualities arising from those grounds. I shall assume, to start with, that Locke had a consistent view even if he was not always as careful as he might have been about expounding it.

I shall approach this by considering the hypothesis that Locke accepted as primary, or at least did not reject, just those qualities that Boyle regarded as primary. As I have pointed out in Chapter 3, Boyle explicitly and consistently regarded shape, size and mobility as the three primary qualities. In the hope of overcoming some initial resistance to my hypothesis I shall just point out here that the lists of qualities in the *Essay* that appear to be, and are often taken to be, intended as lists of primary qualities may not all have been intended as such and that the differences between the lists may be explained in terms of differences in the purpose with which, and the context in which, Locke puts them forward.

We see in III.x.15 that Locke regards *matter* in a similar way to Boyle, for he says

Yet *Matter* and *Body*, stand for two different Conceptions, whereof the one is incomplete, and but a part of the other. For *Body* stands for a solid extended figured Substance, whereof *Matter* is but a partial and more confused Conception, it seeming to me to be used for the Substance and Solidity of Body, without taking in its Extension and Figure: And therefore it is that speaking of *Matter*, we speak of it always as one, because in truth, it expresly contains nothing but the *Idea* of a solid Substance, which is every where the same, every where uniform.

That is, matter is a solid stuff, common to all bodies, whatever their shapes, sizes and other properties, and its solidity is absolute.[2]

I believe that Locke also sees *qualities* as Boyle does, namely as what differentiate bodies from one another and, some of them, allow us to distinguish them by means of ideas caused in us by qualities. He says

our senses failing us in the discovery of the Bulk, Texture, and Figure of the minute parts of Bodies, on which their real Constitutions and Differences depend, we are fain to make use of their secondary Qualities, as the characteristical Notes and Marks, whereby to frame *Ideas* of them in our Minds, and distinguish them one from another. All which secondary Qualities, as has been shewn, are nothing but bare Powers. (II.xxiii.8)

The differences between bodies lie in the bulk, texture and figure of their minute parts. Solidity is not mentioned, no doubt, because it is absolute and so could not be responsible for differences between bodies. If we could directly observe the bulk, figure and texture of the minute parts of bodies that would enable us to distinguish bodies from one another. As we cannot, we are 'fain to make use of' secondary qualities. As I have argued, secondary qualities are textures; so we 'make use of textures' because they have 'effects' on us which do not amount to the perception of textures as they are; they appear to us as colours, etc. To say that we make use of secondary qualities is not to deny that we also make use of the large scale representatives of primary qualities.

Again, in II.viii.26 Locke says

beside those before mentioned *primary Qualities* in Bodies, *viz*. Bulk, Figure, Extension, Number, and Motion of their solid Parts; all the rest, whereby we take notice of Bodies, and distinguish them one from another, are nothing else, but several Powers in them, depending on those primary Qualities; whereby they are fitted ... to produce several different *Ideas* in us...

We distinguish by means of different ideas of secondary qualities, presumably along with ideas of the large scale primary qualities size and shape, but our different ideas of secondary qualities are produced by different arrangements of corpuscles probably having different primary qualities. Again, solidity is not mentioned among the list of qualities; they are said, as so often by Locke, to be qualities of the *solid parts*; solidity is common to the corpuscles, cannot differentiate them and so is not listed as a quality here. Locke often, in this way, treats solidity differently from the qualities, as has often been noticed without explanation.

It must also be remembered that in II.viii.8 Locke says 'the Power to produce any *Idea* in our mind, I call *Quality* of the Subject wherein that Power is'. The power to produce the idea 'round' in us is the roundness of the snowball; the powers to produce the ideas 'white' and 'cold' are certain arrangements of corpuscles; but the idea of absolute solidity is not produced in us in either of these ways. This idea is produced in us by our understanding by augmentation or extrapolation.

Boyle is much clearer than Locke on this question but it is plausible to suppose that Locke follows Boyle in this as he does in so much else. Moreover, from the sense of 'quality' there does not seem much point in regarding as qualities anything in which bodies cannot differ. However, if the evidence is not completely convincing I would ask the reader to entertain it as a hypothesis that Locke takes the same view as Boyle about qualities and solidity and to withhold a final judgement until it is clear what the consequences are and how much appears to become clearer with its help.

It should be said that, as I have put the matter so far, the primary qualities are *determinable*, rather than determinate, shape, size and mobility since what a single corpuscle must have is some shape, some size and some mobility but no particular exemplars of them. So, for accuracy, I must say that the primary qualities are determinables whose determinates are capable of differentiating bodies.

In suggesting that Locke's primary qualities are the same as Boyle's, that is, shape, size and mobility, I have to explain why Locke, in various different lists, appears to include several other qualities among primary qualities. I wish to argue that Locke did not intend these all to be lists merely of primary qualities; that they cannot be in view of his definition of primary qualities discussed in Chapter 5 above; and that he can be defended against the charge of inconsistency by considering what he was doing when he put forward the various lists.

Locke's shortest list is 'Bulk, Figure, or Motion' in II.viii.13 and II.viii.25 where he is clearly talking of corpuscles or parts of bodies. It should perhaps be said that when Locke talks of 'the minute parts', 'the insensible parts' or even 'the parts' of bodies he is almost always referring to corpuscles. At II.viii.9, where he is defining primary qualities, he gives '*Solidity, Extension, Figure* and *Mobility*' and at II.x.6 he gives '*Solidity,*

Extension, Figure, Motion, and *Rest*'. However, in other places he includes, with or without some of these, texture, number, situation and motion of parts. A composite list of all the items that appear in any list that looks like a list of primary qualities is: *solidity, bulk, figure, mobility, situation, number, texture* and *motion of parts*. Thus, if I am to establish my hypothesis I must argue that solidity, texture, situation, number and motion of parts are not to be taken as primary qualities, in spite of appearances.

I have suggested that the definition in II.viii.9 leaves us with '*Solidity, Extension, Figure* and *Mobility*' as essential properties of body, even of a single isolated corpuscle. We might conclude that these are the primary qualities according to his definition. They are, in fact, just those mentioned in the example of the division of a grain of wheat at the end of that paragraph. He does not, however, say unequivocally here that these are the primary qualities. What he does say is that these are inseparable from bodies.

Book II, Chapter viii is clearly central for my argument since it is the main part of the *Essay* in which Locke defines and discusses primary and secondary qualities. In this chapter there are approximately 25 lists that appear to be lists of primary qualities. I say 'approximately' because it is sometimes difficult to decide on the purpose of a particular list. In 9 of these lists Locke includes *number*, in 7 he includes *texture*, in 4 *solidity*, in 3 *solid parts* and in 1 *situation*. Seven lists do not contain any of these. If his definition taken alone seems to point to solidity, shape, size and mobility as the correct list, why does he include texture, number, situation, and motion of parts in some of these lists?

The first step in answering this question is to point out that he puts these lists forward in three different contexts, which may help to explain the differences between the lists. The contexts are

(1) when he is talking of body as such;
(2) when he is talking of observable bodies;
(3) when he is talking of the causes either of ideas in us or of changes in other bodies.

When he is talking of body as such, as when he defines 'primary quality' (II.viii.9), we might expect to find his definitive list and here the first two lists he gives are solidity, extension, figure and mobility, in his example of the division of the grain of wheat. The smallest possible body will be a corpuscle and it must retain these, even in isolation from all other corpuscles. In II.viii.13, where he is again talking of insensible particles he says that each has bulk, figure and motion.

It is important to distinguish mobility from motion. In 14 of the lists 'motion' clearly refers to the motions of the parts, described as 'insensible' or 'solid', of observable bodies, *i.e.* the corpuscles. Body as such, *e.g.* a single corpuscle, need have no parts so motion of parts, according to the definition, cannot be a primary quality; the primary quality is mobility, which occurs in connection with the definition and as 'motion or rest' in other lists that do not mention parts. Even a single corpuscle, alone, has mobility, the ability to be in motion. There are problems here; I discuss them later in this chapter.

Similarly, 'number' occurs along with 'parts', solid or insensible, except in two lists where Locke is clearly talking about the causes of our ideas, and I suggest that it *always* refers to numbers of corpuscles and so is not a primary quality, according to the definition. It would be odd to regard number as a primary quality since before the first corpuscle could be counted it would have to be differentiated from others; so its number cannot be part of what differentiates it, *i.e.* a quality. When Locke discusses number in II.xvi he does not talk of it as a quality.

Any observable body, so the hypothesis goes, is composed of clusters or conventions of corpuscles and so must have, independently of any observer, a number of corpuscles in motion or at rest relatively to one another and the whole must be in motion or at rest relatively to other clusters. So when Locke is discussing observable bodies it is often appropriate to mention number and motion as belonging to the bodies, whether they are observed or not. Although body as such cannot have number and actual motion or rest as essential properties, the observability of bodies depends upon their consisting of numbers of corpuscles in motion or at rest relatively to one another. When one external body acts upon another our information about this must come from changes we observe in them, that is, from ideas caused by them in us. Observable actions between bodies depend upon numbers of corpuscles and their motions, as do ideas caused in us. So it is also appropriate for Locke sometimes to mention these when he is talking of the causes of changes in external bodies and the causes of our ideas.

A single corpuscle, alone in the universe, cannot have *situation.* In II.viii.23, in which occurs the only list in that chapter that includes situation, Locke is again talking of complex, observable bodies. The same is true of what is probably the only other list in the *Essay* in which he includes situation (II.xxiii.9). In both of these motion of parts is included and he appears to be talking of the situations of the corpuscles in the cluster relative to one another. When he discusses 'place', which appears to mean the same as 'situation', in II.xiii.7–10, Locke nowhere suggests that it is a primary quality.

Of course, if Locke believed in absolute space, then place, or situation, would be a primary quality since even a single corpuscle in otherwise empty space would have it. There is, however, evidence that he does not believe in absolute space to be found in his discussion of place. He says

> That our *Idea* of Place, is nothing else, but such a relative Position of any thing, as I have before mentioned, I think, is plain and will be easily admitted, when we consider, that we can have no *Idea* of the Place of the Universe, though we can of all the parts of it; because beyond that, we have not the *Idea* of any fixed, distinct, particular Beings, in reference to which, we can imagine it to have any relation of distance; but all beyond it is one uniform Space or Expansion, wherein the Mind finds no variety, no marks. (II.xiii.10)

Moreover, he goes on,

> to say that the World is somewhere, means no more, than that it does exist; this though a Phrase, borrowed from Place, signifying only its Existence, not Location... (II.xiii.10)

Thus, it seems, a single corpuscle in empty space could not be said to have place, position, situation.

There are two conceptions of absolute space. According to the first, empty space is independent of objects but provides a frame of reference which would allow us to say that a single object in it had a place, or location. I believe that Locke is rejecting this conception in the passages quoted. When he says, in the first of these quotations that 'all beyond it [i.e. the Universe] is one uniform Space or Expansion, wherein the Mind finds no variety, no marks' I believe he is accepting the second and less problematic conception and, at the same time, rejecting the first. According to the second conception, empty space may exist independently of objects but merely as an infinite emptiness, empty even of 'marks' and 'variety', into which objects could be put. If we consider the Universe as one object in empty space it does not make sense to say that it has a place because space would have to have 'marks' by reference to which the universe could have a place. So Locke rejects empty space that provides a frame of reference but not empty space that merely has room for objects. The existence of empty space, in the second sense, is an essential part of atomism. Aristotle and Descartes, among others, had rejected the existence of empty space in both senses.

Some of Locke's lists that appear to be lists of primary qualities may in fact be, rather, lists of what it is in bodies that must figure in causal explanations of our ideas or of interactions between external bodies consisting of clusters of corpuscles. If so, it would be appropriate, though not essential, to include in them qualities of bodies which are 'emergent', that is, which are not essential to body as such, although they depend entirely on body, but which arise when numbers of corpuscles 'convene', as Boyle says. That is, it would be appropriate to include non-primary qualities of bodies. It would be appropriate to mention the solid parts of bodies (i.e. corpuscles) and their arrangements or textures since particular arrangements are thought to be essential to produce particular ideas in us and particular effects upon the arrangements of corpuscles in other bodies. This may be why number and motion of parts, situation and texture appear on some of the lists.

In this connection it is perhaps worth looking in more detail at one or two of the lists in their contexts. Consider, first, Locke's list immediately after his definition of primary qualities. He says

> Take a grain of wheat, divide it into two parts, each part has still *Solidity*, *Extension*, *Figure*, and *Mobility*; divide it again, and it retains still the same qualities; and so divide it on, till the parts become insensible, they must retain still each of them all those qualities. For division ... can never take away either Solidity, Extension, Figure, or Mobility from any Body. (II.viii.9)

And, he might have gone on, if all except one of the corpuscles which are the ultimate products of division were annihilated, that one remaining corpuscle must retain these qualities and could have no others.

At the end of the paragraph, however, he has a list that appears to conflict with the view that this is the correct list, for he says

> These I call *original* or *primary Qualities* of Body, which I think we may observe to produce simple *Ideas* in us, *viz.* Solidity, Extension, Figure, Motion, or Rest, and Number. (II.viii.9)

However, he has here moved from talking about whatever cannot be removed from body to talking about ideas. The final list can be read as a list of ideas produced in us. In any case, he is talking of the production of ideas in us and an explanation of that essentially involves reference to *actual* motions, rather than mobility, and to *numbers* of constituent corpuscles. The sentence in the passage before the one last quoted reads, in full,

> For division (which is all that a Mill, or Pestel, or any other Body, does upon another, in reducing it to insensible parts) can never take away either Solidity, Extension, Figure, or Mobility from any Body, but only makes two, or more distinct separate masses of Matter, of that which was but one before, all which distinct masses, reckon'd as so many distinct Bodies, after division make a certain Number. (II.viii.9)

The whole paragraph, we can now see, may be interpreted as referring to solidity, extension, figure and mobility as the primary qualities, to number and actual motion as arising on division and to solidity, extension, figure, motion and rest, and number as among both the simple ideas caused in us and the causes of those ideas. This involves only the plausible assumptions that 'These' refers back to 'Solidity, Extension, Figure, and Mobility' and that '*viz.*' refers back to '*Ideas*'. One should not assume that Locke's English style is never good.

A second passage allows me to make a similar point and so to reinforce my view. At II.viii.13 Locke says

> After the same manner, that the *Ideas* of these original Qualities are produced in us, we may conceive, that the *Ideas* of *secondary Qualities* are also *produced*, viz. *by the operation of insensible particles on our Senses*. For it being manifest, that there are Bodies, and good store of Bodies, each whereof is so small, that we cannot, by any of our Senses, discover either their bulk, figure, or motion, as is evident in the Particles of the Air and Water, and other extremely smaller than those, perhaps, as much smaller than the Particles of Air, or Water, as the Particles of Air or Water, are smaller than Pease or Hail-stones. Let us suppose at present, that the different Motions and Figures, Bulk, and Number of such Particles, affecting the several Organs of our Senses, produce in us those different Sensations, which we have from the Colours and Smells of Bodies, . . .

Locke here talks of the 'bulk, figure, or motion' of the insensible particles themselves and begins to talk of 'Number' (of such particles) when he comes to talk of bodies affecting our sense-organs and causing ideas in us. Such effects must involve numbers of moving corpuscles, each having in itself only bulk, figure and mobility.

Now let me turn to textures. I have shown that, for Boyle, 'texture' is a technical term referring to an arrangement or pattern of corpuscles, in motion or at rest relatively to one another, the corpuscles individually having only the primary qualities shape, size, and mobility. I have argued that Locke follows Boyle in this. In view of this I believe that the alleged lists of primary qualities when they occur in the *Essay* during discussions of observable objects and their interactions or of the causes of our ideas are in fact lists of primary qualities plus such qualities as emerge when corpuscles are gathered together into groups forming complex bodies. Thus I think the composite list should be regarded as comprising the primary qualities of the solid constituents of any complex body (shape, size and mobility) together with the number, actual relative positions and motions of its constituent corpuscles, the whole of which constitutes the texture of the complex. The observability of bodies, i.e. their power to affect our senses, and the actions of observable bodies upon one another depend upon such textures, which constitute their non-primary qualities. The complete description of a texture would have to mention all the other items in the composite list and different textures would explain the different effects of bodies upon our senses and other external bodies.

The four lists in which Locke includes texture are in sections 10, 14, and 18 of II.viii. For example, in 10, one list is 'Bulk, Figure, Texture, and Motion of Parts'. There are two points about these lists to which I wish to draw attention. The first is that each of them contains 'motion of (insensible) parts', showing that complex bodies are being discussed. The second is that each list is given in the context of a discussion of the causes of ideas of secondary qualities in us. The list in section 18 is in a sentence which does not even use the expression 'primary qualities'. I suggest that these lists could be read in the spirit of my quotation from Boyle (p.123) as 'bulk, figure, the *resulting* texture, and motion of parts'.

However, that looks odd at first sight if the primary qualities are bulk, figure and mobility. The motion or rest of a corpuscle contributes to texture just as its bulk and figure do, so why does 'texture' not come at the end of the list as a resultant of the other three? This, too, can be explained. Motion of parts is not a primary quality since a single corpuscle has no parts. In a complex body the corpuscles (parts) may be in motion relatively to one another or they may not. Their rest or motion contributes to the texture. The simplest case, however, is that in which they are at rest and then texture depends merely upon bulk, figure and their relations. If they are in motion that is an additional factor contributing to the texture but it is not a necessary condition of there *being* a texture.

I have said that Boyle makes it quite clear that there are three primary qualities, shape, size and mobility, and my hypothesis is that Locke accepted this. Boyle refers to solidity as essential to body but not as a primary mood or affection (quality). I have not argued so far that Locke does *not* regard solidity as a primary quality and it indeed figures in lists that appear to be lists of primary qualities. I have, however, given some evidence for thinking that Locke, like Boyle, regards qualities as whatever in bodies may differentiate them from one another; if that is so Locke may, like Boyle, regard solidity as essential to matter but not a *quality* of it.

It is clear from Locke's discussion of solidity (II.iv) that he distinguishes between the absolute solidity of corpuscles and the relative solidity of bodies composed of numbers of corpuscles. (This chapter occurs before Locke has introduced the term 'quality', so no weight can be put on the fact that he does not here *call* it a quality.) Observable bodies are more or less hard, more or less penetrable, but felt hardness is not the same as the absolute solidity of the corpuscles because felt hardness and softness depend upon empty spaces between the corpuscles of a body or the strength of their 'cohesion' (II.iv.4–5). The more space there is, the more compressible, and so soft, the body is. Absolute solidity, Locke says, 'consists in repletion' (II.iv.4) or is 'Impenetrability' (II.iv.1) or is 'the *Idea* belongs to Body whereby we conceive it *to fill space*' (II.iv.2). The only body that completely fills the space within its boundaries is a corpuscle; it is, as it were, whole matter, pure matter, i.e. *solid* matter; and no two portions of pure matter can occupy the same portion of space at the same time. Thus all corpuscles are equally solid because they are absolutely solid so solidity cannot ever be what differentiates, or part of what differentiates, one corpuscle from another; therefore, although it is essential it is not a quality. Solid matter is what ultimately *has*, or is qualified by, the primary qualities; pieces of solid matter may have different shapes, sizes and mobilities but not different degrees of solidity.

We get the idea of absolute solidity through our senses although absolute solidity is not strictly observable: 'The *Idea* of *Solidity* we receive by our Touch...' (II.iv.1). That is, we get the idea of observable bodies' being more or less impenetrable but we arrive at the idea of body as such being absolutely impenetrable or solid by extrapolation. As Locke says

> and though our Senses take no notice of it, but in masses of matter, of a bulk sufficient to cause a Sensation in us; Yet the Mind, having once got this *Idea* from such grosser sensible Bodies, traces it farther; and considers it, as well as Figure, in the minutest Particle of Matter, that can exist; and finds it inseparably inherent in Body, where-ever or however modified. (II.iv.1)

He goes on to distinguish solidity from 'the ordinary Idea of Hardness' (II.iv.3 and 4). Talking of bodies filling space, he says 'This *Idea* of it [*sc.* Solidity], the Bodies which we ordinarily handle sufficiently furnish us with' (II.iv.2: note 'sufficiently') and Solidity, he says, is 'nowhere else to be found or imagin'd, but only in matter' (II.iv.1: note 'matter').

When Locke introduces 'Quality' in II.viii.8, he says it is 'the Power to produce any *Idea* in our mind'. There is a sense in which the solidity of a corpuscle does not have the power to produce any idea in our mind; it does not produce any *particular* idea in our mind but is involved ultimately in the production of *every* idea relating to bodies. The idea of the length of a rod is produced by the lengths of the corpuscles and of the spaces between them and is just the sum of these. The ideas of shape and size produced in us by observable bodies are exactly of the kind we attribute to corpuscles; the idea of solidity produced by them is not. The solidity we attribute to corpuscles is a limiting idea derived from this.

Problems arise for my view, but also some support for it, at the end of Locke's long chapter on Powers. There he says

> And thus I have, in a short draught, given a view of our *original Ideas*, from whence all the rest are derived, and of which they are made up; which if I would consider, as a Philosopher, and examine on what Causes they depend, and of

what they are made, I believe they all might be reduced to these very few primary, and original ones, *viz.*

Extension,
Solidity,
Mobility, or the Power of being moved; which by our Senses we receive from Body:
Perceptivity, or the Power of perception, or thinking;
Motivity, or the Power of moving; which by reflection we receive from our Minds. (II.xxi.73)

It is unfortunate for my argument that the list of ideas received from body includes solidity but excludes shape. However, this is a list of original or primary *ideas*, not qualities and certainly my interpretation takes the idea of solidity to be primary among our ideas of bodies and to be derived, even if indirectly, through our senses. The exclusion of shape may be more apparent than real since the extension of a body, fully specified, would give its shape.

He immediately goes on to say

To which if we add
Existence,
Duration,
Number;
which belong both to the one [*sc.* ideas we receive from bodies], and the other [*sc.* ideas we receive from mind], we have, perhaps, all the Original *Ideas* on which the rest depend. For by these, I imagine, might be explained the nature of Colours, Sounds, Tastes, Smells, and all other *Ideas* we have, if we had but Faculties acute enough to perceive the severally modified Extensions, and Motions, of these minute Bodies, which produce those several Sensations in us. (II.xxi.73)

I suggest that he adds these further ideas because he is moving on to consider explanations of ideas that depend upon large collections of corpuscles which makes it relevant to include, for example, numbers of corpuscles. This, too, explains the longer list at the very end of this section. The closing sentence reads, in part,

it sufficing for my purpose to observe, That Gold, or Saffron, has a power to produce in us the *Idea* of Yellow; and Snow, or Milk, the *Idea* of White; which we can only have by our Sight, without examining the Texture of the Parts of those Bodies, or the particular Figures, or Motion of the Particles, which rebound from them, to cause in us that particular Sensation: though when we go beyond the bare *Ideas* in our Minds, and would enquire into their Causes, we cannot conceive any thing else, to be in any sensible Object, whereby it produces different *Ideas* in us, but the different Bulk, Figure, Number, Texture, and Motion of its insensible Parts. (II.xxi.73)

This last list, as I have suggested earlier, is a list of those qualities of collections of corpuscles needed to explain various ideas and so includes qualities over and above the primary qualities of the individual corpuscles.

The ideas of existence and duration sit oddly in these lists, in view of what Locke says about them elsewhere. Of existence Locke says that it is an idea

> suggested to the Understanding, by every Object without, and every *Idea* within. When *Ideas* are in our Minds, we consider them as being actually there, as well as we consider things to be actually without us; which is that they exist, or have *Existence*. (II.vii.7)

Existence is thus not a *quality* of things because it is suggested by all of them and does not differentiate them. Of duration Locke says that we come by it from the succession of ideas in our minds and

> the distance between any parts of that Succession, or between the appearance of any two *Ideas* in our Minds, is that we call *Duration*. (II.xiv.3)

Thus the idea of duration comes from relations between other ideas and is not, as it were, an idea independent of others. It is therefore strange to find Locke referring to it as an original idea if we regard original ideas as ideas of primary qualities. It seems to me that it is clear that, in this section at least, he is not using the word 'original' in that way. However, if we are explaining the effects of aggregates of corpuscles and this involves referring to actual motions of the corpuscles, duration must enter as fundamental in the explanation. There is much to be explored here but I do not propose to explore it for the present.

It might be thought that I have had to do an unwarranted amount of interpretation in arguing for my hypothesis. I have preferred to assume that Locke was not seriously inconsistent until I cannot sustain that view any longer. I think that the amount of interpretation needed when we read a seventeenth-century text, especially without references to its context, is often underestimated by readers of philosophy. There are several reasons why my interpretation may be necessary.

1. Locke may not have needed, for his purpose, to say precisely what qualities he took to be primary.
2. He may have been genuinely undecided about what to include.
3. Because his main intention was to outline a method of explanation rather than to go into details, he may have been careless in writing down some of his lists, although what he meant was consistent.

The first two of these are both scientific matters: the needs of scientific explanation are what determine which qualities are to be taken as primary. Locke may have thought that Boyle and others had already made it clear what primary qualities were required for the theory at that stage and so there was no need for him (Locke) to go into this in detail. I do not claim to have shown conclusively what Locke's view on this matter was but I think I have given enough evidence to lend considerable weight to my hypothesis that he acccepted or, at least would not dispute, Boyle's list of primary qualities: shape, size and mobility. I have at least shown, I think, that this list would have sufficed for his purpose and does not conflict with his views put forward in the rest of the *Essay*.

However, before this line of retreat is accepted there is another shot to be fired. One thing that is clear in Locke's explicit statements is that he holds that all qualities, primary or non-primary, are in bodies, that bodies really have them. If that is so Locke cannot really have thought that the composite list i.e. *solidity, bulk, figure, mobility, situation, number, texture and motion of parts* contained only primary qualities since there would be nothing left over, in bodies, to be a secondary quality. In my view, it will not do to say that *powers* are left over but the full support for that will be found in Chapter 7. Since the grounds for excluding the qualities I have excluded are similar and based on Locke's definition, there seems to be some justification for my interpretation. There is the added justification that it requires less ingenuity than the alternative interpretation that he did not really think that all qualities are in bodies and it makes better sense.

If the view I have been putting was indeed Locke's, then there are certain problems that arise for him from the idea that a single corpuscle, alone in the universe, could have the primary qualities in question. Indeed, it might well be said that it makes no sense whatever to attribute shape, size or mobility to a corpuscle in isolation from all other material bodies, unless we accept the existence of absolute space which can provide a frame of reference. I believe that neither Boyle[3] nor Locke[4] accepted absolute space in that sense. If these are taken to be, or even to be *among*, the defining properties of matter then the question becomes, I suppose: 'Does it make sense to say that a single corpuscle, alone in the universe, would be material?' Intuitively it seems to make sense if we are prepared to regard matter as a fundamental category at all. Whatever analysis we give of material objects it seems likely to involve fundamental entities of some sort, whether corpuscles or more mysterious sorts of particles or space-time points. It would seem to be contingent that the universe contains the exact number of these that it does and so contingent that it contains more than one. If it contained only one that one would have to have some properties. (One what?) Or, to return to Lockean terms, it seems reasonable to say that if one corpuscle which among its fellows had shape and size were, by some cosmic catastrophe or act of God, left alone it would still have shape and size even if no specification of its shape and size were possible.

It might be thought that since we are considering a universe of many corpuscles between which collisions occur, and so on, their being solid entails their having shape, size and mobility, so if we can show that it makes sense to say that our lonely corpuscle is solid we have done all that is necessary. However, that entailment holds, if it holds at all, I think, only if the stated conditions are satisfied. Someone might claim that the universe is just one infinite solid. Showing that our corpuscle has size, shape and mobility contributes to showing that it makes sense to talk of a universe of many finite corpuscles. If we could show that our corpuscle could intelligibly be said to be solid but not to have shape that would presumably be to show that it could not intelligibly be said to be finite. That, in turn, would perhaps be to show that we could not make sense of our universe in terms of material atoms. So it seems necessary to consider the qualities separately.

I begin with size which seems difficult because of its essentially quantitative and comparative character. For a thing to have size at all it seems essential that there be something else which it is larger than, smaller than or equal to, or by reference to which its size can be precisely specified. Without this we seem to be using the idea of absolute size, which is wicked.

Perhaps a distinction is possible here. The objection to absolute size is that it suggests that a body can have a specific size which is in principle unspecifiable since any specification would be by means of some arbitrary measuring device and, *ex hypothesi*, no arbitrary measuring device can get at the absolute size. However, it may be that to say that our lone corpuscle has size need not involve this. Could we not hold that this is to say that it is the sort of thing for which a size would necessarily be specifiable if other corpuscles were present, that is if we had some standard of measurement? We are talking of a property it now has by virtue of which a certain sort of measurement, under the appropriate conditions, would necessarily give a result. This is not a dispositional property in any ordinary sense because to attribute it is not to say that under certain conditions the corpuscle would behave in a particular way or change in a particular way but merely that a property it now has would appear or be 'revealed' in some way or other depending upon how we looked at it.

We might, alternatively, approach this through solidity. Even though a body in otherwise empty space cannot be said to have a position it can surely be said to occupy a portion of space. Think of space as full of corpuscles. Now think of corpuscles being annihilated until just two are left; there is no difficulty about attributing size to them at least in a rudimentary sense. Now suppose that one of them is annihilated; can we not say that the remaining one occupies a portion of space as it did before? If the body is absolutely solid, some portion of space is not available for further occupancy; if another corpuscle were introduced there would necessarily be two or, since the corpuscles might come into contact, there would be two exemplifications of shape, e.g. two spheres touching. Perhaps we can say that if something has two non-coincident points on it then it has size; if it has some shape then that size is finite.

I now turn to shape. I think it may be possible to make sense of our lone corpuscle's having a particular shape through geometry alone without reference to anything external to the corpuscle. The idea of distance is available since to be solid is to have some volume and we can conceive of points at different distances from one another within any volume. We can also give sense to direction purely internally and geometrically. Then for a corpuscle to have a spherical shape is for there to be a point, x, within it such that in every direction there is a farthest point on the solid such that all these farthest points are equidistant from x. If this condition is not fulfilled then the corpuscle may have some other shape which could be defined along similar lines, although the definition would be more complex. Even this allows us to say that we could give a description of a single-corpuscle universe if that corpuscle is spherical which would differ from its description if the corpuscle were any other shape. This no doubt relates to Putnam's view that distance must be among the 'fundamental magnitudes' of physics since it is an attempt to deal with shape in terms of distance and direction.[5]

I now consider mobility. Boyle says something that may be construed as a reason for his sometimes using 'mobility' instead of 'motion or rest'. It is not essential to matter either that it be in motion or that it be at rest. Matter, he says, 'is as much matter when it rests as when it moves' (*Q.F.Q.* 15). However, in any actual situation in the world, as it is, a body must be in some state of motion (including the zero state). When Boyle and Locke use 'mobility' in this context they must be using it as a technical term since its normal use is to say of something at rest that it could move or be easily moved but we cannot say this of our solitary corpuscle. There appears to be an insuperable difficulty; if to say that it has mobility is to say that it is in some state of motion then it seems impossible to say that it has mobility because there is nothing by reference to which it could be in some state of motion; if to say that it has mobility is to say that it *could* be in some state of motion then it seems impossible to regard mobility as an intrinsic property because in its present isolated state it could not be in some state of motion so its having mobility would seem to depend upon its relations to some other corpuscle. This has seemed an insuperable difficulty at least to most of my critics. I hope to overcome it.

There are signs that the problem worried contemporaries of Boyle. One argument that was used was this. Suppose that there are just two corpuscles in empty space and that they are moving apart. If there is no causal interaction between them then the annihilation of one could make no difference to the other; so if the remaining one was at rest before the annihilation of the other it must still be at rest after the event, and similarly if it was in motion. This clearly will not do – but need I go on?

It might seem that rotary motion was more promising. Could we say that our lone corpuscle was rotating about an axis? For simplicity consider a spherical corpuscle. If there is a point on its surface that is tracing a path relative to some other point, e.g. its geometrical centre, then the sphere is rotating – if not, not. This will not do because the idea of tracing a path depends upon the idea of a change of direction of a line joining two points and that requires an external point of reference; in this case no internal specification of direction will do, if we are dealing with a rigid body, because it will not give a *change* of direction.

It will not do either to attempt to deal with this in a Newtonian way with the help of forces, centrifugal or centripetal. Apart from the fact that this move is not open to Boyle and Locke, it is not clear how this could solve the problem for a rigid body without the assistance of at least one other body to be attracted or repelled by the corpuscle in question.

All this appears to reinforce the view that to make sense of a thing's having mobility we need an external point of reference. However, this may be because it is difficult to avoid thinking of mobility as the ability or capacity of a body to behave in a certain way under certain conditions, that is, in the presence of another body, and that may not be the only way of thinking of it. In trying to find intrinsic properties of matter we are looking for properties that define the kind of stuff it is in the most general sense and that means properties that allow kinds of behaviour, or kinds of properties, each kind being such that specific exemplars of it are inconsistent with one another and such that any particular portion of matter must exhibit some specific exemplar in any circumstances with which we have to deal. In other words, each defining property allows a disjunction of specific properties no two disjuncts of which can be consistently attributed to any one body at the same time. (I am avoiding the terms 'determinable' and 'determinate' for a reason that will emerge.)

Mobility, in the present sense, is a property partly defining a kind of stuff as being, necessarily, at rest or in motion in any situation we encounter. To say that a single corpuscle has mobility is to say that it has the property that ensures that in any actual situation it will be in some state of motion. It would not make sense to say of something that didn't have mobility, say an idea, that it was at rest or that it was in motion. Having mobility is not the same property as either of these; being at rest implies having mobility but having mobility does not imply being at rest, and similarly for being in motion. A body's being in motion is its changing position with respect to something, its being at rest is its not changing position with respect to something; its having mobility is not either or both of these. Its having mobility is its being such that it must either be changing or not changing position if there are positions to change. It is being such that if there are positions it must be in one of them.

We should not think of mobility as if it were comparable to solubility and other dispositional properties. To attribute solubility to something is to say that, certain conditions being fulfilled, it will dissolve; it is not to say that in any actual situation it will either dissolve or not dissolve. To attribute mobility to something is not to say that, certain conditions being fulfilled, it will move; it is to say that in any actual situation it will be either at rest or in motion. To attribute mobility to a lone corpuscle is to say that given any conditions other than that of its being alone it will have certain dispositional properties which will be exhibited under certain further conditions e.g. that of moving when pushed or stopping when resisted. Mobility is more like immersibility in a liquid than it is like solubility. To attribute mobility or immersibility in a liquid to something is part of saying that it is material. To say of something that it is material is to say that it is the sort of thing of which it makes sense to say that it must move or not move, must dissolve or not dissolve, must vaporize or not vaporize, must oxidize or not oxidize, and so on, through a whole range of dispositions. Solubility is a property of some kinds of matter; mobility is a property of all matter whatsoever.

This is not to say something empty. Corpuscles, whether lonely or gathered together in conventions, are neither cheerful nor not cheerful, emotional nor not emotional, ambitious nor unambitious. (It is worth considering why there is something called the materialist theory of mind.) Locke being a convinced dualist, devotes a good deal of space in his letters to Stillingfleet distinguishing between material substance, which is solid, and mental substance, which is unsolid; material things are such things as have shape, size and mobility, mental things are such things as don't have those qualities but do have others. We are here dealing with the most fundamental categories for characterizing stuffs.

The view I have been putting appears to pose difficulties for the idea that mobility is a *quality*, given the conception of quality that I have outlined. Is mobility capable of differentiating one portion of matter, one corpuscle, from another? The contrast with which I am here concerned is with solidity, absolute solidity. There are no degrees of solidity; degrees of empirical hardness are not varying appearances or manifestations of solidity. These are different concepts, solidity involving the absence of empty space from a body, hardness involving its presence. In any situation in which the differentiation of bodies is possible mobility is being manifested in particular states of motion, just as shape and size are being manifested in particular shapes and sizes. That is, the three primary qualities admit, even if a little indirectly, of degrees in any actual situation and this will serve for differentiation.

It might seem appropriate, if I am right, to regard shape, size and mobility, applied to matter as such, as determinables or genera, and particular shapes, sizes and states of motion, manifested in particular bodies, as determinates or species. However, this does not seem right since red, a paradigm of a determinate, is a kind of colour, a paradigm of a determinable; and horse, a paradigm of a species, is a kind of *Equus*, a paradigm of a genus, but a particular state of motion is not a kind of mobility.

Perhaps it would be more appropriate to talk of first-order and second-order properties. I have argued that mobility is a different property from either being in motion or being at rest; a corpuscle's having mobility is its having a property such that under certain conditions it must have the property of being in a specific state of motion. Putnam gives as an example of a second-order property that of having a certain machine table about which he says that it is '*a property of having properties which*...' and this 'although a property of the first level (a property of things), [it] is of "second-order" in the Russell-Whitehead sense, in that its definition involves a quantification over (first-order) physical properties' ('On Properties', p.313). There appears to be a formal similarity here with what I have said about mobility and what I may have to say about size.

(Locke shows signs, in the letters to Stillingfleet, of allowing as qualities anything that differentiates material things from mental things. This would seem to be the idea of second-order qualities.)

However, this exposes another problem. In arguing that it makes sense to attribute mobility to a single corpuscle alone in the universe I have been forced into saying that although it does make sense to attribute mobility it does not make sense to attribute to it the very same property, a particular state of motion, that we attribute to an actual corpuscle and that figures in explanations. How serious a problem is this?

In the first place, one aspect of my problem was the problem whether *any properties at all* could be attributed to the lone corpuscle. If I am right then the answer is that some properties can. I have been covertly asking whether it makes sense to start with material corpuscles at all. In order for this to make sense it seems to me that some properties must be attributable to matter as such, and so to each corpuscle, to distinguish matter from anything else that the universe contains, or might be said to contain. Shape, size and mobility, as explained, appear to suffice for this.

The kind of explanation being considered is in terms of material corpuscles and explanation is called for, and possible, only when there are collections of these, so any explanation must start from some corpuscles among many. In that situation we have, and start from, particular shapes, sizes and states of motion. However, any second-order properties that each of these corpuscles has are such as to ensure the possession by the relevant corpuscles of just those primary qualities, and no others, that figure in any actual explanation.

Having arrived at this point I am struck by the untidiness of the view which I find myself entertaining. The lone corpuscle can be said to have one first-order property, shape, and one second-order property, mobility, and one property, size, which I have left hovering uneasily between the two. However, the whole point of considering the lone corpuscle was to consider what can be said about matter as such. If second-order properties are acceptable at all then presumably there is no objection to saying about our lone corpuscle, or matter as such, that it has the second-order properties shape, size and mobility. To show that it makes sense to say that it has some particular shape (first-order) is not to show that it does *not* make sense to say it has the second-order quality, shape. Nothing needs to have, in order to be material, any particular shape; it needs to have some shape or other. So perhaps I can say that the lone corpuscle has the second-order properties shape, size and mobility, that is, properties such that in any actual situation corpuscles-in-convention would each have the first-order properties, a particular shape, a particular size and a particular state of motion.

The situation, therefore, is this. We understand particular shapes, sizes and states of motion through everyday observation. We apply these without difficulty to the unobservable corpuscles postulated as occurring in collections in actual phenomena. We understand what it is to say that these corpuscles are material by attributing to each the second-order properties shape, size and mobility and these we understand in terms of the corresponding first-order properties. All our explanations start from these first-order properties so these are the properties that are primary for explanation. The second-order properties are primary in the sense that they characterize, in the most general way possible, the kind of stuff we are dealing with and that they indicate what all the phenomena have in common in spite of wide differences between the observed properties of those phenomena.[6]

7

POWERS

Locke says that secondary qualities are 'nothing in the Objects themselves but Powers to produce various Sensations in us' (II.viii.10). This has led commentators to hold that according to Locke secondary qualities are dispositional and relational or mind-dependent or not qualities at all but relations.[1] This poses problems about Locke's consistency if we take seriously his statements about non-primary qualities being *in objects.* I now wish to consider a possible interpretation of the notion of powers that it is plausible to suppose Locke held and that avoids these problems. This interpretation allows one to say that powers of things may be intrinsic properties of them, that is, properties they possess independently of other things.

What, then, is a power? If an object makes me see red or makes a hole in another object then it has the power to make me see red and to make a hole in another object. If *aqua regia* dissolves gold then it has the power to dissolve gold and if a key opens a lock then it has the power to open that lock. These things seem obvious. What is not obvious is the point of saying them. What do they tell us that we didn't know already if we knew the fact of the matter? Do they go any way towards explaining seeing red, making a hole, dissolving gold or opening a lock? Do they add to the description of whatever is said to have the power anything beyond the description that could lead us to attribute the power? Seeing red, the making of holes, the dissolving of gold and the opening of locks seem to be open to our view. Stretching the term a little, they are observable phenomena. To say that something has the power to produce these phenomena looks, at first sight, as if it is to give some sort of explanation of the phenomena or to give a further description of that thing. But very little consideration leads one to question that. Aren't these statements about powers just like the attribution of a *virtus dormitiva* to opium and so empty as explanations and as descriptions?

Boyle and, following him, Locke attacked just such empty explanations and descriptions which were given by the scholastics and the alchemists. To say that a substance is red or gold because of the presence in it of the substantial form of redness or gold is uninformative; to say that a substance is volatile because it contains sulphur, which is just the principle of volatility, is not to explain anything. The aim of natural philosophy, Boyle and Locke thought, is to describe things and substances in detail and in such a way that real and perspicuous explanations could be based on those descriptions and to reject *ad hoc* postulations of occult entities or qualities such as substantial forms and chemical principles. This was perhaps the central aim of the corpuscular hypothesis.

Suppose an object, say a needle, makes a hole in another object, say a cork. Suppose we describe a needle as very hard, thin and with a sharp point and a cork as soft and having a flat surface area of 1 sq.cm. Do we add anything by saying that the needle has the power to make a hole in, or penetrate, the cork? Do we thereby even begin to explain the phenomenon of a slim, sharp, hard body penetrating a soft, flat surface?

Now suppose that we say, with Locke, that the red colour of an object is nothing but a power to produce in us sensations of red. We attribute to the object something other than redness as we see it. But so far we explain nothing. We knew all along that something in the object produced in us the sensation of red; to say that it was a power to do just that is not to explain its doing just that. At most it tells us something about the strict scientific description of the object, but something entirely negative, namely, that such a description is not to include redness. To say that it has this power is a first step towards giving the positive description; it leaves quite open the nature of that description, except to suggest that the description must be different from that we give of something with the power to produce sensations of yellow or holes in other objects. The power statement is like a blank cheque with a specified maximum. It is an indication that further description is needed and an invitation to find it. For explanation we need to fill in the cheque; we need to say what constitutes the power.

The quotation from II.viii.10 with which I began this chapter was incomplete; more fully, it tells us that secondary qualities are

> Such *Qualities*, which in truth are nothing in the Objects themselves, but Powers to produce various Sensations in us by their *primary Qualities, i.e.* by the Bulk, Figure, Texture, and Motion of their insensible parts...

It might be thought that this makes talk of powers, in contrast to talk of 'virtues', explanatory because it indicates what powers depend on. I agree that this is to be taken as pointing to the way in which powers may be elucidated and explanations may be given. I wish to suggest that it may also be seen as pointing towards the elimination of talk of powers, *as distinct from textures*, and that this would be a move favoured by those who regard such a conception of powers as occult. If we can give a detailed description of the fine internal structure of gold and of *aqua regia* which makes it obvious why gold dissolves in *aqua regia* we have explained the phenomenon. It is tempting to say also that we have explained the power and so, in a sense, we have, but there is no longer any need to talk of powers since we can go straight from the internal structure to the phenomenon.

A substance *S* regularly does *X*. We can say, at this stage, that *S* has the power to do *X* but this is mere shorthand if we are considering explanations; if that is what we are doing this is just an invitation to give an explanation. Now if we can say that *S* has the power to do *X* because of internal constitution *C*, we are being unnecessarily wordy. We could say that *S* does *X* regularly because of *C* and lose nothing. So when we get the strict scientific description 'power' talk drops out. We may say that this is because either 'power' was a provisional but empty concept, a mere placeholder for a constitution which would give an explanation, and so is no longer needed or that we have discovered what the power to do *X* really is. To say this is to say that the power is identical with the internal constitution *C*, and an intrinsic property of *S*.

But we have all been well brought up and we bridle at that. If Locke meant *that*, we are inclined to say, he was wrong, which is not surprising because our upbringing has also led us to expect Locke to be often wrong. The power of *aqua regia* to dissolve gold necessarily involves reference to gold as well as *aqua regia* so how can it be an intrinsic property of *aqua regia*? And that's that. But is it that? I shall adopt a roundabout route to the conclusion that it is not that. In the course of it I want to consider not what *we* should mean if we talked of qualities as powers but what it is probable that *Locke* meant.

Reginald Jackson, in an over-celebrated article,[2] argued that Locke's distinction between primary and secondary qualities was really a distinction between qualities and powers; that primary qualities really are qualities and secondary qualities are powers *and not qualities*. This flies in the face of several references by Locke to 'three sorts of qualities in bodies' (e.g. II.viii.23) and his statements 'the Power to produce any *Idea* in our mind, I call *Quality* of the Subject wherein that power is' and 'a Snow-ball having the power to produce in us the *Ideas* of *White, Cold* and *Round*, the powers to produce those *Ideas* in us, as they are in the Snow-ball, I call *Qualities*' (II.viii.8). 'Round' is shortly to be called a primary quality. These statements appear to allow primary as well as secondary qualities to be powers. Jackson is rather cavalier with these statements but then he is clear from the beginning that Locke is full of inconsistencies. But let us fight against our upbringing and accept Jackson's own principle that 'there seems something to be said for selecting where possible, as the normal meaning of a term [in Locke] what Locke says he means by it' (pp.53–4).

As an aside, I may say that Jackson claims to find in Boyle some things that I cannot find, including some of the faults I cannot find in Locke either. It is true that he gives page references to Boyle but it is difficult to see what on those pages he takes to support his views (pp.55–7). However, I do not need to grind that axe here.

According to Jackson, an important reason why Locke needed the distinction between primary and secondary qualities is that 'he is anxious to preserve an absolute distinction between qualities and relations' (p.58) and, he says, 'Knowledge of what a given body is in itself is knowledge, not of the relations in which it stands, but of the qualities which it has' (pp.58–9). However, 'many apparent qualities turn out ... to be relations' (p.59). So it is important for Locke 'to smell out pseudo-qualities'. 'Now' says Jackson, 'there is one kind of relation which he thinks especially likely to be mistaken for a quality, the kind of relation which, according to his wider use of the term, he calls a "power"' and 'according to Locke a body can have a power only by means of a quality, and as long as we know only the power we are ignorant of what the body is in itself; while if we know the quality, we do not need to mention the power in a statement of what the body is in itself' (p.59).

I think that the last of these statements is true but I think also that Jackson failed to see all its implications. He might have added that we do not need to mention the power in a statement of what the body will do if we know in detail what the body's constitution is.

Now Jackson goes on to say that 'the distinction between Primary and Secondary qualities is a special case of the distinction between Qualities and Powers' (p.60), where powers are relations rather than qualities. However, if Locke saw this distinction as preserving an absolute distinction between qualities and relations and if he thought that secondary qualities are relations and *not* qualities and if this was of such great importance to him, it is surely odd that he did not say these things, or even hint at them. It is noticeable, but not surprising, that Jackson gives no reference to Locke in the pages in which he is urging this part of his view (pp.57–60). It is not surprising because Locke never seems explicitly to connect secondary qualities with relations in the way this suggests. It is true that at II.xxi.3 Locke says '*I confess power includes in it some kind of relation*' but this is in the course of an argument for regarding ideas of powers as simple ideas and he says that in this respect they do not differ from primary qualities. In II.viii, where Locke makes the distinction, he does not appear to be concerned about a body's relations to other bodies except in so far as the effects of a body on others can be explained in terms of the qualities of the body. He refers to three sorts of qualities *in bodies* and, as Jackson himself points out, he is anxious to say that secondary qualities are not 'in the mind' but are in bodies. When Locke says that non-primary qualities are 'only powers to act differently upon other things' he adds 'which powers result from the different modifications of these primary qualities' (II.viii.23). Jackson says (p.66) that Locke 'uses the term secondary qualities only "to comply with the common way of speaking"' (II.viii.10), but Jackson has supplied the 'only'; Locke does not use it, so he does not imply that the common way of speaking is wrong in the relevant respect.

In II.xxxi.2 Locke, talking of light and heat from a fire, says

For though Fire be call'd painful to the Touch, whereby is signified the power of producing in us the *Idea* of Pain; yet it is denominated also Light, and Hot; as if Light and Heat, were really something in the Fire, more than a power to excite these *Ideas* in us; and therefore are called *Qualities* in, or of the Fire. But these being nothing, in truth, but powers to excite such *Ideas* in us, I must, in that sense be understood, when I speak of secondary *Qualities*, as being in Things; or of their *Ideas*, as being in the Objects, that excite them in us. Such ways of speaking, though accommodated to the vulgar Notions, without which, one cannot be well understood; yet truly signify nothing, but those Powers which are in Things, to excite certain Sensations or *Ideas* in us. Since were there no fit Organs to receive the impressions Fire makes on the Sight and Touch; nor a Mind joined to those Organs to receive the *Ideas* of Light and Heat, by those impressions from the Fire, or the Sun, there would yet be no more Light, or Heat in the World, than there would be Pain if there were no sensible Creature to feel it...

He is not here saying, I believe, that secondary qualities are *not* qualities *but* powers. He is saying that they are *both* qualities and powers. It is heat and light, as felt and seen, the ideas, that are neither qualities nor powers, and he is asking us to remember this if he should talk of the sun as being hot and light. Ideas are not qualities but qualities, whether primary or secondary, are. (See next chapter.)

Is not my case spoilt by the words 'result from' in the quotation from section 23 above? I think not. If we talk about powers in this context we do have to talk of their 'resulting from' modifications of matter just because such expressions as 'the power to dissolve gold' are interim, schematic, blank cheque expressions. Such expressions naturally carry with them the idea of dependence upon an internal structure and they drop out when the full description of the structure leaves nothing more to be described in providing an explanation. At this point the power is identified with the structure. I shall return to this.

Consider again the clause 'which powers result from the different modifications of those primary qualities'. 'Modifications' can mean either the action of making changes or the result of such changes. If we think of 'modifications' in the active sense we can see this fits in with my view. The primary qualities of the corpuscles are *brought into* various combinations and arrangements which form various internal constitutions or textures. These constitutions or textures may then be identical with powers and can be said to result from the modifications (the bringing into those patterns) of the primary qualities. The internal constitutions need not be said to be the grounds of, rather than identical with, the powers.

Locke does not appear to me to say anything in support of Jackson's view in his four chapters on relations, where one might expect to find support (II.xxv–xxviii). In the first of these, in listing four points about relations in general, Locke says

Fourthly, That *Relation* being the considering of one thing with another, which is extrinsical to it, it is evident, that all Words that necessarily lead the Mind to any other *Ideas*, than are supposed really to exist in that thing, to which the Word is applied, are *relative Words. v.g. A Man Black, Merry, Thoughtful, Thirsty, Angry, Extended;* these, and the like, are all absolute, because they neither signify nor intimate any thing, but what does, or is supposed really to exist in the Man thus denominated; But *Father, Brother, King, Husband, Blacker, Merrier, etc.* are Words, which, together with the thing they denominate, imply also something else separate, and exterior to the existence of that thing. (II.xxv.10)

It is important to note that in this passage he includes 'black' as well as 'extended' in the list of absolute words. It is true that what Locke says leaves open the possibility that 'extended' may signify something that 'does really' exist in the man while 'black' may signify something only 'supposed really to exist in the man', but falsely. However, in the light of chapter viii, as I have interpreted it, it seems that what is falsely supposed so to exist is the *idea* of black and not the correlative internal constitution. The point being made here, I believe, is that 'blacker' involves two similar internal constitutions which are compared, whereas black does not.

In the final section of chapter xxv Locke says he proposes to show 'how all the *Ideas* we have of *Relation*, are made up, ... only of simple *Ideas*...'. Thus, the idea of power being a simple idea may be thought of not as an idea of relation but an idea that goes to make up certain ideas of relations. This may seem to involve Locke in contradiction since I have said that for him power talk is relational. However, the way in which he may avoid contradiction is already implicit in what I have said. Unanalysed talk about the powers of objects always involves references to other objects but when we discover what powers are as they really are in objects we see them as intrinsic properties of the objects said to have the powers. The scientific description of the world shows what powers really are.

Jackson finds a number of problems connected with Locke's distinction. Is knowledge of a body as it is in itself knowledge of an individual persisting through perishing states or knowledge only of those perishing states (p.61)? Jackson sees part of the trouble as arising from Locke's failure to distinguish between an *individual*, 'something that persists through passing states', and a *particular*, 'something to which every quality, that belongs to it at all, is essential' (p.62).[3] He thinks that to avoid the danger of being put in the position of having to say that we know only perishing states of an individual and not the individual itself Locke defines primary qualities in such a way as to enable them to qualify an individual body throughout its different states. But, says Jackson, 'he can do this only by taking qualities abstractly and abstracting from just those determinations of qualities which on his view distinguish one body from another. The determinations which, if taken as always qualifying the individual, are fatal to its persistence, are indispensable to its distinguishability from other individuals' (p.61). Because a piece of wax may be spherical and at rest at one moment and cubical and in motion at another 'Locke is unwilling to include its determinate shape and its determinate motion-and-rest among its primary qualities' (p.61). Jackson goes a long way round to extricate Locke from these difficulties.

I believe that this rests upon a misunderstanding of Locke. He defines primary qualities as

> such as are utterly inseparable from the Body, in what estate soever it be; such as in all the alterations and changes it suffers, all the force can be used upon it, it constantly keeps; and such as Sense constantly finds in every particle of Matter, which has bulk enough to be perceived, and the Mind finds inseparable from every particle of Matter, though less than to make it self singly be perceived by our Senses. (II.viii.9)

Note that the primary qualities are defined in relation to particles of matter i.e., ultimately, corpuscles.

The first point to be made is that, in a passage to which Locke probably owes much, Boyle clearly has in mind determinate shape, size and mobility as primary qualities. He says these are

> three essential properties of each intire or undivided, though insensible part of matter; namely, magnitude (by which I mean not quantity in general, but a determined quantity, which we in *English* often times call the size of a body) shape and either motion or rest... (*O.F.Q.* 16, *S.* 20)

I have said that Locke follows Boyle in this and that he regards qualities, just as Boyle does, as what may differentiate bodies from one another and sometimes enable us to distinguish them.

When he uses the expression 'such as' in defining primary qualities Locke means qualities *of a sort* that are inseparable, whether to reason or to sense, i.e. shape, size and mobility. Our conception of body depends upon both the mind and the senses, the senses giving us information which enables us to see how it is plausible to conceive body as such.

The primary qualities, as determinables, are inseparable from all bodies and endure through all observable change. However, the primary qualities of the corpuscles, their determinate primary qualities, are also inseparable from them. They, as determinates, are unchanging and endure through all observable and unobservable changes, which are changes of arrangement. If we fully understood the observable changes of bodies we should know the persisting elements of their internal constitution, their determinate qualities. A piece of gold, in order to be gold, need not have a particular determinate shape and size but it must be composed of corpuscles having determinate shapes and sizes and arranged in a limited range of determinate patterns. This surely is the force of the example of dividing a grain of wheat which immediately follows the definition of 'primary quality'. At each stage of the division, each part has some determinate shape, size and motion or rest, and at the beginning of the division these can be observed. If we imagine it to be divided into its smallest actual particles, the corpuscles, each of these considered separately must be conceived of as having a determinate shape, size and mobility which is both essential to it and unchanging. The determinate shape, size and motion or rest of the corpuscles of a piece of gold are its ultimate primary qualities; if we knew these we should know bodies as they are in themselves and not merely changing states of them. To know what gold is in itself we should have to know these plus the range of patterns of corpuscles emerging from their convention. To know a piece of gold at a given time we should have to know what we might call its gross primary qualities, its observable shape and size, but this is not part of knowing that it is gold. The secondary qualities of gold are essential to gold since they are patterns or arrangements of corpuscles each having only primary qualities. They are also determinate; they help to differentiate gold from, say, silver. They may change, as when gold is melted, but presumably only within certain limits as long as the gold remains gold.

The identity of a piece of gold depends upon its shape, size, motion or rest and continuous history. The identity of the substance gold depends upon the arrangement of its constituent corpuscles and their primary qualities; so it depends upon both primary and secondary qualities. The identity of one corpuscle among many depends upon its primary qualities plus its situation and continuous history. Its situation depends upon its relation to other corpuscles. Is that, then, a non-primary quality? I think not. The distance between two corpuscles is just an instance of extension. It is a property of a portion of space, exactly like the diameter of a corpuscle. It is a quality of the same kind as the diameter of a corpuscle.

The determinate primary qualities of complex objects, including observable ones, depend upon the determinate qualities of the corpuscles composing them and their arrangements. The determinate primary and secondary qualities of complex objects may alter with alterations in the arrangements of their corpuscles. The determinate primary qualities of the corpuscles do not alter; it is these in terms of which the qualities and interactions of complex bodies are to be explained. What we know of a body through sense-experience is its perishing states but if we come to understand these states fully through scientific investigation and inference we shall come to know what persists through those perishing states and what both explain the states and figure in the explanation of their perishing. The spherical shape of a piece of wax at a given moment is not inseparable from the piece of wax but neither would it figure in the explanation of the colour it appears to have; the determinate shape of its constituent particles is inseparable from it since without this it would not be wax and this shape would figure in the explanation of its appearing yellow to us. We might say that the primary qualities of the corpuscles are primary, fundamentally, but the primary qualities of a complex collection of them are primary, only derivatively. They are primary because they are of the same *kind* as the fundamental ones.

I now return to my contention that for Locke secondary qualities are both qualities and powers and that powers may be intrinsic properties of things. I start with a quotation from Mackie which gives a more orthodox view.[4]

When he [Locke] is about to introduce the distinction between primary and secondary qualities, he first distinguishes between ideas – the word must here be used in a broad sense – 'as they are ideas or perceptions in our minds' and 'as they are modifications of matter in the bodies that cause such perceptions in us', and proposes to call only the former *ideas*. That is, he is introducing a narrow sense of 'ideas' in which these are to be 'perceptions in our minds' and not 'modifications of matter'. One would expect him here to call only the latter *qualities*, but in fact this is not what he proposes to do; rather he says he will give the name *qualities* to the various powers of objects to produce ideas in us. But immediately afterwards his usage is partly inconsistent with this proposal, for what he identifies as *primary qualities* are solidity, extension, figure, motion or rest, and number, and these are not powers: rather they are intrinsic properties of things which may be the grounds or bases of powers, and they are 'modifications of matter in the bodies...'

I have already pointed out that Locke heads sections 7–8 'Ideas *in the Mind, Qualities in Bodies*' and section 23 '*Three sorts of Qualities in Bodies*'. Nowhere, as far as I can see, does Locke distinguish between qualities and powers in the way that one would expect if Mackie and Jackson were right. Locke says that the third sort of qualities, which are 'barely' powers, are 'as much real qualities in the subject' as secondary qualities and he refers to the power of fire to melt wax as 'as much a quality in fire as the power it has to produce in me a new *idea* or sensation of warmth or burning' (II.viii.10). To suppose that he was really saying that these powers are not qualities at all would merely be to credit Locke with an extraordinary insensitivity to the English language; he was as well in command of the expression 'as little' as he was of the expression 'as much'. Of course, these passages may be read in either way but if the point had been as important to Locke as Mackie and Jackson suggest one would expect him to leave the matter less open.

However, there are more substantial points to be made about Mackie's paragraph than this. He is prepared to saddle Locke with the inconsistency of saying that he is going to give the name *qualities* to various powers and then immediately going on to give as examples of qualities things that are not powers, in order to save Locke from the more serious philosophical howler of identifying some powers with intrinsic qualities of things and to allow him an expression, 'primary qualities', for 'the grounds or bases of powers'. I think it highly implausible that Locke should allow himself, or have missed, so obvious an inconsistency within the space of two paragraphs in a passage in which he is setting out the central ideas of his system. I think it more likely that Mackie is being anachronistic and that Locke did not regard the alleged howler as a howler at all. I believe that Locke can be read with less strain and saved from more inconsistencies, in this and other passages, if he is regarded as explicitly identifying qualities and powers, though not as identifying all qualities with powers to produce ideas *directly* in us.

Mackie appears to think that 'solidity, extension, figure, motion or rest, and number' cannot be powers because they are intrinsic properties of things and powers cannot be intrinsic properties of a body because describing them always involves reference to another body. When we attribute a power to a body or substance such as the sun or *aqua regia* we always have to use such expressions as 'the power of the sun to melt *wax*' or 'the power of *aqua regia* to dissolve *gold*'. As Jackson has it, statements attributing powers to a body are always relational statements, relating that body to others. So powers cannot be intrinsic properties of bodies.

This appears to be the orthodoxy but it depends upon a certain way of thinking about powers which there is reason to believe was not Locke's way. (I have already quoted from Boyle in support of the kind of view I am putting about Locke.) Consider a single corpuscle. It must be solid, it may be spherical and have a diameter of x units. According to Mackie, solidity, shape and extension are primary qualities but not powers. If we were able to observe it by some means these primary qualities are what we should find and a complete list of its primary qualities would be a list of all that was there to be found. This description would not include, for example, its power to deflect another corpuscle on collision. It *has* this power but we could not find it by observation of it alone. We could discover it only by observing a collision with another corpuscle and this would entitle us to say that it has the power. But even then, we don't observe the power, as Hume saw. We observe only *that* the corpuscles collide and *that* one is deflected or both are deflected.

Locke and Boyle were suspicious of 'occult qualities' and I believe they regarded powers as occult if they were conceived as properties of corpuscles over and above their primary qualities. We can describe a corpuscle completely in terms of its primary qualities. (I say this for convenience here. There are reservations to be made but they don't affect my present argument.) There is nothing more in the corpuscle to describe. When it interacts with others we can describe the effects it has on those others. But this, of course is not to describe its powers; it is to describe its relations to other corpuscles and changes it produces in them. Still we can say that it has the powers to cause these changes. Is this a further description of the corpuscle? Locke does not appear to think that it is merely an inductive generalization of the singular statement describing the effect in question. He appears to attach greater explanatory force to the notion of power than this would allow, which is probably why Hume objected.

So what can a statement that a corpuscle has a certain power mean? I suggest that it means nothing for Locke, or for Boyle, unless the power is *given* in the description of its primary qualities; unless, that is, the appropriate power statements are entailed by that description. This explains how there can be, for Locke, necessary connections in nature. Consider the description 'solid, spherical and x units in diameter'. This entails that anything so described will fit into a circular hole of $x + \delta x$ units in diameter. That is, that it has the power or the ability to fit into such a hole. This is not a further description of the corpuscle and if it is a description of its relations to other corpuscles it is one without which we cannot understand the original description. To understand that description fully *is* to understand that anything falling under it will fit into holes of certain shapes and sizes. Primary qualities are powers: the sphericity of a corpuscle *is, inter alia*, its power to fit neatly only into spherical holes and its size *is, inter alia*, its power to fit only into holes above a certain size. To this extent there is no distinction between what it is and what it can do.

Primary qualities and powers are intrinsic properties of things. Any relational element in this notion of powers is not empirical.

Now consider a collection of corpuscles making a recognizable substance, say gold. The corpuscles have specific shapes and sizes and are in some state of motion or rest relatively to one another. They are certain distances apart or they move in such a way as to keep the distances between them within certain limits. That is, they are arranged in a stable pattern. A certain number of corpuscles sufficient to form that pattern is necessary in order that the collection be the substance gold. A complete description of gold can be given in terms of just this pattern, mentioning just these properties. (By a 'complete' description I mean one that is necessary and sufficient to explain all the properties, appearances and activities of gold.) Suppose that part of this description is that the corpuscles have certain specified shapes, are *x* units in diameter, at rest and *y* units apart. Then its solubility might be explained by the complete and regular intermingling of this pattern with the patterns of various liquids having, for example, spaces between corpuscles greater than *x* units and equal to *y* units, retaining features of the original patterns but forming new stable patterns, thus,

```
                 X X X X X    X X X X X
0 0 0 0                         0 0 0 0
                 X X X X X    X X X X X
0 0 0 0                         0 0 0 0
                 X X X X X    X X X X X
0 0 0 0                         0 0 0 0
                 X X X X X    X X X X X
  Gold             Solvent      Solution
```

Now, if we have the full description of gold we can say, from that description alone, that it has the power or ability to dissolve in any substance which has one of the range of patterns having certain features. We do not need to know of the existence of any such substance or, more importantly, to be able to test empirically for this. If solubility is just a matter of such fitting, then we can know, in advance of experience, the conditions for solubility. To understand the description of gold in these terms is to understand some necessary features of the description of any substance that will dissolve it. If we could discover the inner constitution of gold in these terms and the inner constitution of *aqua regia* we should know without trial that gold is soluble in *aqua regia*. As we cannot do this at least in the present state of science, we have to conduct trials. But that is a merely human and possibly temporary limitation. God would not need to conduct trials.

Similarly for other properties. The felt hardness of a substance and its resistance to penetration may depend simply on the shape of its corpuscles and their distances apart. Solids, liquids, and gases may differ in these respects. Ductility might depend upon the extent of homogeneity in shape and size of the corpuscles of the substance in question. And so on.

Colours, sounds, tastes, etc. depend also upon intrinsic properties. A lump of gold has intrinsic properties, other than the primary qualities of its corpuscles, i.e. properties that make it gold and that lump of gold in particular. Given the full description of the patterns of the corpuscles composing its surface we can tell from that alone that if there is light having certain patterns of corpuscles then it would all be reflected or absorbed, or partially reflected and partially absorbed, again without trial. If we knew enough about the corpuscular patterns of light and of the human sense organs we should be able to tell, without trial, under what conditions it would look yellow. That, at least, is the ideal but, as Locke points out in Book IV, what would also be needed would be an account of the conversion of brain-states into ideas.

Boyle's celebrated passage about the lock and the key (*O.F.Q.*, 18) quoted in Chapter 3 is of considerable relevance to this. This passage and others that follow it closely are strikingly similar to the following passage from Locke. He says

> I doubt not but if we could discover the Figure, Size, Texture, and Motion of the minute Constituent parts of any two Bodies, we should know without Trial several of the Operations one upon another, as we do now the Properties of a Square, or a Triangle. Did we know the Mechanical affections of the Particles, of *Rhubarb, Hemlock, Opium,* and a *Man*, as a Watchmaker does those of a Watch, whereby it performs its Operations, and of a File which by rubbing on them will alter the Figure of any of the Wheels, we should be able to tell before Hand that *Rhubarb* will purge, *Hemlock* kill, and *Opium* make a Man sleep; ... The dissolving of Silver in *aqua fortis*, and Gold in *aqua regia*, and not *vice versa*, would be then, perhaps, no more difficult to know, than it is to a Smith to understand, why the turning of one Key will open a Lock, and not the turning of another. (IV.iii.25)

The view I am putting is supported by many other passages in which Locke is clearly asserting the existence of necessary connections in nature. He says that he doubts that any hypothesis other than the corpuscularian one 'will afford us a fuller and clearer discovery of the necessary Connexion, and *Co-existence*, of the Powers, which are to be observed united' in different sorts of things (IV.iii.16). He asks: 'What is that Texture of Parts, that real *Essence*, that makes Lead, and Antimony fusible; Wood and Stones not? What makes Lead, and Iron malleable; Antimony and

Stones not?' (III.vi.9) and repeatedly writes as if there were answers to these questions, even though we do not know them.

Because we do not know the real essence of gold, from which all its properties flow, the best we can do is to presume a real essence (II.xxxi.6); that is why we have to rely upon observation and experiment rather than deduction for the discovery of the properties of things. If we knew the inner constitution of gold 'it would be no more necessary, that *Gold* should exist, and that we should make Experiments upon it, than it is necessary, for the knowing the Properties of a Triangle, that a Triangle should exist in any Matter' (IV.vi.11). This comparison of the relation between inner constitutions and properties of substances, on the one hand, and that between the definition of a geometrical figure and its properties, on the other, is adverted to again and again. For example, Locke says 'it is as impossible that two Things, partaking exactly of the same real *Essence*, should have different Properties, as that two Figures partaking in the same real *Essence* of a Circle, should have different Properties' (III.iii.17 and see also III.vi.8). Again, he says, that if we had a complex idea of the real essence of any substance

> then the Properties we discover in that Body, would depend on that complex *Idea*, and be deducible from it, and their necessary connexion with it be known; as all Properties of a Triangle depend on, and as far as they are discoverable, are deducible from the complex *Idea* of three Lines, including a space. (III.xxxi.6)

The account I have given of powers and the interaction of substances appears to me to depend upon the most plausible way of supposing that powers were understood in order that the corpuscular hypothesis would allow the theoretical explanation of the necessary connections that Locke regarded as holding in nature.[5]

It may be wondered why, in this account, there is no mention of forces, especially as the Newtonian theory is supposed to have influenced Locke. I believe that both Boyle and Locke were suspicious of forces, regarding them as occult and, therefore, too similar to the occult explanatory devices which they were attacking in the work of the scholastics and alchemists. They were largely interested in chemical interactions but they had no such idea of chemical action as was to be developed later and they were suspicious of the earlier notions of affinities and repugnancies just as they were suspicious of 'mechanical forces'. They wished to develop a science, including a chemical science, that was material and mechanical and without action at a distance. So they thought of chemical reactions in geometrical and kinematic terms. Their corpuscular theory was not dynamic. The only way in which we can conceive bodies acting is 'by impulse' (II.viii.11).

The first edition of Locke's *Essay* was published in 1690 but was sent to the printer in 1689[6] and Newton's *Principia* was published in 1687. Between the first and the third and fourth editions of the *Essay* Locke noted that as a result of reading the *Principia* he would have to make some alterations in future editions of the *Essay*. This was in his third letter to Stillingfleet. The relevant passage is this.

It is true, I say, 'that bodies operate by impulse, and nothing else [II.viii.11]'. And so I thought when I writ it, and can yet conceive no other way of their operation. But I am since convinced by the judicious Mr. Newton's incomparable book, that it is too bold a presumption to limit God's power, in this point, by my narrow conceptions. The gravitation of matter towards matter, by ways inconceivable to me, is not only a demonstration that God can, if he pleases, put into bodies powers and ways of operation above what can be derived from our idea of body, or can be explained by what we know of matter, but also an unquestionable and every where visible instance, that he has done so. And therefore in the next edition of my book I shall take care to have that passage rectified. (*L.S.* 467–8)

The changes Locke made in II.viii.11 and 12 were minimal and did not show a clear change in his convictions. In the first, second and third editions, section 11 reads

The next thing to be consider'd, is how *Bodies* operate one upon another, and that is manifestly *by impulse*, and nothing else. It being impossible to conceive, that Body should operate on what it does not touch, (which is all one as to imagine it can operate where it is not) or when it does touch, operate any other way than by Motion.[7]

In the fourth edition this became

The next thing to be consider'd, is how *Bodies* produce *Ideas* in us, and that is manifestly by *impulse*, the only way which we can conceive Bodies operate in.

This is surely no more than a softening of the original contention.

The next section (12) begins, in the first, second and third editions thus

If then Bodies cannot operate at a distance; if external Objects be not united to our Minds, when they produce *Ideas* in it;...

In the fourth edition Locke merely deleted 'then Bodies cannot operate at a distance; if' and inserted 'then' between 'If' and 'external Objects'.[8]

It is worth mentioning, also, that Descartes was well known at the time to have argued for a geometrical conception of physics. Thus the last item in Pt. II of his *Principles* reads

LXIV That I do not accept or desire any other principle in Physics than in Geometry or abstract Mathematics, because all the phenomena of nature may be explained by their means, and sure demonstration can be given of them.[9]

Thus I think that for Locke there is not, as Mackie holds, a distinction between powers and intrinsic properties or between powers and qualities. Powers just are more or less complex qualities, primary or non-primary.

When Locke says that he will use 'qualities' for powers to produce ideas in us (II.viii.8) he means it. He cannot, of course, be giving a definition of 'qualities' if this is taken to mean the power to produce ideas in us *directly*. A corpuscle can be said to be spherical although it never does or can produce in us the idea of sphericity, or *its* sphericity. However, all the ideas produced in us by external objects depend ultimately upon the primary qualities of the corpuscles involved, so the sphericity of the corpuscles contributes to the producing in us of the idea of sphericity and the idea of yellow when we see a golden ball; so it, along with other qualities of corpuscles, produces ideas in us indirectly. I think that here Locke intended to say only enough about qualities and powers to enable him to get on with his argument about primary and secondary qualities. In any case, I think he wants to define 'qualities' independently of our ideas of them, ontologically rather than epistemologically, as whatever is capable of differentiating bodies. They may also, some of them, enable us to distinguish bodies from one another, but that is a further, epistemological, point.

There is another way of putting all this. We must be careful to distinguish between Locke's ontological and epistemological statements. There is a difference between the power of a body to produce ideas in us or changes in other bodies and the ideas of such powers that we derive merely from experience. Locke says different things about powers according to whether he is or is not considering epistemological questions.

We discover the qualities of bodies in the first place by observing them and their reactions with one another. Thus we become aware of their powers to affect us and one another and the power-statements we make on this basis must be relational. In this context, to assert the power of *aqua regia* to dissolve gold necessarily involves mention of gold, *aqua regia* and, as far as observation can tell us, a contingent relation between them. Such statements cannot attribute powers as intrinsic, non-relational qualities to bodies. Thus Locke, talking of the possibly incurable inadequacy of our complex ideas of substance says

> The simple *Ideas* whereof we make our complex ones of Substances, are all of them (bating only the Figure and Bulk of some sorts) Powers; which being Relations to other Substances, we can never be sure that we know all the Powers, that are in any one Body, till we have tried what Changes it is fitted to give to, or receive from other Substances, in their several ways of application... (II.xxxi.8)

This passage appears to support Jackson, at least by suggesting that secondary qualities are powers and powers are relations. However, it also seems to say that primary qualities are powers. The sense of the bracketed portion is not to except all primary qualities; it just means that observed figure and bulk, for example, are not ideas that go to make up our complex ideas of all sorts or species. The idea of the sort 'elephant' includes the idea of the particular shape of an observable object but the idea of the sort 'gold' does not. The particular shape of an elephant is a power to produce in us the idea of 'elephant-shaped', which is part of our idea of the sort 'elephant', but the particular shape of a particular lump of gold is not a power to produce in us part of our complex idea of the sort 'gold'. However, this does not bear on my main point here.

The immediate point is that Locke is here talking of powers as we know them through observation, and he frequently denies that when we discover in this way the powers of substances we are discovering parts of the real essences of those substances. For example, shortly after this he says

> since the Powers, or Qualities, that are observable by us, are not the real Essence of that Substance, but depend on it, and flow from it, any Collection whatsoever of these Qualities, cannot be the real Essence of that Thing. (II.xxxi.13)

In contrast, when Locke is talking of qualities and powers at II.viii he is, most of the time, talking of things as they are, that is, what he later calls the 'real essences' of things, rather than things as they appear to us in observation. There, he gives no indication that powers are relations rather than intrinsic properties and he leaves open the possibility that different things have to be said about powers as they are in things and as they appear in observation. Ontologically, I think he means, powers are intrinsic and non-relational; in so far as they are accessible to observation they are relational and non-intrinsic. Ontologically, they are 'real qualities in the subject' (II.viii.10).

If we rely merely on observation, because we are limited to observing the qualities of bodies only through their actions on us and other bodies we necessarily have to describe those qualities relationally and what we are describing are the nominal essences of things. Ideally, if scientific investigation could somehow give us access to their real essences, their intrinsic qualities and their inner constitutions, we should be able to describe their powers non-relationally in the way I have suggested. That is why certain things could then be known about the natural world 'without trial', even things that we are now forced to discover through trials and that we now describe relationally, such as the power of *aqua regia* to dissolve gold or of a file to smooth another piece of metal. This is surely indicated shortly after the passage from II.xxxi.8 just quoted when Locke says

So that *all our complex* Ideas *of Substances are* imperfect and *inadequate*. Which would be so also in mathematical Figures, if we were to have our complex *Ideas* of them, only by collecting their Properties, in reference to other Figures ... Whereas, having in our plain *Idea*, the whole Essence of that Figure [*sc.* an ellipse], we from thence discover those Properties, and demonstratively see how they flow, and are inseparable from it. (II.xxxi.11)

Mathematical figures are '*figures* that the mind has the power to make' and are *simple modes* of space (II.xiii.6). Our ideas of them are complex (II.xxxi.3). Since they are made by the mind their '*real* and nominal essence is the same' (III.v.14). As things stand, if we could have proper scientific knowledge of bodies we should be in principle able to come to knowledge of their real essences and we should be able to 'from thence discover those properties [*sc.* those we now have to discover by observation], and demonstratively see how they flow, and are inseparable' from them.

8

WHAT ARE SECONDARY QUALITIES?

It is quite usual for commentators on Locke to say that his primary qualities are 'Solidity, Extension, Figure, and Mobility' and his secondary qualities are 'Colours, Sounds, Tasts', etc. These lists are quoted from *Essay* II.viii.9–10, since this is probably the main source for the idea, but their actual composition is not crucial for my argument in this chapter. I intend here to argue more fully for the view already mentioned that colours, sounds, tastes, etc. are not Locke's secondary qualities; my hypothesis is, as I have said, that he used such words as 'colour' and 'red' for *ideas* of secondary qualities rather than for secondary qualities themselves. This is not a trivial hypothesis depending merely upon the view that he holds that *all* words stand for ideas. He clearly means 'extension' and 'figure' to be names of primary qualities as well as of ideas of them, as in II.viii.18 for example; my contention is that it is part of the distinction between primary and secondary qualities that, in contrast, neither 'colour' and 'odour' nor 'red' and 'sour' are names of secondary qualities.

The establishment of this hypothesis would, I believe, be an important step because it would remove some apparent contradictions which have troubled even Locke's most sympathetic critics.[1] The most important of these is that he appears to say *both* that *all* qualities are in bodies independently of perceivers, being powers to produce effects upon us or other bodies (e.g. II.viii.8–10 and II.viii.23) *and* that in a world lacking perceivers there would be no secondary qualities (as, allegedly, at II.viii.17).

We may here leave aside the view that says bluntly that, for Locke, secondary qualities are 'subjective' or 'mind-dependent' or 'in the minds of perceivers' whereas primary qualities are not. However, the price to be paid for this has been regarded by some to be the admission of an inconsistency in Locke since, in spite of his official doctrine that secondary qualities are just powers in bodies to produce ideas in us, when talking of fire and snow he says

But *Light*, *Heat*, *Whiteness*, or *Coldness*, are *no more really in them, than Sickness or Pain is in* Manna. Take away the Sensation of them; let not the Eyes see Light, or Colours, nor the Ears hear Sounds; let the Palate not Taste, nor the Nose Smell, and all Colours, Tastes, Odors, and Sounds, as they are such particular *Ideas*, vanish and cease, and are reduced to their Causes, *i.e.* Bulk, Figure, and Motion of Parts. (II.viii.17)

About porphyry, he says

Hinder light but from striking on it, and its Colours Vanish; it no longer produces any such *Ideas* in us: Upon the return of Light, it produces these appearances on us again. Can any one think any real alterations are made in the *Porphyre*, by the presence or absence of Light; and that those *Ideas* of whiteness and redness, are really in *Porphyre* in the light, when 'tis plain *it has no colour in the dark*? (II.viii.19)

Clearly, these passages are inconsistent with the official doctrine only if 'colours', 'tastes', 'odours', 'sounds', 'whiteness', and 'redness' are taken to name secondary qualities or classes of secondary qualities. I believe that the inconsistency is not Locke's. He did not mean to say that in a world lacking perceivers there would be no *secondary qualities*; if he is read carefully he does not even appear to say it. In these two passages he is talking of *ideas* being removed when the conditions for perceiving are not satisfied. For him, the fact that without perceivers bodies would have no colours, tastes, odours or sounds does not entail that they would have no secondary qualities. The mistake of his interpreters can be exposed by a careful reading of the very passage upon which it is based, when it becomes clear that Locke adhered consistently to the maxim: 'Ideas *in the Mind, Qualities in Bodies*'.

I believe that there is no passage in the *Essay* in which Locke unequivocally indicates that he is using such words as 'red' and 'colour' as the names of secondary qualities or classes of them. On the contrary, the most natural reading of the relevant passages supports my view.

In what follows, it is of great importance to bear in mind what I have said about Boyle's and Locke's use of 'texture' as a technical term. A consequence of this is that secondary qualities are those textures of bodies that produce sensations of colours, etc. in us. This is part of the attempt of both Boyle and Locke to avoid explanations that rely on occult qualities; if causes of sensations were something over and above these textures they would be occult. On the other hand, if we could describe a texture in complete detail we should understand how it would act upon other possible and fully described textures.

I have argued that, in spite of appearances, Locke does not intend texture to be a primary quality. It is noteworthy that Locke discusses in some detail all the items that appear in the lists that seem to be lists of primary qualities except textures. If, as I say, 'texture' is used as a technical term is it not surprising that he did not think it important to give it the separate discussion he gave to the other items? There are two possible replies, not exclusive of one another. The fact that he did not give it individual discussion suggests, first, that Locke did, like Boyle, regard texture as derived from primary qualities and, second, that he took the word 'texture' to be familiar to his readers in his sense, and so not to need special explanation. This is supported by the *Oxford English Dictionary* where, under 'texture' we find

> 4. In extended use: the constitution, structure or substance of any thing, with regard to its constituent or formative elements.
> a. Of organic bodies and their parts. 1665 Boyle *Occas. Med.* IV.iv *The Leaves ... of a Tree ... are of a more solid Texture and a more durable Nature than the Blossoms...*
> b. Of inorganic substances as stones, soil, etc.: Physical (not chemical) constitution; the structure or minute moulding (of a surface). 1660 Boyle *New Expt[s]. Phys. & Mech.* xxii, 165 *Air is ... endowed with an Elastical power that probably proceeds from its Texture.*

The quotations from Boyle are the first in each list of examples.

My suggestion, then is that secondary qualities, being qualities of bodies, are textures in the sense explained. However, Locke holds that secondary qualities are *powers* in bodies to produce sensations, these powers residing in patterns of the groups of corpuscles which constitute bodies. Am I then identifying powers with textures? I am at least entertaining the hypothesis that Boyle and Locke did identify them. However, Locke and Boyle say such things as that the powers of bodies 'depend upon' or reside in the patterns of primary qualities. Locke, in II.viii.19 talks of 'such a texture, that hath the power to produce such a sensation in us'.

Concerning the first point, I have said that there was little idea at the time of chemical action; what we think of as chemical action was then thought of as mechanical. I have suggested in the last chapter that powers are to be conceived of in a mechanical and, especially, a *geometrical* way. To say, for example, that one body has the power to penetrate another is just to say that the corpuscles of the two bodies are so arranged that the second has interstices of such a shape and size that projections on the first will *fit* them. To describe the texture of the first body completely is to describe something that fits certain other textures. The texture of gold is such that the particles will fit neatly, in some regular fashion, into the interstices in the texture of *aqua regia* and that is what dissolving is. To describe in terms of its shape and size a cone-shaped corpuscle is to describe its *ability* to fit a cone-shaped cavity of a certain size; the ability to fit is not something more in the corpuscle to be described.

As for what Boyle and Locke say about powers *depending upon* textures, consider this: to say that one body has the power to penetrate another is to describe from the point of view of gross observation an ability which can be explained in terms of, is obvious from, a complete description of the unobserved textures of the two bodies. It is a matter of observation that gold dissolves in *aqua regia* but it does not help to explain it merely to say that *aqua regia* has the power to dissolve gold. What is needed, and what Boyle and Locke envisage, is an analysis of such powers that will explain why *aqua regia* has that power, what accounts for that power. In II.viii.15 Locke says 'what is Sweet, Blue, or Warm in *Idea*, is but the certain Bulk, Figure and Motion of the insensible Parts in the Bodies themselves, which we call so'. Boyle says that the power of a key to turn a lock is nothing distinct from its shape and size, which fits the shape and size of the inwards of the lock and that the attributes of being soluble in *aqua regia* and insoluble in *aqua fortis* 'are not in the gold any thing distinct from its peculiar texture' (*O.F.Q.* 18, *S.* 24). I have quoted the important passage in IV.iii.25 where Locke echoes this and uses just the examples used by Boyle.

I must now discuss an ambiguity I ignored in introducing the hypothesis of this Chapter. Consider again my hypothesis: Locke uses such words as 'colour' and 'red' for *ideas* of secondary qualities, or classes of them rather than for secondary qualities themselves. However, since he holds that '*Words, in their primary or immediate Signification, stand for nothing, but the* Ideas *in the Mind of him that uses them* ...' (III.ii.2), it would be trivially inconsistent for him to say that 'red' stands primarily or immediately for a secondary quality. So, can a secondary quality, as distinct from an idea of a secondary quality, be named at all? I think it can, as will emerge. What is clear is that the *primary or immediate* signification of words 'standing for secondary qualities', if there can be such, must be *ideas* of secondary qualities.

At this point we must recognize that the expression 'idea of a secondary quality' is ambiguous; we must take into account the fact that Locke uses the word 'idea' for both 'percept' and 'concept'. He usually uses it, in this context, to mean a percept; thus 'red', I allege, is the name of an *idea* of a secondary quality in this sense. More fully, it is the name of a percept caused in us by a secondary quality. He calls this percept the idea of a secondary quality because the only way in which the secondary quality, i.e. a particular texture, can *appear* to us is as that percept.

At present we have neither names for secondary qualities nor descriptions of the specific textures which produce specific sensations in us. Locke appears to think that we may eventually obtain limited knowledge of these textures but not universal and perfect knowledge (IV.iii *passim* and IV.xii *passim*). If we could arrive at descriptions of textures it would be by scientific inference. The only descriptions we have of secondary qualities are of the form 'the power to produce the idea of red in us' but these, as I have argued, are uninformatively of the *virtus dormitiva* kind. Descriptions of textures would be informative descriptions of secondary qualities, qualities of objects. We should then have what I am calling 'concepts of secondary qualities' and not just percepts caused by secondary qualities.

My hypothesis, then, is that Locke intends such words as 'red' to be used, in philosophical contexts, as names of ideas (percepts) caused by secondary qualities but not for ideas (concepts) of the cause of those percepts, i.e. secondary qualities. This concerns primary or immediate signification.

Can we say that descriptions of textures could, for Locke 'stand for' secondary qualities or must we say that they could 'stand for' only ideas (concepts) of secondary qualities? Here, Locke's resemblance thesis becomes important. However we analyse this thesis, it seems clear that it is meant to catch whatever we mean by 'having an accurate idea of' something. One consequence of the thesis is that the names of ideas (percepts) of primary qualities will also serve as the names of primary qualities; 'extension' names both an idea and a quality. One aim of natural science is to provide descriptions of natural phenomena and this involves arriving at accurate ideas of them. So we can say that one aim of science is to get ideas (concepts) of particular secondary qualities which are accurate. Then descriptions of these would also serve as descriptions of secondary qualities. If we then invented names for various textures, or secondary qualities, they would also serve as names for our ideas (concepts) of them. The ideal is to bring our descriptions of the world in terms of secondary qualities to the same level as our descriptions of it in terms of primary qualities.

Is this consistent with Locke's view that words stand for ideas? As has been pointed out from time to time, his statements such as '*Words in their primary or immediate Signification, stand for nothing, but the* Ideas *in the Mind of him that uses them* ...' (III.ii.2) do not entail that words stand for *nothing but* ideas.[2] The reservation '*in their primary or immediate Signification*', which Locke italicizes, must be taken seriously. Locke thinks that men are *right* when they suppose that words stand also for ideas in the minds of others and for the reality of things, because he *does* think that if words did not so stand we should talk in vain, not be understood and not talk of things as they really are (III.ii.4 and 5). (This will be discussed more fully in a later chapter.) There cannot, then, be any objection, on grounds of inconsistency, to saying that names of ideas, whether percepts or concepts, of qualities can be *also* the names of those qualities.

If I am right, colours, tastes, odours and sounds are not, for Locke secondary qualities but ideas; secondary qualities are colourless, tasteless, odourless and soundless textures of bodies. However, I believe that my negative thesis can be established independently of my positive thesis. I now turn to a more detailed consideration of my hypothesis.

Chapter viii of Book II must surely be the most relevant to this hypothesis since it is here that he distinguishes ideas from their causes, qualities from ideas and secondary from primary qualities. Within this chapter, sections 9 and 10 are probably of the greatest importance since they contain the first and, as I think is generally agreed, the best statement of the distinction between primary and secondary qualities. The passage that might seem to work most strongly against my hypothesis is that in which he first explains secondary qualities. He says

> 2dly, Such *Qualities*, which in truth are nothing in the Objects themselves, but Powers to produce various Sensations in us by their *primary Qualities*, *i.e.* by the Bulk, Figure, Texture, and Motion of their insensible parts, as Colours, Sounds, Tasts, *etc.* These I call *secondary Qualities*. (II.viii.10)

The form of this passage has misled readers in two ways, most notoriously and inexcusably because some have stopped paying attention after the word 'themselves' and so failed to notice that 'nothing' goes with 'but'. Locke is not, of course, saying that secondary qualities are nothing in the objects themselves; we cannot conclude from that sentence that he is saying they are mind-dependent. He is here beginning to say just what it is in objects that constitutes secondary qualities.

The second error in interpreting this passage is less vulgar and obvious. The word 'These' in the last sentence has been taken to refer back to 'Colours, Sounds, Tasts'; I believe that it refers back to 'Qualities' in its first appearance or to 'Powers to produce various Sensations in us', or both. Qualities are not sensations and the phrase 'as Colours, Sounds, Tasts, *etc.*' lists sensations or ideas. In other words the passage should be read as if it were written

> Secondly, such qualities which, in objects themselves, are nothing but powers to produce in us various sensations, such as colours, sounds, tastes, etc., by their primary qualities, i.e. by the bulk, figure, texture, and motion of their insensible parts. These qualities (i.e. powers) I call secondary qualities.

When Locke returns to his definitions in section 23 he heads the section '*Three sorts of Qualities in Bodies*' and in between he has sections headed '*How primary Qualities produce their* Ideas' and 'How secondary' and he several times says that secondary qualities are 'merely', 'barely', or 'only' powers. Qualities are related to ideas as causes to effects. Colours, sounds, etc. are not powers; they are ideas, the effects of powers on us. In view of all this, my reading of the above passage seems more natural and plausible than one that takes 'Colours, Sounds, Tasts, *etc.*' to be a list of secondary qualities. It also saves Locke from such infelicities as the view that ideas, in his strict sense, can be in objects.

It also has the advantage that the contrast between primary and secondary qualities, as Locke states it, is clear and obvious by enabling us to see the importance of inseparability. The thought is not that whereas extension, for example, is inseparable from body, colours, for example, are not. Of course, colours as we perceive them are separable since they are ideas. Shape, size and mobility are inseparable from body, *as such*, and what is being contrasted with this is the patterns or textures formed by conglomeration of corpuscles. These are in bodies, just as much as primary qualities but they are separable because they are not in *all* bodies. That is, they are separable from bodies as such because single corpuscles do not have them but not separable from complex bodies consisting of numbers of corpuscles.

The paragraph defining secondary qualities clearly refers to complex bodies because it refers to the motion of their insensible parts. Bodies *as such* do not have insensible parts; the smallest actual bodies are those that are the insensible parts of complex bodies. The definition of primary qualities must admit only qualities that are inseparable from anything that is a body; the definition of secondary qualities must admit qualities that only complex bodies have and so are separable from bodies, because they are not in the simplest bodies. Secondary qualities *are* textures and ideas of secondary qualities such as colours, sounds, tastes, etc. are all to be explained by reference to textures. A single corpuscle cannot be coloured since colour is an idea; it cannot cause colour-ideas because it is too small and too simple. It does not have the texture that can reflect, and selectively reflect, light. Presumably one light corpuscle could bounce off some other variety of corpuscle but the reflected corpuscle could not enable us to see anything and could not be 'coloured'. What is seen as coloured is a stream of corpuscles with a particular texture. (Section 10 is also the first in which textures are mentioned.)

I have already (in Chapter 7) said enough to remove what looks like another telling and ready-to-hand objection to my interpretation. I refer to the passage in section 8 about the snowball. As I have argued, we must pay attention to the development of Locke's argument in this chapter. If we see him as distinguishing, within what are often called 'ideas', between sensations or perceptions, which he proposes to call '*Ideas*', and the corresponding powers in bodies to produce those ideas, which he proposes to call 'Qualities', then this passage no longer suggests the confusions of which Locke is accused.

When Locke again describes the distinction between primary and secondary qualities, in II.viii.23, he says, in describing secondary qualities

> Secondly, The *Power* that is in any Body, *by* Reason of *its* insensible *primary Qualities*, to operate after a peculiar manner on any of our Senses, and thereby *produce in us* the *different Ideas* of several Colours, Sounds, Smells, Tasts, *etc.* These are usually called sensible Qualities.

Here he appears to distinguish the powers of bodies to operate upon our senses from the ideas thereby produced, *i.e.* colours, sounds, smells, and tastes. He is defining secondary qualities and he seems to be contrasting them with what 'are usually called sensible Qualities', which are ideas. He has clearly said, in the previous paragraph, that we can discover by observation the primary qualities of middle-sized bodies so he does not mean to say, here, that *all* primary qualities are insensible, in contrast to secondary qualities, all of which are sensible. It might be objected that Locke's italics in this passage constitute some evidence against my interpretation. I suggest, however, that in the phrase '*its* insensible *primary Qualities*', the word 'insensible' is to be stressed and so is left unitalicized in an italicized context; the *insensible* primary qualities are just primary qualities of the individual corpuscles, which contribute to textures which are, and cause ideas of, secondary qualities. We may take what are 'usually called sensible qualities' to be colours, sounds, smells, tastes, etc. which are *ideas*. Locke is, I suggest, objecting to this usual way of speaking since he has already said that these are not qualities. The argument goes on in section 23; this way of talking misleads people into thinking that ideas are qualities and that all ideas are *real* qualities. I think that he is again proposing a way of talking different from the usual way: colours, sounds, smells, tastes, etc., should be called not 'sensible qualities' but 'ideas of secondary qualities', which qualities really are qualities of bodies, though they are insensible because they are the fine texture of bodies composed of corpuscles. It is not absurd to suppose that Boyle would have accepted this amendment.

In II.xxxi.3 he says 'And sensible Qualities as Colours and Smells, etc., what are they but the *Powers* of different bodies in relation to our perception, etc.?' He is surely here saying that the so called sensible 'Qualities' are not really qualities but ideas that arise from the interaction of the powers or qualities of bodies and our sense organs.

In II.viii.14 and 15, which deal with resemblances, I believe that Locke is, *inter alia*, contrasting what are usually called sensible qualities with what he calls secondary qualities. In section 14 he says

What I have said concerning *Colours* and *Smells*, may be understood also of *Tastes* and *Sounds, and other the like sensible Qualities*; which, whatever reality we, by mistake, attribute to them, are in truth nothing in the Objects themselves, but Powers to produce various Sensations in us...

If I am correct, for Locke what the man-in-the-street means by 'sensible qualities' must be secondary qualities *as sensed*, i.e. ideas.

I cannot rely too heavily on this point because Locke does not appear to be entirely consistent in his use of 'sensible qualities'. However, when in II.xxiii.9 and IV.iii.28, for example, he appears to contradict this by talking of 'sensible secondary Qualities' he may be using 'sensible' not in this vulgar sense but in the learned sense to mean 'able directly to cause sensations'.

Both Locke and Boyle were worried by the uncertain and conflicting ways in which the word 'quality' was commonly used at the time. Boyle says that he is not going to waste time enquiring into 'all the several significations of the word *quality*, which is used in such various senses, as to make it ambiguous enough...'[3]

Locke from time to time speaks, with signs of unease, about vulgar ways of talking in general. In III.iv.16 he says

... the general term *Quality*, in its ordinary acception, comprehends Colours, Sounds, Tastes, Smells, and tangible Qualities with distinction from Extension, Number, Motion, Pleasure, and Pain...

and he clearly does not accept this when he defines 'quality' in Book II. In II.xxxi.2 he says

For though Fire be call'd painful to the Touch, whereby is signified the power of producing in us the *Idea* of Pain; yet it is denominated also Light and Hot; as if Light and Heat, were really something in the Fire, more than a power to excite these *Ideas* in us; and therefore are called *Qualities* in, or of the Fire. But these being nothing, in truth, but powers to excite such *Ideas* in us, I must, in that sense, be understood, when I speak of secondary *Qualities*, as being in Things; or of their *Ideas*, as being in the Objects, that excite them in us. Such ways of speaking, though accommodated to the vulgar Notions, without which, one cannot be well understood; yet truly signify nothing, but those Powers, which are in Things, to excite certain Sensations or *Ideas* in us.

That is, Locke is rejecting the vulgar way of calling heat and light qualities, and so secondary qualities; when he uses 'secondary qualities' he means not heat and light but the powers in objects to give us those ideas.

All this is connected with Locke's calling secondary qualities, in II.viii.22, 'imputed qualities'. We never strictly perceive secondary qualities because they are textures responsible for our perceiving colours, smells, sounds and tastes. The textures fit to produce these ideas in us must be *imputed* to bodies on the basis of scientific investigation and inference. Primary or real qualities, on the other hand, we sometimes, at least, perceive as they really are in bodies. The passage reads, in part,

> it being necessary in our present Enquiry, to distinguish the *primary* and *real Qualities* of Bodies, which are always in them, (*viz.* Solidity, Extension, Figure, Number and Motion, or Rest; and are sometimes perceived by us, *viz.* when Bodies they are in, are big enough singly to be discerned) from those *secondary* and *imputed Qualities*, which are but the Powers of several Combinations of those primary ones, when they operate, without being distinctly discerned ... (II.viii.22)

It is noticeable that he gives a list of primary qualities but not of secondary qualities; there are no handy names for secondary qualities and the shortest way of describing them in general is by calling them powers. Moreover, he says distinctly that primary qualities are sometimes perceivable but that secondary qualities are not.

Light is thrown on this by II.xxx.2, where Locke says that even though whiteness and coldness are not in snow yet these are 'real *Ideas* in us whereby we distinguish the Qualities that are really in things themselves' because these ideas are either 'constant effects' or 'exact resemblances' of those qualities. 'And thus' he says 'our simple *Ideas* are all real and true, because they answer and agree to those Powers of Things, which produce them in our Minds, that being all that is requisite to make them real and not fictions at Pleasure'. Our simple ideas enable us to distinguish objects from one another because differences in ideas correspond to differences in their causes in objects. All simple ideas are real and not 'fantastical'. When Locke calls primary qualities real and secondary qualities imputed he is not using the contrast of real with fictional, for in that sense, the causes of *all* simple ideas are real. The contrast of 'real' with 'imputed' concerns not the nature of qualities but the basis on which we attribute qualities to objects.

Why, then, does Locke call the powers of bodies to affect us 'secondary qualities immediately perceivable' and their powers to affect other bodies 'secondary qualities mediately perceivable' if, as I say, no secondary qualities are strictly perceivable? Locke does not mean to say that some secondary qualities are perceivable in the way in which the shape and colour of a piece of porphyry are. He explains what he means in II.viii.26: secondary qualities immediately perceivable are those powers of bodies whereby they are fitted 'by immediately operating on our Bodies to produce several different *Ideas* in us' and secondary qualities mediately perceivable are those powers whereby they are fitted 'by operating on other Bodies, so to change their primary Qualities, as to render them capable of producing *Ideas* in us, different from what they did before'. Perceiving is having ideas caused by qualities of bodies; secondary qualities produce ideas in us either by operating immediately on our senses or by operating mediately on our senses through other bodies whose secondary qualities are changed and so produce different ideas in us. The contrast is between the immediate and mediate operation on our bodies of the corpuscles of external bodies.

It seems to me, then, that a consistent account of II.viii can be given on the hypothesis that Locke used such words as 'red' and 'colour' for ideas of secondary qualities, or classes of them, rather than for secondary qualities themselves. I do not think that anything fundamental in the rest of the *Essay* conflicts with this interpretation. If anything Locke says does need explaining away it will involve less serious modifications and more plausible interpretation than supposing that he did not accept his own definitions in II.viii.

There is a long section in Book IV (IV.vi.11) which seems to weigh against much of what I have said in interpreting Locke; this is a convenient point at which to attempt to show that it does not.

Locke begins the section by putting his by now familiar point about the relationship between the inner constitutions or real essences of substances, their qualities and the ideas they cause in us. He says

> Had we such *Ideas* of Substances, as to know what real Constitutions produce those sensible Qualities we find in them, and how those Qualities flowed from thence, we could, by the specifick *Ideas* of their real Essences in our own Minds, more certainly find out their Properties, and discover what Qualities they had, or had not, than we can now by our Senses: and to know the Properties of *Gold*, it would be no more necessary, that *Gold* should exist, and that we should make Experiments upon it, than it is necessary for the knowing the Properties of a Triangle, that a Triangle should exist in any matter, the *Idea* in our Minds would serve for the one, as well as the other. (IV.vi.11)

If we knew the inner structure of, say, gold in all its details we should know how it would react with other specified inner structures without experiment, even if gold or, presumably, those other structures, did not exist.

However, Locke now proceeds to talk in a way that appears, on the face of it, to conflict with this; as if all the qualities of things or substances were dispositional and dependent upon other things or substances rather than intrinsic as the above passage suggests. We are inclined, he says, to think of substances as possessing their qualities independently of other substances and so we overlook the many surrounding bodies and substances that affect them to cause changes that we observe. He refers to

> the Operations of those invisible Fluids, they are encompassed with; and upon whose Motions and operations depend the greatest part of those qualities which are taken notice of in them, and are made by us the inherent marks of Distinction, whereby we know and denominate them.

However, he continues,

> Put a piece of *Gold* any where by itself, separate from the reach and influence of all other bodies, it will immediately lose all its Colour and Weight, and perhaps Malleableness too; which, for ought I know, would be changed into a perfect Friability. *Water*, in which to us *Fluidity* is an essential Quality, left to it self, would cease to be fluid.

Now, I believe that Locke is not saying that all the qualities that he seems to have said are intrinsic to substances are, after all, dispositional and relational. He is not saying that the piece of gold would lose that quality or inner constitution that makes it look yellow in its normal setting or have weight in relation to other bodies or bend rather than crumble when struck by a hammer. I believe he is talking about the *ideas* of its colour, weight and malleability produced in me by a specific inner constitution, but only produced in the presence of other bodies such as light corpuscles, and so on. What disappears when other bodies are removed, Locke says, is

> the greatest part of those qualities *which are taken notice of in them*, and are *made by us the inherent marks of Distinction*, whereby *we know and denominate them*. (IV.vi.11, my italics)

The 'qualities which are taken notice of' are qualities *as sensed*, that is *ideas*, and so are caused by intrinsic qualities; what Locke sometimes calls 'sensible qualities'. Were it not for the surrounding bodies we should not be able to 'take notice' of these qualities because they would not produce ideas in us, or alterations in other bodies which would in turn produce ideas in us. Although in the absence of light gold would not look yellow it would still have the quality (texture) that necessarily makes it look yellow in normal conditions.

Throughout the rest of this section Locke puts in expressions that indicate that he is talking about what we observe rather than about the nature of things apart from our observation.

Thus in the next sentence he says

> if inanimate Bodies owe so much of their present state to other Bodies without them, *that they would not be what they appear to us*, were those Bodies that environ them removed, it is yet more so in *Vegetables* ... (my italics)

and again

> if we look a little nearer into the state of *Animals*, we shall find, that their Dependence as to Life, Motion, and the most considerable Qualities *to be observed in them*, is so wholly on extrinsecal Causes and Qualities of other Bodies, that make no part of them, that they cannot subsist a moment without them: though yet those Bodies on which they depend, are little taken notice of, and make no part of the complex *Ideas*, we frame of those Animals. (my italics)

Our complex ideas of animals are, of course, collections of simple ideas we gain from observation. If we take away air from living creatures they soon lose sense, life and motion; we all depend upon the sun as a source of heat; we all depend upon environments of which much is unknown to us. These things are

> absolutely necessary to make them [animals] be, *what they appear to us*, and to preserve those Qualities, *by which we know*, and distinguish them (my italics)

and

> We are then quite out of the way, when we think, that Things contain within themselves the Qualities, *that appear to us in them*. (my italics)

All this is consistent with the view that qualities that bodies *appear to have* depend upon other bodies but the real qualities of bodies do not.

Finally, Locke says about bodies

> Their *observable* Qualities, Actions, and Powers, are owing to something without them; and there is not so complete and perfect a part, *that we know*, of Nature, which does not owe the Being it has, and the Excellencies of it, to its Neighbours; and we must not confine our thoughts within the surface of any body, *but look a great deal farther*, to comprehend, perfectly those Qualities that are in it. (IV.vi.11, my italics)

Note the last phrase 'to comprehend perfectly those Qualities that are in it'; Locke is again stressing that there are qualities in it, intrinsic qualities, and he has just reminded us, as he has often done before, that we should not suppose that all the apparent qualities in things are really in them. He is here making mainly epistemological points about bodies as they appear to us. The properties of a substance include both its actions on other substances and their actions on it. Gold is soluble in *aqua regia*, animals and plants need air to survive. If we could discover directly their intrinsic properties, by discovering their inner structures, we should know all their properties but we cannot do this in the present state of the art. So, in our present state of ignorance, we have to observe and experiment, to try the effects of bodies upon one another; and when we observe them, their properties present themselves as dispositional and relational. The only sign *we* can get that gold is soluble in *aqua regia* is that it dissolves; the only sign that animals need air is that they die if we deprive them of it. These occurrences depend upon other bodies.

However, the dissolving of a sample of a substance in *aqua regia* is not identical with that substance's being soluble in *aqua regia*; an animal's dying when air is removed is not identical with its being the sort of thing that needs air for its survival or proper functioning. A gas stove when connected to the gas main and turned on may or may not light; if we knew in every detail its inner constitution we should be able to say for certain, in advance, whether it would or would not light under those conditions. The *solubility* of gold and an animal's *being the sort of thing that needs air* are intrinsic properties of the gold and the animal, respectively, on which the actual dissolving and the actual dying when deprived depend.

That is the point of the passage first quoted above from this section. The solubility of gold in *aqua regia* is a structural property and, therefore, an intrinsic property of gold. If we knew that structure we could say that it must necessarily dissolve in something having certain other structures, whether or not that structure existed, whether or not we had discovered anything with that structure and whether or not we had tried the experiment and found gold to dissolve. Locke says that we could know the properties of gold even if gold did not exist. This may be interpreted in the most abstract and theoretical way conceivable. It would be possible to start with the conception of various sorts of corpuscles with all their properties specified, 'construct', on paper or in the head, a certain texture, whose properties resulted from those of the corpuscles, out of some selection of them and then by considering the properties of this structure to 'construct' another such that the first would necessarily dissolve in it. The first could be the structure that gold actually has and the second the structure that *aqua regia* actually has even though neither gold nor *aqua regia* existed or was known.

If gold and *aqua regia* exist then if we could fully investigate their structures directly and separately from one another we should be able to say, without putting them together, that one would dissolve in the other. We are not, so far able to do this; our only way of finding out that gold does dissolve in *aqua regia*, as things stand, is to try the experiment.

To say that an animal needs air for survival is also to refer indirectly to a particular inner constitution such that certain things would not happen within the animal, things we associate with animal life and behaviour, if something with the constitution of air were not available to it. If we knew the inner constitution of the animal we should be able to say with certainty 'That is the constitution of an air-needing animal', without trying the effect of withholding air from it.

In conclusion, then, the section I have been discussing appears to conflict with my interpretation of Locke only if we confuse what Locke says about ontological and epistemological matters. A world containing necessarily connected structures must present itself to us, when we have to rely on observation and induction for information about it, as a world full of things with relational properties; only different methods of discovering its details would be capable of giving us a picture of it as it really is.

9

OBSERVABILITY

Locke is frequently taken to hold that secondary qualities are perceivable and primary qualities not. However, it follows from my interpretation that primary qualities are perceivable and secondary qualities not. It seems to me possible to hold the former view only if one holds that colours, tastes, odours and sounds are secondary qualities and I have argued that they are not. There are various passages that appear to conflict with my view. Perhaps the most important of them is that in which Locke says

> secondary Qualities are those, which in most [substances] serve principally to distinguish Substances one from another... For our Senses failing us, in the discovery of the Bulk, Texture, and Figure of the minute parts of Bodies, on which their real Constitutions and Differences depend, we are fain to make use of their secondary Qualities, as the characteristical Notes and Marks, whereby to frame *Ideas* of them in our Minds, and distinguish them one from another. (II.xxiii.8)

I have argued that secondary qualities are textures and not colours, odours, etc., which are ideas produced by them, and textures are unobservable as Locke says in this passage. Locke puts the matter in the way he does here because, I believe, he thinks that we are really capable of distinguishing bodies from one another and that our ability to do so depends partly upon qualities actually possessed by the bodies. When one body appears red and another blue to us that is not *merely* a matter of their appearing differently to us; if we have been careful they really are different in respect of the textures responsible for their red and blue appearances. These textures being secondary qualities we are in fact making use of secondary qualities in the only way we are able to, by making use of ideas caused by them in us, by taking notice of their appearances to us. We are not observing secondary qualities directly, which would enable us to describe those secondary qualities, textures, as they are in the bodies, but indirectly through their effects, which enables us to describe only the effects and to say there are real differences between bodies corresponding to them.

One must distinguish also between the primary qualities of the corpuscles and the primary qualities of observable bodies. The textures of bodies, in the fundamental sense of that word, are unobservable if only because they involve relations between corpuscles, the corpuscles are 'too small to be singly sensible' and the primary qualities *of the corpuscles* are unobservable. However, the primary qualities of observable bodies, their shapes and sizes, are observable and these are qualities of the same sort as the primary qualities of the corpuscles. Indeed, it is only our ideas of primary qualities obtained through observation that accurately represent qualities of bodies.

Of course, we would not be able to observe these primary qualities if it were not for secondary qualities; more generally, an object must have secondary qualities in order to be observable at all. This is, in the first place, because it must be above a certain size and, in the second place, because we would not be able to distinguish one body from another visually if we could not distinguish it by colour from its surroundings, or by any other sense, if it did not have some secondary qualities. It might be thought that when I distinguish by touch in the dark, say, a cube from the table on which it stands I do so by means of primary qualities alone but this cannot be so because in order to detect the cube and the table in the first place I need their felt hardness and the felt hardness is an idea caused by a secondary quality, as I have already argued.

It follows that to be observable a body must have a texture; to be large enough it must consist of a collection of corpuscles and to be distinguishable from its surroundings it must be capable of producing ideas of secondary qualities in me. On this view, it follows that single corpuscles are in principle unobservable because a single corpuscle does not have a texture in the relevant sense. It cannot look coloured, feel hard, have a smell or taste or make audible sounds. Of course, a single corpuscle must be solid, in the absolute sense and it might be thought that this would form the basis of observability. So it does but only by contributing to a texture: the theory is that it is the interaction of textures of bodies and sense organs, sometimes through the intervention of light rays, that results in perception. A single corpuscle is just too small and too simple for its solidity to have a detectable effect on our sense organs.

There are passages in the *Essay*, especially in II.xxiii, which have led some critics to suppose that Locke takes the corpuscles to be in principle observable. For example, John Yolton[1] argues, mainly on the basis of these passages, against Maurice Mandelbaum's thesis[2] that they are in principle unobservable. I do not propose to enter into a detailed discussion of their arguments but merely to present an interpretation of these passages, based on my account so far, which favours the view that the corpuscles are in principle unobservable, according to Locke. These passages seem to me to be quite clearly consistent with the view for which I have argued.

Locke says

> Had we Senses acute enough to discern the minute particles of Bodies, and the real Constitution on which their sensible Qualities depend, I doubt not but they would produce quite different *Ideas* in us; and that which is now the yellow Colour of Gold, would then disappear, and instead of it we should see an admirable Texture of parts of a certain Size and Figure. This Microscopes plainly discover to us: for what to our naked Eyes produces a certain Colour, is by thus augmenting the acuteness of our Senses, discovered to be quite a different thing; and the thus altering, as it were, the proportion of the Bulk of the minute parts of a coloured Object to our usual Sight, produces different *Ideas*, from what it did before. (II.xxiii.11)

The first sentence is what most strongly suggests that the corpuscles are in principle observable. Locke seems to be saying that if we could see the corpuscles and the real constitutions of bodies they would appear very differently to us and this can hardly be denied since we cannot even conceive what it would be like to see things without colours except that it would be very different from seeing things as we do. This can be said sensibly and truly without implying that it is in principle possible that we should be able to see corpuscles. Perhaps it was misleading of Locke to use the word 'acute'; we should need not merely senses that were more acute but senses that were quite different. Locke mentions this possibility later when he says

> what other simple *Ideas* 'tis possible the Creatures in other parts of the Universe may have, by the Assistance of Senses and Faculties more or perfecter, than we have, or different from ours, 'tis not for us to determine. But to say or think there are no such, because we conceive nothing of them, is no better an argument, than if a blind Man should be positive in it, that there was no such thing as Sight and Colours, because he has no manner of *Idea*, of any such thing, nor could by any means frame to himself any Notions about Seeing... What Faculties therefore other Species of Creatures have to penetrate into the Nature, and inmost Constitutions of Things; what *Ideas* they may receive of them, far different from ours, we know not. (IV.iii.23)

We can conceive of other creatures having senses different from ours so we are not entitled to deny the possibility of it; but because we cannot conceive what it would be like to have such senses we are unable to say what those perceptions would be like. When we detect something by any of our senses we are able to do so only because for each sense there are means of discriminating that thing from its surroundings. We can conceive of a creature's having a sense which discriminates by some means different from any means available to us but we cannot conceive what that means might be or what it would be like to have it available. But we do know that we do not have such extra senses so nothing we can correctly say about this can alter the fact that the corpuscles are in principle unobservable by us.

The second sentence of the first passage quoted above appears to say merely that the microscope shows us that variations in the acuteness of our senses make a great difference to what we see. It nowhere suggests that we could see uncoloured corpuscles but merely that we should see different colours; and this is intended to continue the theme of the first sentence. The examples that Locke goes on to use, sand, hair and blood, are seen under the microscope as coloured, but differently coloured. Thus he says

> Blood to the naked Eye appears all red; but by a good Microscope, wherein its lesser parts appear, shews only some few Globules of Red, swimming in a pellucid Liquor; and how these red Globules would appear, if Glasses could be found, that yet could magnify them 1000, or 10 000 times more, is uncertain. (II.xxiii.11)

If that is uncertain, what and how we could possibly see with a microscope powerful enough to reveal the corpuscular structure is beyond our comprehension.

Shortly after this Locke says

> ...if that most instructive of our Senses, Seeing, were in any Man 1000, or 10 000 times more acute than it is now by the best Microscope, things several millions of times less than the smallest Object of his sight now, would then be visible to his naked Eyes, and so he would come nearer the Discovery of the Texture and Motion of the minute Parts of corporeal things; and in many of them, probably get Ideas of their internal Constitutions... (II.xxiii.12)

That a man so equipped could 'come nearer' the discovery of the inner constitution of bodies and even 'get *Ideas*' of them does not imply that he would be able to see them. There are other sorts of ideas than ideas of sensation and if much more detailed observation were possible it might well be possible to construct better hypotheses about the description of inner constitutions without actually seeing them. We must remember that our naked eyes, as they are, give us no ideas whatsoever of inner constitutions through observation. Nevertheless the corpuscular hypothesis gives us *some* idea, reasonably plausible, of what they are like.

In these sections Locke appears to think that the microscope could be indefinitely improved and, at that time, there was little reason to think otherwise. If he did, and if he thought that eventually we might be able to see textures and individual corpuscles then it is difficult to explain why in Book IV, he is so pessimistic about the future possibility of coming to a knowledge of real essences, or inner constitutions. He even at times appears to say that this is in principle impossible. I shall discuss this later.

It might be objected that in consequence of my contention that simple ideas of sensation are qualities as sensed I ought to say that having such ideas just is observing qualities so no distinction can be drawn in respect of observability between primary and non-primary qualities. However that is to miss part of the point of the resemblance doctrine. Ideas of primary qualities are qualities as sensed which give us some information about the character of the qualities as they are in objects; ideas of secondary qualities are qualities as sensed but they give us no information about the character of the qualities as they are in objects. So there is some point in distinguishing the primary qualities as observable in contrast to the non-primary qualities.

There has been much discussion of Locke's alleged representative theory of perception and the idea that this view implies that we are cut off from the world by an impenetrable 'veil of perception'.[3] Locke says at IV.iv.3 that the mind 'perceives nothing but its own *Ideas*'. I am inclined to sympathize with the view put by John Yolton[4] and Ian Tipton[5] that Locke has been misread and misinterpreted on this matter.

I shall not deal in detail with the various arguments that have been put for and against the 'veil of perception' view but merely show how my interpretation of Locke so far enables us to give an account that, at least, does not entail an objectionable form of representative perception and perhaps entails no form of it.

Locke, as I have said, is following the views of some contemporary natural philosophers and is sympathetic to their explanations. He starts from the assumption of the existence of an external world which is independent of our perceiving it. He speaks often as if we perceive this external world (see e.g. IV.xi.3) but through the medium of ideas. Of course we must have percepts if we are to perceive: that is, there must, in some sense, be something we perceive, our perceptions must have some content. This does not, however, commit us to regarding percepts as objects or entities that 'stand in' for or replace or obscure external objects. As I have argued, ideas are simply the appearance of these objects to us. The scientific account is intended to tell us what happens when we perceive external objects and it would be just mistaken if it concluded that we *don't* perceive external objects because it would be denying the very facts it set out to explain. It aims to show how external objects produce in us perceptions of them. It goes like this for visual perception: objects react with light in various ways and light of a certain character consequently impinges on our eyes where it sets up a train of physical occurrences ending in the brain at which point the mind, mysteriously, perceives the objects.[6] Perceiving is an event which Locke calls 'having an idea'; it is a mental event and the content of the idea is just how the object is perceived, the appearance of it.

If one finds such arguments unconvincing and thinks that Locke is committed to the 'veil of perception' view, would there remain any sense in which qualities of any sort could be regarded as observable? If one accepts the view that the existence of external objects, the causal relation between qualities and ideas and the resemblance thesis are all part of a grand hypothesis which is up for confirmation such a possibility does remain, although the line between the observable and the unobservable might then be drawn differently from the way I have drawn it. It would be part of the hypothesis that some qualities operate directly on us to produce ideas; others, the third sort, operate only indirectly through changes in secondary qualities. This might be a basis for regarding both primary and secondary qualities as observable and the third sort as unobservable or, because of the resemblance thesis, for regarding primary qualities as observable and non-primary qualities as unobservable. The argument does not appear to be settled by the acceptance of the 'veil of perception' view.

10

PATTERNS AND RESEMBLANCE

Locke says that ideas of primary qualities resemble primary qualities but ideas of secondary qualities do not resemble secondary qualities. I have urged that it is of great importance to understand that Locke does not use the idea of resemblance in making his original distinction between primary and secondary qualities and he certainly never regards it as a criterion for distinguishing between them.

The first mention of resemblance occurs in II.viii.7. The word 'idea' is used in a vulgar or vague sense. Among the things referred to when it is so used, Locke proposes to distinguish ideas properly so-called from modifications of matter in bodies. He says

> that so we *may not* think (as perhaps usually is done) that they [*sc.* ideas] are exactly the Images and *Resemblances* of something inherent in the subject; most of those of Sensation being in the Mind no more the likeness of something existing without us, than the Names, that stand for them, are the likeness of our *Ideas*, which yet upon hearing, they are apt to excite in us. (II.viii.7)

Having made the point he now leaves this notion alone while he defines primary and secondary qualities (9–10) in the way I have discussed, with no mention of resemblance. He mentions resemblance again in section 13 merely to say that it is not inconceivable that God should 'annex' ideas of the colour and smell of a flower to motions of minute particles 'with which they have no similitude'. This is because secondary qualities have been defined in terms of the motions of insensible parts of bodies and yet ideas of secondary qualities are quite different, being colours, sounds, tastes and smells.

Sections 11–14 are taken up with a discussion of the way in which bodies produce ideas in us and this is in fact a brief exposition of the corpuscular theory leading us from the discussion of the way things are, in the distinction between primary and secondary qualities, to an account of how we might know the way things are. The ideas of primary and secondary qualities are both, according to the theory, produced in us by *impulse* originating in the sizes, shapes and motions of the constituent corpuscle of bodies.

Then section 15 begins

From whence I think it is easie to draw this Observation, That the *Ideas of primary Qualities* of Bodies, *are Resemblances* of them, and their Patterns do really exist in the Bodies themselves; but the *Ideas, produced* in us *by* these *Secondary Qualities, have no resemblance* of them at all. There is nothing like our *Ideas,* existing in the Bodies themselves. (II.viii.15)

This is the first time that the resemblance/non-resemblance distinction has been put to work in relation to primary/secondary quality distinction. It is clearly intended to be a conclusion drawn from what has gone before. The clue to 'whence' Locke thinks 'it is easie to draw this distinction' lies in the clause 'their Patterns do really exist in the Bodies themselves'. The resemblance of primary qualities and the non-resemblance of secondary qualities to their causes is not a consequence merely of the definitions of primary and secondary qualities. It is a consequence of the definitions together with other features of the theory of which the definitions are a part, the corpuscular theory.

Bodies produce ideas in us 'by *impulse*'. Since primary qualities of bodies are perceivable *at a distance* there must be some 'singly imperceptible Bodies' that come from external bodies to our sense-organs which transmit their motion to 'our Nerves, or animal spirits, by some part of our Bodies, to the Brain or the seat of Sensation, there *to produce in our Minds the particular* Ideas *we have of them*' (II.viii.12). Ideas of secondary qualities are produced in us 'after the same manner'. Since secondary qualities in bodies are analysable in terms of primary qualities but appear to us as colours, sounds, and so on, it follows that there is no resemblance between our ideas and the corresponding secondary qualities.

It might be thought that the definition of secondary qualities would be sufficient for the conclusion that ideas of them do not resemble them since they are

but Powers to produce various Sensations in us by their *primary Qualities, i.e.* by the Bulk, Figure, Texture, and Motion of their insensible parts...

That is not so, however. A cubical object has the power to produce in us an idea that resembles it; it looks cubical. An object might be, independently of us, coloured because a certain 'Bulk, Figure, Texture, and Motion' of insensible parts produced colour, in the object, which then produced a resembling idea in us just as a certain other combination of these might make a body soluble in certain solvents. It is just because the theory aims to explain certain *apparent* qualities in terms of others that the non-resemblance of certain qualities and ideas follows. The importance of resemblance relates to our knowledge of the external world. If sense-experience is to be able to give us information about bodies, and sense-experience is explained in terms of ideas, then it is important that at least some ideas of the qualities of things should be accurate. This would be ensured if they resembled the qualities in question.

There is a distinction to be made between representation and resemblance and Locke relied on it. One thing may represent another without resembling it. I might, for example, play a war-game using toy soldiers to represent real soldiers, which they resemble, or using tiddly-winks to represent real soldiers, which they do not resemble. Toy soldiers normally have limbs, heads, and so on, resembling those of real men; they wear uniforms and carry standards and weapons of colours and designs resembling those of particular regiments. Even using these we may rely on representation which is not resemblance as when we allow one toy soldier to stand for ten or a hundred real soldiers. When we use tiddly-winks we are not using objects that resemble real soldiers in design and we need not use objects that resemble in colour the uniforms of the real soldiers they represent; we may use green and yellow tiddly-winks to represent, respectively, soldiers with red and blue uniforms.

We may take this further as I believe Locke does. A cheque for £50 represents that amount of currency but there are obvious and important ways in which it does not resemble it. More relevantly, however, if an event of a certain kind is uniquely caused by a specific set of conditions, the event may be said to represent those conditions although it does not resemble them. The crucial kind of example for Locke is that of colour; if a red colour, and no other, is caused by a specific arrangement of non-coloured corpuscles and a blue colour, and no other, is caused by a different arrangement of non-coloured corpuscles, then a red colour represents the first arrangement and a blue colour represents the second, although neither red nor blue, as seen, resembles an arrangement of corpuscles. On the other hand, the length of an object represents one aspect of an arrangement of corpuscles and also resembles it, being the sum of lengths of corpuscles and distances between them. Resemblance implies representation but not *vice versa*; one may have representation with or without resemblance. In Locke's view all simple ideas of sensation represent qualities of objects but only some of them also resemble those qualities.

What are transmitted from external bodies to our brains is the patterns or textures in which the corpuscles are arranged, or parts of these patterns. A pattern is reproduced in the corpuscles of any medium between the body and the appropriate sense-organ and in the corpuscles of the nerves and the brain. By the brain, ideas are produced, somehow, in the mind, and certain of these ideas resemble, somehow, the primary but not the secondary qualities of the body. It must be concluded that ideas may sensibly be spoken of as having patterns which resemble, or fail to resemble, the patterns of objects.

In his chapters 'Of Perception' and 'Of Retention' Locke makes much use of such words as 'impression', 'trace', 'imprinting', 'stamping' and 'inscription'. These are used both metaphorically and non-metaphorically, or almost so, depending on whether the mind or the body is being considered. It is basically a mechanical conception: patterns are impressed on the body as a pattern is impressed on wax by a seal, or inscribed as a point inscribes a pattern on a wax tablet, and so on. Thus there may be impressions 'made on the outward parts' which, if not taken notice of, result in no perceptions (II.ix.3). There are 'sensitive plants' which move as the result of external stimuli because patterns are transmitted but no ideas produced so 'it is all bare Mechanism' (II.ix.11). Concerning memory, recall appears to be explained by physical impressions or traces in the brain which may or may not be taken notice of and which may also wear out, resulting in total forgetting; 'the Inscriptions are effaced by time'. This is compared to 'Characters drawn on [the Brain] like Marble', or free-stone or sand. (II.x.3–5). This is the non-metaphorical or less metaphorical use. However, Locke also talks of ideas being imprinted on the mind (II.ix.4,7 and 8) and of 'original Characters impressed on' the mind (II.ix.6) which appears to be a fully metaphorical use.

The corpuscular philosophy is committed to attempting to explain perception and memory. The account appears to go like this. Bodies cause alterations in sense organs, nerves and brain. The process is conceived of as series of patterns of events structurally similar to the patterns in the bodies. Patterns are handed on without the transference of matter from one series, such as light rays, to the next, such as those in the eyes. These patterns that are transmitted are, interpreted according to the corpuscular philosophy, to be found in the textures of the objects in which the whole process originated and so may themselves be called 'textures'.

Locke, like other philosophers of the period, talks occasionally of 'animal spirits' in our bodies that are responsible for the transmission of motions or patterns from, for example, our sense-organs to our brains, that is, through our nervous systems (e.g. at II.viii.4 and 12, II.x.5 and II.xxiii.6). Animal spirits were regarded as a subtle fluid which on the corpuscular philosophy could be regarded as being corpuscular and so having textures and transmitting patterns. The conception was involved in attempts to allow mechanical explanations of perception and memory and sometimes, mistakenly, to make the mind/body problem more tractable, as by Descartes in his talk of the pineal gland. Locke talks of sensations being produced in us 'only by different degrees and modes of

Motion in our animal spirits, variously agitated by external Objects' but he shies away from detailed discussion of the idea, perhaps because he was aware of, and baffled by, the difficulties, both empirical and logical, in the use that was made of it. He says most when he is discussing memory and the association of ideas (II.x.5 and II.xxxiii.6).

His view about the association of ideas is that we sometimes associate, voluntarily or by chance, ideas that are not naturally related but have been presented to us together on a number of occasions. If we have heard the notes of a particular tune often enough we remember it and, on hearing the first few notes, recall the whole tune. In this connection he talks of 'Trains of Motion in the Animal Spirits' which wear 'a smooth path' in our bodies. This can be interpreted in terms of the corpuscular philosophy or, at least, that was what was hoped. The successive textures of the air originating from each note sounded on a musical instrument and impinging on our ears cause a series of modifications in the texture of our animal spirits and brains on successive hearings until their paths are well worn. Later, on hearing the first few notes of the tune, the paths of the rest are reactivated so that we recall the whole tune. That is, a series of textures is re-established in the animal spirits and transmitted to the brain, upon which we have the appropriate ideas. This, of course, gives only a sketchy idea of the way the explanation would work and is not very plausible until the details of the mechanism by which the paths are reactivated are filled in; it was perhaps because he was not clear how to fill in these details that Locke avoids discussion of the precise mechanism.

All these accounts of perception and memory conclude that when the pattern reaches the brain ideas may be produced in the mind; this last step is mysterious and raises a problem which haunted Locke throughout the *Essay*, that is, the mind/body problem which is, Locke repeatedly suggests, likely to remain forever beyond the reach of a corpuscular, or any other, explanation (IV.iii.28). So the corpuscular theory can at best hope to explain the mechanical parts of the whole process.

Now although Locke recognizes our inability to say how a resembling pattern is produced in the mind by the brain or even to conceive how it might happen or what it might mean, he does think that it makes sense to talk of the resulting ideas as, some of them, having patterns which resemble the patterns in the original bodies. We must consider how this can make sense.

It is important to realize that since we do not see, or otherwise perceive, the texture, or corpuscular structure, of bodies but only resultant overall structures of large collections of corpuscles, the patterns of ideas must be different in detail from the patterns reproduced in our nerves and brain. What this means is that the most we need to attribute to patterns in ideas is, for example, overall sizes and shapes which correspond to overall sizes and shapes of collections of corpuscles constituting bodies. When we see a six-inch cube and a three-inch sphere, for example, our perceptual ideas are not of collections of corpuscles but of a six-inch cube and a three-inch sphere. This is how we see them, in the right conditions. They may also look red and blue respectively and this is because some features of the corpuscular structures have certain characteristics which can be transmitted but which are not colours, and which, again, we do not see.

The idea of resemblance that is in Locke's mind is indicated at II.viii.11, though in a negative way. Bodies, he says, are commonly thought to have qualities such as *hot* and *white* 'which qualities are commonly thought to be the same in those Bodies, that those *Ideas* are in us, they are the perfect resemblance of the other, *as they are in a Mirror*' (my italics). This Locke denies. I think that, in the absence of evidence to the contrary, Locke must be taken to be asserting of ideas of primary qualities just what he is here denying of ideas of secondary qualities, *i.e.* that an idea of a primary quality resembles the primary quality as a mirror image resembles its object. At II.viii.18 he says that when a moving piece of manna produces ideas in us 'This *Idea* of Motion represents it as it really is in the *Manna* moving: A Circle or Square are the same, whether in *Idea* or Existence; in the Mind, or in the Manna.' I shall, shortly, discuss the implications of the use here of the analogy with mirror images.

My account so far is supported by Locke's apology for his 'little Excursion into Natural Philosophy' (II.viii.22), which must refer at least to sections 8 to 21 since he clearly refers to the distinction between primary and secondary qualities as part of that excursion. Moreover, he indicates that through it we 'may come to know what *Ideas* are, and what are not, resemblances of something really existing in the Bodies, we denominate from them'. This supports the view that resemblance is a necessity of the explanatory theory and not something drawn from our perceptual experiences.

In the penultimate section of Chapter viii Locke explains why we take bodies to have qualities resembling our ideas of secondary qualities. Ideas of colours and sounds 'containing nothing at all in them, of Bulk, Figure, or Motion' do not suggest that they can be explained in terms of the bodies' primary qualities. So we think of colours and sounds as qualities of bodies because nothing in our perceptions suggests otherwise; indeed, they look to be in the bodies. We contrast this with our observation of, for instance, the sun melting wax, when we perceive the sun and the wax but nothing in the sun resembling the melting of the wax; so we conclude that the cause is different from the effect and can contemplate the explanation of the melting in terms of qualities, *primary* qualities ultimately, which do not resemble it. We have not the same reasons, *from experience*, for entertaining similar explanations of colours and sounds; *that* requires an argument with theoretical backing.

There occurs another important reference to resemblance at II.xxx.2 where Locke says that our ideas serve to distinguish bodies from one another since they are either 'constant effects or else exact Resemblances of something in the things themselves' and adds, significantly, 'But whether they answer to those constitutions [*sc.* of things], as to Causes, or Patterns, it matters not; it suffices that they are constantly produced by them.' Our simple ideas enable us to distinguish between bodies because they are caused by qualities of bodies and we can spot difference of cause indirectly by difference of effect; under given conditions, differences between ideas correspond to differences between qualities of bodies. We should thus be able to distinguish bodies from one another, correctly, even if none of our ideas resembled the qualities of bodies. What resemblance enables us to do, however, is something beyond this: to conceive the world as it really is, at least in some respects, that is, as composed of bodies describable in terms that relate to some of the features we meet with in sense-experience.

Locke's reliance on resemblance has been regarded by most critics as open to serious objections in two main ways. *First*, it is asked how we could ever know that ideas of primary qualities resemble those qualities and ideas of secondary qualities do not since in perception we are aware only of the ideas and never, independently, of the corresponding qualities so we are never able to do the necessary comparison between them on which a judgement of resemblance or non-resemblance must rest. *Second*, it is said that the notion of resemblance between ideas and objects is incoherent because ideas are mental and objects are material so they cannot share kinds of properties in respect of which they could be conceived of as resembling one another. I believe that both these criticisms can be answered.

The answer to the first objection is implicit in the account I have already given. The resemblances or non-resemblances between ideas and qualities are not *discovered* in sense-experience, or claimed to be; there is no question of doing, or needing to do, 'the necessary comparison' between ideas and qualities. These relations between ideas and qualities are *postulated* as part of the corpuscular theory; the resemblance of primary qualities and ideas explains how sense-perception can give us information about the world; the non-resemblance of secondary qualities and ideas is a consequence of the way in which secondary qualities are analysed, and their causing ideas in us explained, according to the theory. The resemblance/non-resemblance is not something that could be known through experience but something postulated, *inter alia*, to explain knowledge through experience. If this has the consequence that we cannot strictly *know* about the resemblance then it is in no worse a position than any other hypothesis relying on unobservables by means of which we attempt to explain what we observe; the probability of the conjecture depends upon the success of the explanations.

Hence Locke's talk about resemblance in his attempt to give an account of the way in which some of our ideas may be accurate and give us information about the world. He is aware of ways in which he may be misunderstood and aware, in particular, that he may be criticized in the way I am considering. In his discussion of knowledge in Book IV he says

> 'Tis evident, the Mind knows not Things immediately, but only by the intervention of the *Ideas* it has of them. *Our Knowledge* therefore is *real*, only so far as there is a conformity between our *Ideas* and the reality of Things. But what shall be here the Criterion? How shall the Mind, when it perceives nothing but its own *Ideas*, know that they agree with Things themselves? (IV.iv.3)

Ideas, he has already said, are real if they have a 'Foundation in Nature', have a 'Conformity with the real Being, and Existence of Things' although that does not mean that they have to be 'Images, or Representations' of things; it is enough that they be *caused by* them (II.xxx.1 and 2).

It is noticeable that when, after the above passage from Book IV, Locke goes on to discuss the criterion he does not consider resemblance as a possible criterion. Indeed, one of the points of this passage is that we cannot use resemblance, even for ideas of primary qualities, as a criterion for distinguishing 'real' from 'fantastical' knowledge. Not only is this impossible but if it were possible it would draw the distinction in the wrong place and make knowledge of secondary qualities 'fantastical', which he does not want.

In some passages I shall discuss later he goes on to say that there are two sorts of ideas 'that, we may be assured, agree with Things', namely, simple ideas and complex ideas other than those of substances. We are concerned here with simple ideas and in arguing for the reality of these Locke uses theoretical considerations. They are produced by 'Things operating on the Mind in a natural way', they are 'the natural and regular productions of Things without us, really operating on us' and 'they represent to us Things under those appearances which they are fitted to produce' (IV.iv.4). This is a general defence of the reliability of simple ideas in that they, treated carefully, are sufficient for real knowledge. The criterion is not to be found in the experiences but in the theory.

He is clearly not using resemblance in talking of the conformity of ideas to things. The 'natural and regular production' of ideas by things is a causal notion and makes essential references only to representation, not to resemblance. Ideas of primary qualities have the additional features that they show us directly, if to a limited extent, what the world is really like. If the aim of the corpuscular theory were realized we should be able to specify how a thing that looks blue differs in inner constitution from a thing that looks red, under the same conditions; we are already in a position to say that two things that look different in length, from the same point of view, *are* proportionately different in length.

I now come to the second objection. How is it possible that an idea, which is mental, can be sensibly said to resemble a physical thing, even in shape and size? How can something mental have a shape or a size? Berkeley appears to have this problem in mind when he says that 'nothing can be like an idea but an idea'.[1]

In *Dialogues* I[2] Philonous says

> But how can that which is sensible be like that which is insensible? Can a real thing in itself *invisible* be like a *colour*; or a real thing which is not *audible*, be like a *sound*? In a word, can anything be like a sensation or idea, but another sensation or idea?

Of course, Locke did not hold that a colour is like something that is invisible and Berkeley may not have intended this as a direct criticism of Locke; it may rather be a criticism of anyone who holds that objects are, independently of ideas, coloured, where that word is applied univocally to both ideas and things. However, Berkeley's last sentence generalizes, so it does become a criticism of Locke's view.

This is clear when, in the *Principles*,[3] Berkeley says

> But say you, though the ideas themselves do not exist without the mind, yet there may be things like them whereof they are copies or resemblances, which things exist without the mind, in an unthinking substance. I answer, an idea can be like nothing but an idea; a colour, or figure, can be like nothing but another colour or figure... Again, I ask whether those supposed originals or external things, of which our ideas are the pictures or representations, be themselves perceivable or no? If they are, then they are ideas, and we have gained our point; but if you say they are not, I appeal to anyone whether it be sense, to assert a colour is like something which is invisible; hard or soft, like something which is intangible; and so of the rest.

Berkeley's principle that 'an idea can be like nothing but an idea', figures importantly in his argument against the primary/secondary quality distinction. He appears almost to regard the principle as self-evident. Locke would clearly not accept it but nevertheless it poses a question with which anyone who champions Locke must deal. It is important to explain just how an idea can be like something that is not an idea. How does it make sense to say that an idea can be extended or shaped, as it appears that they must be if we can say that they resemble those qualities in objects? This would perhaps have seemed an even more acute problem in the context of seventeenth century dualism than it does to us. If material and mental substances are regarded as being independently definable, in terms, say, of extension and thinking, respectively, then nothing mental could be extended since that would make it material.

The problem is to give an unobjectionable account of what it is to have an accurate idea of something in terms of the resemblance between an idea and a quality. E. M. Curley[4] suggests a causal interpretation which goes well with the kind of account I have been giving. The causal account of perception is different for primary qualities from the causal account for secondary qualities. The account of our seeing an object as elliptical involves reference to that object's having some shape, though not necessarily an elliptical shape. On the other hand, the account of our seeing an object as blue does not involve reference to the object as having some colour, blue or any other; indeed, this is explicitly ruled out by the theory and the account involves reference only to the same primary qualities as the account of our seeing the object as elliptical. Thus

> while the causal explanation of our perception of the shape of an object enables us to understand why we perceive it as having a shape (since its having some shape is essential to the explanation) nothing in the causal explanation of our perception of color really makes it intelligible that we should see the object as colored (since only primary qualities are involved in the explanation).

Resemblance lies in the fact that the same cluster of terms is used in explaining the perception causally as in describing it for primary qualities but not for secondary qualities.

In my paper 'Boyle and Locke on primary and secondary qualities'[5] I made a related suggestion. When we distinguish in language between our ideas we use such descriptions as 'idea of a circle', 'idea of red', if we are Lockeans. Primary qualities are such that the words we use in describing our ideas of them are also the appropriate words for describing the corresponding qualities; but this is not so for ideas of secondary qualities. The resemblance is between the description of the idea and the description of the corresponding quality.

It now seems to me that there is a weakness in both these accounts. Both of them rely on describing the idea, or perception, in terms, for example, of shape; but if that description is to be correct then the idea must have shape and whether something mental can be sensibly said to have shape is the very point at issue. I want to argue that such things can be sensibly said and that Curley's account and my own are not so much wrong as incomplete; they require that argument. The argument depends upon the passage quoted from Locke earlier in this chapter on mirror images and upon what I said in my chapter about ideas (Chapter 4).

First, however, it must be pointed out that such expressions as 'an idea of that shape' and 'an idea of a given shape' are, on the face of it, doubly ambiguous. In the first place, to talk of having 'an idea of a given shape' may be to mean that the idea has that shape but it is consistent with holding that the idea is not itself shaped but is merely *of an object which has that shape*. In the second place, in Lockean terms there is a difference between having an idea of a given shape while I am seeing it and having an idea of a given shape when I am not seeing or having an image of it. The difference corresponds to that between seeing, say, a right-angled triangle and thinking about a right-angled triangle, without an image. The thinking could be carried out in words, the seeing could not. The idea of resemblance with which I am here concerned is involved in the seeing but not the thinking of the object. I suggest that in talking of 'an idea of a given shape' in connection with seeing Locke holds that it does make sense to say that ideas can have shapes.

It might be wondered whether I could apply these considerations to solidity. Can it be said that with respect to ideas of touch Locke could hold that it makes sense to say that ideas can be solid? I think it can. I have distinguished between absolute solidity and relative hardness and softness and argued that ideas of sensation include ideas of relative hardness and softness but not ideas of absolute solidity. If I touch a piece of wood I get an idea of hardness; if I touch a piece of foam rubber I get an idea of softness. An idea is an appearance and the wood appears hard, the rubber soft. If we think of ideas as in the mind and think of this as meaning that they are mental images or pictures then it seems absurd to say that ideas can be hard or soft; but if we think of 'in the mind' as meaning something different perhaps it does not seem absurd. I can say, touching the wood, 'This is hard' and, touching the rubber, 'This is soft' and regard 'this' as the present appearance to me, the idea I am having. That seems no more absurd than allowing that, in some sense, an idea can have shape. Perhaps this sense will become clearer and more acceptable as I proceed.

Consider again Locke's passage at II.viii.16, quoted above, in which he says that bodies are usually thought to have such qualities as hot and white 'which qualities are commonly thought to be the same in those Bodies, that those *Ideas* are in us, the one the perfect resemblance of the other, as they are in a Mirror'. I suggested that Locke is affirming of primary qualities and their ideas just what he is denying of secondary qualities and their ideas: that they resemble one another as an object and its mirror image do.

It is worth asking what the analogy with a mirror and its images amounts to. Consider an object and its image in a plane mirror. If the object is one foot in front of the mirror its image is, or appears to be, one foot behind it, that is, it has a position in space; the image has the shape and size of the object; if the object is moved, the image moves; both the object and the image can be indicated by pointing. These resemblances, I suggest, are the kinds of resemblance Locke is talking about. Nothing he says suggests that the analogy is to be pushed any further; nothing about the object and its mirror image corresponds to the distinction between physical and mental, except the conception of an object and its appearance and perhaps the image's being 'in the mirror' as the idea is 'in the mind'. If Locke means what I take him to mean then in respect of ideas of primary qualities similar things must be said: the object and the idea (its appearance) resemble in shape and size; they both have a position in physical space, this time, usually the *same* position; if the object is moved its appearance moves; we can point at the object and its appearance but this time it amounts usually to pointing at the same position. This, I suggest, is the extent of what Locke is using the analogy to do.

It is perhaps also important to point out that although mirror images have colours, they do not reproduce the smell or taste or ability to make sounds that their objects have; in a somewhat similar way, the smells, sounds, tastes and colours of appearances are not *reproductions* of qualities in the objects. It may also be significant that at II.i.25 Locke talks of the 'Images or *Ideas*' produced in mirrors.

One conclusion that I think may be drawn from all this is that ideas are not like what is usually meant by a mental image. The more I think about it the less clear I am about what mental images *are* like but at least they are often talked about as if they were pictures 'in the head' or on a private cinema screen. This analogy has often been used. But if an idea is like a mirror image only in the respects I have just mentioned then Locke does not seem to be treating ideas as mental images in the way that he is often thought to be.

What I have said about mirror images reinforces and supplements what I earlier said about ideas. If ideas of sensation are like mirror images in the respects I have mentioned then those ideas can be said to be the appearances of things to us, as I argued in Chapter 4. What I have called their 'minimal publicity' is analogous to the publicity of mirror images: they can both be pointed at. These ideas disappear, usually, when the object is removed just as mirror images do. They have sizes and shapes just as mirror images do. They are 'in the mind' as mirror images, we say, are 'in the mirror' and just as we don't point at the mirror but at a position behind it if we point to an image so we don't point at the mind, whatever that might mean, or at ourselves but at a position external to us if we are pointing at appearances or ideas. The expression 'in the mirror' does not, I suggest, indicate a location but gives an indication of the way in which the image is generated; neither does it indicate privacy. The location of the image is at a point the same distance behind the mirror as the object is in front of it. Of course, there is a sense in which it is not 'really' there; we don't find it if we look behind the mirror but only if we look through it or in it; the image is an appearance but *ipso facto* not private to the mirror. Similarly, I suggest, the expression 'in the mind' when used of ideas of sensation does not for Locke indicate a location but gives an indication of the way in which the idea, or appearance, is generated; neither does it indicate privacy. The idea is mind-dependent in that the ideas or appearances only arise from the interaction of objects and minds. The difference between an idea and a mirror image is that the idea is usually where the object is; the object usually *appears* to be where it is.

The main difference, then, is that we don't see both the object and its appearance as we see both the object and its mirror image. This presents no problems for the doctrine of resemblance if resemblance is taken to be theoretical rather than empirical and if we do not suppose that it is a *criterion* for recognizing accurate ideas. To regard it as a criterion would be to attempt to push the analogy too far. To say that ideas resemble qualities is to say that things appear to us as, in certain respects, they really are; what we see is extended just as what causes us to see it is. We see both an object and its mirror image and we see that they are both extended so we see the resemblance; we see the appearance of an object and we see it as extended but the theory tells us that in this respect it resembles the object of which it is an appearance. To say that the idea is 'in the mind' is to say that it is mind-dependent just as to say that the image is in the mirror is to say that it is mirror-dependent; neither expression indicates position.

It might be thought that when we point at a mirror image we are pointing at the mirror. However, consider the experiments that beginning students of optics do with pins to find the position of the image; if the mirror is small enough to do the experiment we shall point past it to the position of the pin behind it to indicate the position of the image. Next consider a mirror, that is too large for that. We can point to the image 'in', or as I prefer to say 'through' the mirror. Make a mark on the surface of the mirror on the straight line between the eyes and the image; pointing at the image is not the same as pointing at that mark. Pointing at that mark would not indicate the position of the image except by a lucky chance. Consider pointing at a house on a distant hill through some intervening telegraph wires.

These ways of talking may sound odd to us but that is mainly because we have never really grasped Locke's technical use of 'idea' in these contexts. The terminology of 'sensa' and 'sense-data' was perhaps in part an attempt to make such talk appear less odd but it spawned a great deal of mythology that Locke did not need. By seeing just what Locke did need we can avoid the mythology. When I see a billiard ball I see *it* but as it looks to me. When I point at a billiard ball I point at *it* as it looks to me. Locke sometimes talks of perceiving ideas. Perceiving ideas happens whenever I perceive things. Things-as-perceived-by-me are ideas. An idea of sensation is not in the mind in the full-blooded way in which an imaginary adventure of mine in Monte Carlo in 1905 is; but neither is it just in the world as something independent of me. It is an appearance of an object to me and so mind-dependent. If we say it is 'mental' we cannot mean that it is mental in the way in which my worry about my income tax is mental.

What Locke requires is that it be sensible to say that, in respect of primary qualities, the way a thing looks is, at least roughly, the way it is, while, in respect of secondary qualities, the way it looks is not even roughly the way it is. However, we must allow for such facts as that a circular coin may look elliptical; we must restrict the first statement to determinables. If a thing looks circular or elliptical and ten inches wide then it looks as if it has shape and extension and it does indeed have shape and extension; if it looks red or blue then it looks as if it has colour, in the same sense as it has shape, but in fact it does not. A perceptual idea is part of the look of a thing, which may or may not be like the thing itself.

Now this may seem an empty claim, in respect of primary qualities at least, because anything we see looks to be extended. But this is to miss the point. The primary/secondary quality distinction and the notion of resemblance are, as I have previously said, all part of an explanatory theory relating the way the world is to our experience and knowledge of it. The whole point of this is that by distinguishing primary qualities and explaining secondary qualities in terms of them, by distinguishing ideas of both and identifying their respective common characteristics and by introducing the idea of resemblance it is hoped that it will be possible to explain our perceptions and decide the extent to which they can give us knowledge of the world.

Of course, when we come to describe our experiences and, we hope, the external world, we are concerned about determinate shapes, extensions, colours and so on, and *real* shapes, extensions, colours, and so on, but this is a project within the general theory of our relations with the external world and our experience of it. The ways in which we take ourselves to be able to decide upon the real shapes, etc. of things does not directly involve Locke's doctrine of resemblance between some ideas and some qualities.

The *real* shapes and colours of things are matters about which experience itself gives us many clues, but we rely also on an explanatory theory concerning the relations *between* our experiences. Experience shows us, for example, that a penny that feels circular and looks circular from certain points of view also looks variously elliptical and even rectangular from certain other points of view; theoretical considerations, of geometry and mechanics, for instance, show us that the various ways in which the penny looks can be most simply and economically explained on the assumption that the real shape of the penny is circular, with parallel faces. That a penny is larger than a halfpenny and smaller than a book or a church tower is just what we would expect from the way they look at various distances from one another and from an observer; again theoretical considerations about relative sizes allow us to explain why these things look as they do. I am not concerned here to decide how informative this talk of real shapes and sizes is but merely to say that in some such ways as these Locke can accommodate our everyday beliefs about real shapes. Then, bringing into play the general theory and relying on the idea of resemblance, we are able, on his view, to give some sense to the idea of the real shapes and sizes of things in the world, independently of our perceptions. Locke's empiricism is not of the extreme variety that disallows the theoretical and explanatory considerations I have mentioned. Without such considerations we should be in danger of having to say either that the conception of the real shape of something is unintelligible or that the real shape of something is always the shape that we see, which is absurd. Both these possibilities appear to me to spring from a faulty conception of shape and Locke shows no signs of saying either of them. The theory he espouses is not constructed merely from our experiences and the way things look to us. Resemblance does not, and was not intended to, allow us to infer directly from the mere look of a thing on one occasion, or even from many looks of it on many different occasions, to how in detail it really is.

11
SUBSTANCE-IN-GENERAL

I

Locke said, in the course of a discussion of substance with Edward Stillingfleet, Bishop of Worcester,

> Men's notions of a thing may be laughed at by those whose principles establish the certainty of the thing itself. (*L.S.* 449)

The conception of substance is that for which Locke is regarded by many as most radically mistaken, imperceptive, inconsistent and foolish. However, I believe, and wish to argue, that his conception of it is more plausible, more consistent with the rest of his philosophy and more complex than was dreamt of by conventional critics of Locke. Before we laugh at his notion of it, or indeed, at anyone's notion of anything, if we think it worth mentioning, we should try to get it right.

There are, I suppose, three basic questions to be asked: (1) what did Locke mean by 'pure substance in general' (his expression at II.xxiii.2 and elsewhere)? (2) did he assert its existence? and (3) how could he think himself justified, on his own principles, in asserting its existence, if he did? I believe that he did assert its existence and that my account of what he thought he was asserting shows how he could think himself justified in asserting it.

It is perhaps as well at this stage to say in a general way what rôles I consider that substance was intended to play. I believe there are three: (1) to provide something for qualities to qualify; (2) to account for the existence of stable collections of qualities, such as objects or what Locke calls 'bodies'; and (3) to provide, in the Aristotelian manner, something enduring underlying change which would allow the differentiation of something changing into something else from something being substituted for something else.

II

Locke, of course, distinguishes between particular sorts of substances, such as lead or gold, and substance-in-general which is, crudely, what would be left if all the qualities that distinguish particular sorts of substances from one another could be stripped away, that is a qualityless substratum for the qualities.

We may find some support for regarding Locke as denying altogether the existence of substance as a support for qualities. However, this conflicts with other statements, some quite explicit, to the contrary; it would have left Locke, in his own eyes, faced with insuperable logical problems; and it would lay him open to being accused of a cowardly vacillation about facing the consequences of this denial. We may also find support for regarding him as asserting the existence of substance-in-general conceived as something we know not what, an absolutely indeterminate and undescribable somewhat underlying everything of which, on his own ground, we could have experience or anything approaching knowledge. This is the more usual, conventional view and, of course, if this is what he held then he has deserved everything he has got in the way of criticism, from Berkeley onwards, since it leaves him with insuperable problems of epistemology and meaning. He has, after all, been described as 'the father of British Empiricism' although I am not sure that he would welcome the title if he could examine his alleged issue.

It would be as well to have before us some of the main passages that have led to these interpretations so I propose to quote these from the *Essay* in the first place. I quote them, on the whole, in the order in which they occur because I believe that, here as elsewhere, we are faced with an argument, or at least a view, that is *developing* as the work proceeds.

Locke's first important reference to substance-in-general is, I believe, this

> there is another *Idea*, which would be of general use for Mankind to have, as it is of general talk as if they had it; and that is the *Idea of Substance*, which we neither have, nor can have, by *Sensation* or *Reflection*... We have no such *clear Idea* at all, and therefore signify nothing by the word *Substance*, but only an uncertain supposition of we know not what (*i.e.* of something whereof we have no particular distinct positive) *Idea*, which we take to be the *substratum*, or support, of those *Ideas* we do know. (I.iv.18)

This is when he is considering various ideas that might be claimed to be innate and, at first sight, it looks damning for the view that Locke accepted the existence of substance-in-general. However, I think it is not; it concerns not substance itself but the idea of substance and the word 'substance'. The passage makes a suggestion that I believe runs through Locke's later discussion of the topic: since our idea of substance is not clear we really attach no meaning to the word but to say that we signify nothing by it is not to say that there is nothing corresponding to the idea but only that we do not know what it is, how to describe it. There is evidence that Locke wished to guard against the suggestion that we have *no* idea of substance since in this passage he originally talked of 'something whereof we have no *Idea*' and changed this to 'something whereof we have no particular distinct positive *Idea*' in the 4th Edition, that is, after the correspondence with Stillingfleet. The importance of that will emerge. Why does Locke say that it would be 'of general use to mankind' to have a clear idea of substance? Because *there is* substance-in-general and if we had a *clear* idea of it we should know not only that it exists, because it must, but how it works. My contention is that Locke later becomes less pessimistic about saying what substance *is*.

The next important passage is

> The *Ideas* of *Substances* are such combinations of simple Ideas, as are taken to represent distinct particular things subsisting by themselves; in which the supposed, or confused *Idea* of Substance, such as it is, is always the first and chief. (II.xii.6)

I suggest that this is neutral as between the assertion and the denial of substance; it leaves open the possibility that just as combinations of simple ideas *do* represent particular things so our ideas of substance represent an existent, but not clearly.

I leave the next important passage (II.xiii.18), which requires much discussion, until I can use it more effectively and move to the first passage about the poor Indian, who does not know a hawk from a handsaw, and his European colleagues, who are no clearer in spite of their good opinion of themselves.

> They who first ran into the Notion of *Accidents*, as a sort of real *Beings*, that needed something to inhere in, were forced to find out the word *Substance*, to support them. Had the poor *Indian* Philosopher (who imagined that the Earth also wanted something to bear it up) but thought of this word *Substance*, he needed not to have been at the trouble to find an Elephant to support it, and a Tortoise to support his Elephant: The word *Substance* would have done it effectually. And he that enquired, might have taken it for as good an Answer from an *Indian* Philosopher, That *Substance*, without knowing what it is, is that which supports the Earth, as we take it for a sufficient Answer, and good Doctrine, from our *European* Philosophers, That *Substance*, without knowing what it is, is that which supports *Accidents*. So that of *Substance*, we have no *Idea* of what it is, but only a confused obscure one of what it does. (II.xiii.19)

Here Locke is inviting us to laugh, I think, not at the poor Indian so much as at the European philosophers who think that their doctrine of substance as a support for accidents is superior to that of the elephant and the tortoise as supports for the earth; they think they have described and *explained* something when they have invented the word 'substance' to stand for the support that accidents need, even though they can say nothing more about it. It is left open whether substance exists and does something although our idea of how it does it is as obscure as our idea of what it is; so the European philosophers may have been right to postulate substance-in-general but wrong in some of their views about it. I think that Locke later accepts this.

It is worth noting that in the next section (II.xiii.20) Locke says that it would not be of much use to a foreigner ('an intelligent American') who wanted to know about our architecture if he were told that a pillar is what is supported by a basis and a basis is what supports a pillar. 'But' he goes on, 'were the Latin words *Inhaerentia* and *Substantia*, put into the plain English ones that answer them, and were called *Sticking-on* and *Under-propping*, they would better discover to us the very great clearness there is in the Doctrine of *Substance and Accidents*, and show of what use they are in deciding of Questions in Philosophy.'

Is Locke being ironical? Yes, and no. I think he is saying that the Latin words suggest a description of the support whereas their English equivalents merely indicate the job to be done by substance without saying what substance is. The noun 'substance' suggests that we know the kind of thing we are talking about and any irony is directed at philosophers who suppose that we do; but 'under-propping' suggests, more accurately, that all we can say is that substance, whatever it is, supports. This fits in with the last sentence of the previous section, already quoted. To be told that a basis supports a pillar is not to be told what a basis is but only what it does. To be told that substance supports the Indian's tortoise is not to be told what does the supporting whereas to be told that the tortoise supports the elephant is to be told what does the supporting. In either case we are told something: we are clear enough about the idea of *supporting* from our everyday experience of such things as tables supporting cups.

He introduces substance at the beginning of his long chapter on our complex ideas of substances by saying

> a certain number of these simple *Ideas* go constantly together; which being presumed to belong to one thing, and Words being suited to common apprehensions, and made use of for quick dispatch, are called so united in one subject, by one name; which by inadvertency we are apt afterward to talk of and consider as one simple *Idea*, which indeed is a complication of many *Ideas* together; Because, as I have said, not imagining how these simple *Ideas* can subsist by themselves, we accustom ourselves to suppose some *Substratum*, wherein they do subsist, and from which they do result, which therefore we call *Substance*. (II.xxiii.1)

It is important to note that here, and elsewhere, there is a possibility of being misled by Locke's form of words. He talks as if substance were required to support *ideas* and their going constantly together and yet he is clearly talking in general about *material* substance as a support for *material* qualities and accidents. If *ideas* need a substance to support them one would expect them to need *mental* substance. I must refer again to the passage in II.viii.8 which I have previously said must be taken seriously. Having said that by '*Ideas*' he means 'sensations or perceptions' he adds 'which *Ideas*, if I speak of sometimes as in the things themselves, I would be understood to mean those qualities in the objects which produce them [*sc.* ideas] in us'. Thus when he says such things as 'the power of drawing iron is one of the *Ideas* of the complex one of that substance we call a Loadstone' (II.xxiii.7) he is to be understood as saying *not* that a power in a *substance* is an idea but that the idea of the power is part of the idea of a loadstone. The power is what *causes* the idea in us. So also, in the present passage, he is to be regarded as talking of substance as a support for qualities in objects corresponding to ideas in us.

So in this passage I take Locke to be saying that we cannot imagine how groups of qualities, represented by groups of simple ideas, can subsist by themselves; qualities need something to qualify. Whatever they *all* qualify we call 'substance'. It should be noted that Locke does not say here that we are *wrong* when we accustom ourselves to suppose a substratum.

This passage is shortly followed by the second reference to the poor Indian.

> if any one will examine himself concerning his *Notion of pure Substance in general*, he will find he has no other *Idea* of it at all, but only a Supposition of he knows not what support of such Qualities, which are capable of producing simple Ideas in us; which Qualities are commonly called Accidents. If any one should be asked, what is the subject wherein Colour or Weight inheres, he would have nothing to say, but the solid extended parts: And if he were demanded, what is it, that that Solidity and Extension inhere in, he would not be in a much better case, than the *Indian* before mentioned; who, saying that the World was supported by a great Elephant, was asked, what the Elephant rested on; to which his answer was, a great Tortoise; But being again pressed to know what gave support to the broad-back'd Tortoise, replied, something, he knew not what. (II.xxiii.2)

Locke here talks about substance as a support for qualities and not merely the idea of substance as relating ideas of qualities and he mentions solidity and extension, I think for the first time, as if they were qualities requiring support; I shall have more to say about this later.

In the same section he says

> The *Idea* then we have, to which we give the general name Substance, being nothing, but the supposed, but unknown support of those Qualities, we find existing, which we imagine cannot subsist, *sine re substante*, without something to support them, we call that Support *Substantia*; which, according to the true import of the Word, is in plain *English, standing under*, or *upholding*. (II.xxiii.2)

There may be no objection here, as in previous quotations, to reading the statement that we cannot imagine qualities existing without support as making a logical, rather than a psychological point, that is, as the assertion that it is *inconceivable* because logically impossible that qualities should exist without support. Indeed, shortly after this Locke says

Hence when we talk or think of any particular sort of corporeal Substances, as *Horse, Stone, etc.* though the *Idea*, we have of either of them, be but the Complication, or Collection of those several simple *Ideas* of sensible Qualities, which we use to find united in the things called *Horse* or *Stone*, yet because we cannot conceive, how they should subsist alone, nor one in another, we suppose them existing in, and supported by some common subject; *which support we denote by the name Substance*, though it be certain, we have no clear, or distinct *Idea* of that *thing* we suppose a Support. (II.xxiii.4)

If Locke is saying, as he seems to be, that we cannot conceive of qualities with nothing to qualify, or 'unsupported', and if he is criticizing the European philosophers for supposing that their idea is clear or that their *unclear* idea has explanatory value we should expect him to go on to urge *either* that we can, after all, conceive of qualities as subsisting alone, or one in another, *or* that we can have a clearer idea of substance than at first appears. I believe that he does not do the first but does do the second.

The passages I have quoted so far have concerned material objects and substance-in-general in relation to their qualities. Locke goes on to say, rather more briefly, similar things about mental or spiritual things. Thus he talks of our supposing

a Substance, wherein *Thinking, Knowing, Doubting*, and a power of *Moving, etc.* do subsist, *We have as clear a Notion of the Substance of Spirit, as we have of Body;* the one being supposed to be (without knowing what it is) the *Substratum* to those simple *ideas* we have from without; and the other supposed (with a like ignorance of what it is) to be the *Substratum* to those Operations, which we experiment [*sc.* experience] in our selves within. (II.xxiii.5)

It may be significant that he here says that we have 'as clear a Notion', rather than 'as unclear a Notion', of the substance of spirit as we have of the substance of Body, thus suggesting that the idea of substance for mental things is to be retained along with the idea of substance for material things rather than that both are to be rejected.

In general he says,

Whatever therefore be the secret and abstract Nature of *Substance* in general all *the* Ideas *we have of particular distinct sorts of Substance*, are nothing but several Combinations of simple *Ideas*, co-existing in such, though unknown, Cause of their Union, as makes the whole subsist of itself. (II.xxiii.6)

III

Was Locke asserting the existence of substance-in-general? It might be thought, as some features of these passages admittedly suggest, that he was ridiculing the whole idea of substance and denying the necessity of postulating it; that he was saying that we are mistaken in thinking that anything in the world corresponds to our confused and obscure idea of it. Among the first to suggest this was Edward Stillingfleet, Bishop of Worcester, with whom Locke had a voluminous, and sometimes tedious, correspondence in 1696–8.[1] Stillingfleet accuses Locke[2] of having 'almost discarded substance out of the reasonable part of the world' (*L.S.* 5). Locke again and again denies that this was his intention. The correspondence took place between the 3rd and 4th Editions of the *Essay* and Locke made a number of emendations in the 4th Edition which appear to be intended to obviate the misleading impression he has created.

In his first letter to Stillingfleet Locke says, in direct reply to the Bishop's accusation, 'as long as there is any such thing as body or spirit in the world, I have done nothing towards the discarding substance out of the reasonable part of the world' (*L.S.* 7) and here he is clearly thinking about 'the confused idea of something' in which the simple ideas making up the complex ideas of substances inhere, something that 'supports accidents' (*L.S.* 7). He quotes Stillingfleet as saying 'we find we have no true conceptions of any modes or accidents (no matter which) but we must conceive a substratum, or subject wherein they are' and says that he agrees with this (*L.S.* 12–13). He says that our idea of substance-in-general is founded in this: 'that we cannot conceive how modes or accidents can subsist by themselves' (*L.S.* 19) and that 'the mind perceives their [*sc.* modes' and accidents'] necessary connection with inherence or being supported' (*L.S.* 21). In the *Essay* he has talked of accustoming 'ourselves to suppose some Substratum' which the Bishop takes to mean that we have a mere supposition, unsupported by evidence or reason, of substance-in-general. Locke here replies 'I grounded not the being, but the idea of substance, on our accustoming ourselves to suppose some substratum' (*L.S.* 18) which I think makes it clear that he was saying how we came to have the idea; the existence of substance is undoubted, and undoubtedly independent of this supposition, 'the being of things' as he says 'depending not on our ideas'.

If this is not enough, Locke returns, much later in the correspondence, to this question and says

> For I held we might be certain of the truth of this proposition, that there was substance in the world, though we have but an obscure and confused idea of substance. (*L.S.* 236)

This occurs, in fact, in the course of another argument to the effect that to say that our idea of substance is obscure and confused is *not* to say that substance does not exist. Elsewhere he says

But in some cases we may have certainty about obscure ideas; *v.g.* by the clear idea of thinking in me, I find the agreement of the clear idea of existence, and the obscure idea of a substance in me, because I perceive the necessary idea [agreement] of thinking, and the relative idea of a support; which support, without having any clear and distinct idea of what it is, beyond this relative one of a support, I call substance. (*L.S.* 42. The first edition has 'agreement' instead of 'idea' as indicated.)

Although the idea of what is the character of the support is not clear, the idea of the necessity of a support *is* clear. Some confirmation of this interpretation, beyond what I have already given, may be gleaned from a much later passage when Locke returns to 'the poor Indian'. He says

Your lordship's next words are to tell the world, that my simile about the elephant and the tortoise 'is to ridicule the notion of substance, and the European philosophers for asserting it'. But if your lordship please to turn again to my *Essay* [Ref. given: II.xix.13, should be II.xiii.19] you will find those passages were not intended to ridicule the notion of substance, or those who asserted it, whatever that 'it' signifies, but to show, that though substance did support accidents, yet philosophers, who had found such a support necessary, had no more a clear idea of what that support was, than the Indian had of that which supported his tortoise, though sure he was it was something. (*L.S.* 448)

But even if, Locke continues, these passages were intended to ridicule some European philosophers' notions of substance

it will by no means follow from thence, 'that upon my principles we cannot come to any certainty of reason, that there is any such thing as substance in the world'. Men's notions of a thing may be laughed at by those whose principles establish the certainty of the thing itself... (*L.S.* 449)

Finally, Locke says that he agrees with Stillingfleet that 'it is a repugnance to our first conception of things that modes or accidents should subsist by themselves' and that 'however imperfect and obscure our notion may be, yet we are as certain that substances are and must be, as that there are any beings in the world' (*L.S.* 452). Although the last passage refers to 'substanc*es*' which, in the plural, is usually used by Locke to refer to particular substances rather than substances-in-general, taken with what has gone before and the context of this passage, we can interpret him *either* as saying that since substanc*es* have qualities and qualities depend upon substance-in-general, the existence of substances implies the existence of substance-in-general *or*, as we shall see is likely, that there exist at least two substances-in-general.

I think, therefore, that we may safely conclude that Locke accepts the existence of substance-in-general and that the main reason for his (admitted) certainty about this is the impossibility of conceiving qualities (*i.e.* accidents) or modes or stable collections of qualities as existing without support. Qualities, accidents, modes must inhere in something, be qualities, accidents, or modes *of* something, this something is what he calls 'substance' even if he cannot say anything else about it.

'So much the worse' you may say 'for Locke's view' since to assert the existence of something of whose character we can say nothing is to make no contribution to knowledge, to explain nothing. I would agree, if that were all there was to it. I believe, however, that Locke does not in the end think that he cannot say anything more about it. Before I attempt to establish that I must get one or two things out of the way.

The term 'accidents' has led to some confusion. Locke does not often use it in the *Essay* but when he does I think it is interchangeable with 'qualities' as it is in Boyle's work. Locke talks of 'such Qualities, which are capable of producing simple *Ideas* in us; which Qualities are commonly called accidents' (II.xxiii.2). Qualities capable of producing simple ideas in us are all the primary and secondary qualities, at least, whether they are observable or unobservable; the idea of roundness is produced in us by the roundness of a ball, the idea of yellow by a particular unobservable texture or quality in an object that looks yellow. Boyle explains his use of the term 'accident' when he calls primary qualities 'inseparable accidents'; shape is inseparable because a body must have *some* shape but an accident because a body need not have *some one particular and unalterable* shape (*O.F.Q.* 16, *S.* 20). Thus the solidity of the corpuscles cannot be an accident since it is absolute and the same in every corpuscle.

Shortly after this Boyle points out that the term 'accident' is used in two senses by logicians and philosophers. The first sense, called *accidens praedicabile* in the schools, is that in which it refers to a non-essential, as opposed to an essential, property; that is, it 'may be present or absent without the destruction of the subject'. The second sense, called *accidens praedicamentale*, is the sense in which it refers to what is contrasted with substance in Aristotle's list of categories; substance is one category and the others are accidents, needing 'the existence of some substance or other, in which they may be, as in their subject of inhesion' (*O.F.Q.* 16–17, *S.* 21). It is clear that Boyle's term 'inseparable accidents' does not use 'accidents' in its first sense in which it applies only to non-essential properties, since 'inseparable accidents' would then be self-contradictory. He is using it in its second sense to apply to either essential or non-essential qualities, although, incidentally, he warns us against taking substance as capable of existing without accidents as, he says, some schoolmen did. I think we may take it, then, that accidents, separable or inseparable, are just qualities, primary, secondary or of the third sort. Locke talks of accidents much more frequently in the correspondence with Stillingfleet than in the *Essay* and I think we may take the word to be used in Boyle's sense. In the *Essay* he uses 'accident' in contrast to 'substance' at II.xiii.17, II.xiii.19, II.xxiii.24 and II.xxiii.15 when he talks about substance supporting accidents. I take 'accidents' to refer to any qualities, whether essential to body or not and whether observable or not.

The most usual view, historically, has probably been that Locke accepted the existence of substance-in-general but that since he admitted that it was a 'something we know not what', he had no philosophical justification for accepting it. The problem is well-known. In the ordinary way we describe things in terms of their qualities but substance-in-general cannot *have* qualities, in the sense in which a lump of lead *has* qualities, it merely supports them; I suppose we may say that it is what enables a lump of lead to have qualities held together in a stable collection. The relation of supporting qualities must be conceived of as different from the relation of having qualities. Substance-in-general, together with particular qualities, make up a particular substance which we can describe in terms of those qualities. Substance-in-general is, as far as we can know, a mysterious indeterminate stuff which is *logically* indescribable except that we can say that it supports qualities. However, it was postulated just because it was thought that qualities needed support, so the hypothesis of substance is a paradigm *ad hoc* hypothesis: since the only thing we can say about substance is that it does what it was invented to do, there can be no independent evidence for the existence or the nature of substance.

This view appears to be supported by a number of passages in which Locke says that substance is common to everything or is the same everywhere. For example, at II.xxiii.6 he refers to simple ideas, making up our ideas of bodies, being supposed by us 'to rest in, and be, as it were, adherent to that unknown common subject, which inheres not in anything else,' and at II.xxiii.4 he talks of particular ideas as 'existing in, and supported by some common subject; *which Support we denote by the name Substance*', and at II.xxiii.14 he says that the ideas we attribute to particular substances 'all terminate in sensible simple *Ideas*, all united in one common subject'. At III.x.15 he refers to 'the *Idea* of a solid substance, which is every where the same, every where uniform'. In the correspondence with Stillingfleet, on the subject of spiritual substance, Locke talks of 'the general idea of substance being the same every where, the modification of thinking, or the power of thinking joined to it...' (*L.S.* 33) and he quotes Stillingfleet with approval as saying of matter that '"it consists in a solid substance, every where the same"' (*L.S.* 410). If substance-in-general is the same in everything and if it is a support for all qualities, it appears that there is no possibility of knowing anything more about it than that. As Locke says, 'we have no *Idea* of what it is, but only a confused, obscure one of what it does' (II.xiii.19).

It seems clear that Locke did not intend to reject substance-in-general altogether but that he was certainly criticizing some views of substance, for example, in the passages about the Indian. In view of his, and Boyle's, suspicion of occult entities and qualities it seems to me implausible to suppose that he was asserting the existence of an absolutely unknowable, absolutely undifferentiated stuff postulated solely on the grounds that qualities logically require some support. So it seems worth searching for an alternative interpretation.

It is worth noting that the standard criticism of Locke often fails to make a distinction that may be important. Locke appears not to doubt the *existence* of a support for qualities and if he was relying on the argument that this is a *logical* necessity then, supposing this argument to be sound, there is no need for independent evidence for its existence and it is no criticism of Locke that he did not provide it. What rightly worried Locke, in so far as he was an empiricist, was the fact that this involves postulating an entity with an important function to perform but gives us no inkling of the nature of this entity and how it performs its function.

What precisely was Locke attacking? I suggest that he was attacking supposed scholastic conceptions of substance according to which it was not only unknowable but was also a *thing* capable of existing by itself, *i.e.* without the qualities it was postulated to support. Part of my evidence for this view can be given here but part of it will more conveniently come later.

The elephant and the tortoise postulated by the Indian as supports for the earth are *things* that can exist, relatively, by themselves and to these Locke compares some European philosophers' conceptions of substance-in-general. Just after the second passage about the Indian Locke says

'Tis the ordinary Qualities, observable in Iron, or a Diamond, put together, that makes the true complex *Idea* of those Substances, which a Smith, or a Jeweller, commonly knows better than a Philosopher; who, whatever substantial forms he may talk of, has no other *Idea* of those Substances, than what is framed by a collection of those simple *Ideas* which are to be found in them; only we must take notice, that our complex *Ideas* of Substances, besides all these simple *Ideas* they are made up of, have always the confused *Idea* of *something* to which they belong, and in which they subsist: and therefore when we speak of any sort of Substance, we say it is a *thing* having such or such Qualities, as Body is a *thing* that is extended, figured and capable of Motion; a Spirit a *thing* capable of thinking; and so Hardness, Friability, and Power to draw Iron, we say, are Qualities to be found in a Loadstone. These, and the like fashions of speaking intimate, that the Substance is supposed always *something* besides the Extension, Figure, Solidity, Motion, Thinking, or other observable *Ideas*, though we know not what it is. (II.xxiii.3)

Notice, first, that Locke refers to *substantial forms*, a conception which Boyle attacked as being occult and as being held, by some, to be of something capable of existing without matter (*O.F.Q.* 17, *S.* 22).

This lends support to the view that Locke is here attacking the move from saying '*something* must support qualities' to saying 'some *thing* must support qualities'. The mistake depends upon a false analogy; we say that a loadstone is a thing with qualities and, by analogy, that substance-in-general is a thing with, or supporting, qualities. The use of the word '*thing*', italicized by Locke, suggests the possibility of the existence of substance-in-general without qualities and something of which no further description can be given. Later, as we shall see, Locke may be seen explicitly attacking this notion of independent existence and showing how the mistake arises.

In the next section Locke says

yet, because we cannot conceive, how [the qualities of a Horse or Stone] should subsist alone, nor one in another, we suppose them existing in, and supported by some common subject; *which Support we denote by the name Substance*, though it be certain, we have no clear, or distinct *Idea* of that *thing* we suppose a Support. (II.xxiii.4)

I suggest that Locke is arguing that the Indian's European colleagues were right to postulate a support for qualities but wrong in thinking of it as a *thing* that could exist independently of what it supports and wrong in thinking that they had a clear idea of what it is.

IV

If Locke was, as I suggest, satisfied with the logical argument for the *existence* of substance-in-general but was unhappy about the inability of this to establish anything about its nature, we might expect him to attempt to show how it might be describable, at least in principle, in some terms other than 'a support for qualities'. One possibility would be to identify substance with real essence; even though it may not be possible in the present state of our knowledge and investigative techniques to describe the real essences of things the possibility of their eventual description is not excluded on logical grounds. That is surely the point of his talk, at IV.iii.13, of a *more and a less* 'incurable ignorance'.

If it is correct to regard Locke as holding that *observable* qualities give us the basis for the nominal essences of things, that this is all we at present know of external objects and that these qualities, and so those essences, depend upon the real essences of things, then it may seem plausible to suppose that these inner constitutions are identical with substance-in-general regarded as a support for observable qualities. This is a view which Stillingfleet, in his haphazard way, argues at one point. More recently, Maurice Mandelbaum has argued for a much more subtle and persuasive version of the identification.[3] However, I hope it will become obvious that close attention to the correspondence makes his view look less plausible than it does if it is based mainly on the *Essay*.

The view has difficulties of its own in relation to Locke's text and probably goes too far in the opposite direction to the standard view although it is in the spirit of Boyle's mechanism and his attack on scholastic doctrines of prime matter, substantial forms and real qualities. A central problem is that Locke is most naturally read as saying that in material things, at least, there is a *common* substratum; if substance were identical with real essence, substance in gold would not be the same as substance in silver; what makes them particular substances is their *different* real essences or inner constitutions. M. R. Ayers, in an article[4] that I find often illuminating but finally baffling, arguing against the idea that Locke saw substance as 'a mysterious, undifferentiated substrate, the same in everything', distinguishes general or determinable substance from determinate substances and holds that Locke accepted both but that our only access to the former is through the latter. In the course of this, Ayers puts an argument that might be used to support the identification of substance and real essence.[5] Of one important passage in the correspondence in which Locke talks of 'the general idea of substance being the same every where' Ayers says that Locke means strictly that the *idea* is the same everywhere, that is, 'the idea is equally lacking in positive content wherever it occurs' and is not to be taken as asserting that 'there is a mysterious undifferentiated substrate, the same in everything'.

If Locke had meant simply this in all such passages then one of the main obstacles to the view that he identified substance and real essence would be removed. Some passages I have quoted certainly appear to suggest that he may have meant this, but others suggest that he did not mean *only* this. However, allowing for some laxity of expression on Locke's part, as one often must, a determined interpreter might find considerable support for the view.

I propose next to show why I think this view cannot ultimately be sustained and then to argue for a very different view of the reasons for which Locke thought he was justified in thinking that substance was not a mysterious undifferentiated substrate about which nothing could be said.

V

If Locke intended to identify substance-in-general and real essence then it is odd that such an important identification is not mentioned in the long discussion of real and nominal essences in III.vi. There is a passage in Book II which some may think conflicts with my view; it would give me no trouble were it not for a particular interpretation put upon it by recent scholars. At the beginning of II.xxiii Locke talks about the way in which we acquire the idea of substance-in-general. This, of course, comes in the first place from observable qualities. Thus

> If any one should be asked, what is the subject wherein Colour or Weight inheres, he would have nothing to say, but the solid extended parts: And if he were demanded, what is it, that that Solidity and Extension inhere in, he would not be in a much better case, than the *Indian* before mentioned... (II.xxiii.2)

The expression 'the solid extended parts' has been taken by Yolton and Ayers[6] to mean the solid extended *observable* parts of a body and not the corpuscles of which it is composed. Consequently they take the passage to concern the explanation of merely *observable* solidity and extension. This might lead one who thinks that Locke intended to identify substance and real essence to suppose that the references to the Indian philosopher were all concerned with the question of what supports *observable* qualities and an acceptable answer to that would be substance-in-general conceived as real essence. This, however, seems implausible to me if only because anyone giving that answer *would be*, in Locke's eyes, in a better case than the Indian.

Moreover, the Indian takes the matter farther than that. He says that an elephant, presumably unobservable, supports the earth and a tortoise, presumably also unobservable, supports the elephant and then, in Locke's development of the view, something, he knows not what, supports the tortoise. This suggests that, even if the original question concerns what supports observable qualities and the answer to that is the real essences of particular substances, the questioning goes beyond that and concerns what supports qualities making up these real essences, and the answer to that is substance-in-general. After all, if qualities are logically in need of support that must apply to unobservable as well as observable qualities. It is as sensible to ask about the support for the shape of the corpuscle as it is to ask about the support for the shape of an orange. Qualities *as such* have a 'necessary connection with inherence or being supported'.

That Locke really did accept this becomes clearer in the correspondence because he is replying to an accusation that he has rejected substance-in-general altogether. He quotes Stillingfleet as saying 'we find, that we can have no true conceptions of any modes or accidents (no matter which) but we must conceive a substratum, or subject wherein they are' (*L.S.* 12–13) and says that he agrees. In a famous passage Locke says 'I ground not the being, but the idea of substance, on our accustoming ourselves to suppose some substratum; for it is of the idea alone I speak there, and not of the being of substance' (*L.S.* 18). Accustoming ourselves to suppose a substratum concerns our acquisition of the ideas of qualities, inherence and substance-in-general, but the necessity for its existence is a logical one for, Locke says, 'the mind perceives their [*sc.* modes' and accidents'] necessary connection with inherence or being supported' and 'the mind frames the correlative idea of a support' (*L.S.* 21) and, much later, he says that 'sensible qualities imply a substratum to exist in' (*L.S.* 447).

There are several fairly direct rejections of the identification of substance and essence in the correspondence since Stillingfleet suggests it. He objects to Locke's putting weight on the derivation of the word 'substance' and points out that Cicero and Quintilian 'take substance for the same as essence...; and so the Greek word imports'. Locke replies

> though Cicero and Quintilian take *substantia* for the same with essence, as your lordship says; or for riches and estate, as I think they also do; yet I suppose it will be true, that *substantia* is derived *à substando*, and that that shows the original import of the word. (*L.S.* 23)

Locke clearly appears to be rejecting the identification. He seems to be saying that whatever earlier writers have said about substance the basic sense upon which he is relying comes from its original derivation, meaning supporting or standing under. We can ask about essence what supports it. It seems likely to me that Locke is thinking of the Greek word '*hypokeimenon*' whereas the Bishop is thinking of the word '*ousia*', as equivalents for '*substantia*'. See also *Essay* II.xiii.20 when Locke seems to be making the point that the word 'substance' is more misleading, perhaps because it is a noun, than the words 'standing under', emphasizing the relation.

Locke is perhaps even clearer in the correspondence when, in various places, Stillingfleet identifies substance with both essence and nature. Locke says

> Your lordship, in this paragraph, gives us two significations of the word nature: 1. That it is sometimes taken for essential properties, which I easily admit. 2. That sometimes it is taken for the thing itself in which these properties are, and consequently for substance itself. And this your lordship proves out of Aristotle. (*L.S.* 70)

I take Locke to be agreeing that it is so used. He says 'that your lordship thinks fit to call substance nature, is evident'. Shortly afterwards he says

> I must then confess ... that I do not clearly understand whether your lordship, in these two paragraphs, speaks of nature, as standing for essential properties; or of nature, as standing for substance. (*L.S.* 73)

This certainly suggests that Locke wishes to distinguish between substance-in-general and essential properties, or real essence.

Later again (*L.S.* 81), when the Bishop has said that 'we cannot comprehend the internal frame or constitution of things, nor in what manner they do flow from the substance' Locke says 'I do not take them [*sc.* essences] to flow from the substance in any created being, but to be in every thing that internal constitution, or frame, or modification of the substance, which God in his wisdom and good pleasure thinks fit to give to every particular creature, when he gives [it] a being' (*L.S.* 82).

This suggests that substance-in-general is a kind of raw stuff underlying and common to different real essences which God creates, that of which different real essences are different modifications. Of course, *different* real essences could not *flow from* (be caused by) *the same* substance.

In his third letter to Stillingfleet, Locke returns to the question of substance, nature and essence, and the derivation of the word 'substance' because Stillingfleet has returned to it. This time, no doubt knowing the influence this is likely to have on Locke, Stillingfleet uses the authority of Boyle for identifying substance and real essence. Locke says

> Indeed your lordship brings a proof from an authority that is proper in the case, and would go a great way in it; for it is of an Englishman, who, writing of nature, gives an account of the signification of the word nature in English. But the mischief is, that among eight significations of the word nature, which he gives, that is not to be found, which you quote him for, and had need of. For he says not that nature in English is used for substance, which is the sense your lordship has used it in, and would justify by the authority of that ingenious and honourable person; and to make it out, you tell us, 'Mr. Boyle says the word essence is of great affinity to nature, if not of an adequate import;' to which your lordship adds, 'but the real essence of a thing is a substance'. So that, in fine, the authority of this excellent person and philosopher amounts to thus much, that *he* says that nature and essence are two terms that have a great affinity; and *you* say, that nature and substance are two terms that have a great affinity. For the learned Mr. Boyle says no such thing, nor can it appear that he ever thought so, till it can be shown, that he has said that essence and substance have the same signification. (*L.S.* 364, my italics)

Moreover, Locke says, Boyle does not even use the word 'substance' in the passages in question, and he is right.[7]

Again, Locke complains that Stillingfleet does not give 'a very tolerable account' of Locke's idea of substance because 'the account you give over and over again of my idea of substance is, that "it is nothing but a complex idea of accidents"' (*L.S.* 456). He then quotes *Essay* II.xii.6 in which substance-in-general is contrasted with particular substances.

Finally, perhaps an important passage for our purpose occurs when Locke, in reply to Stillingfleet, returns to Cicero and Aristotle. This concerns the question whether the soul is an immaterial substance and Locke denies that Cicero held this. Cicero, he says, takes the human body to be distinguished from the soul and, indeed, says it is 'the prison of the soul'. But he does not distinguish matter in general from the soul. So he appears to distinguish between body and matter, as Locke does. Cicero concludes that the soul is 'not like other things here below, made up of a composition of the [Aristotelian] elements' (*L.S.* 484). He excludes the 'gross elements' earth and water, but 'beyond this he is uncertain'. He argues, however, that there are reasons for thinking the soul might be composed of air or fire and so is not immaterial. But in the end he inclines to the view that 'the soul was not at all elementary [*i.e.* composed of one or more of the four elements], but was of the same substance with the heavens' (*L.S.* 484). That is the '*quinta essentia*', the quintessence or fifth element, of which heavenly bodies, according to Aristotle, are composed. This, however, still means that the soul is not immaterial.

All this is in the course of an argument about whether matter might think but it is not this that concerns me here. Earlier in the discussion Locke has said

> The idea of matter is an extended solid substance; wherever there is such a substance there is matter, and the essence of matter, whatever other qualities, not contained in that essence, it shall please God to superadd to it. (*L.S.* 460)

This appears to mean that the 'essence of matter' is to be distinguished from the real essence of particular substances and this is the conclusion I take the argument about Cicero, very indirectly, to support.

Although Locke does not accept Cicero's account in terms of the Aristotelian elements he appears to look with approval upon certain formal features of that account. Cicero may be regarded as saying that a person is made up of body and soul. The body has a structure or inner constitution since it is composed of some combination of the four elements. But matter is to be distinguished from body because the four elements, and the *quinta essentia*, are material but do not have structures or inner constitutions in the way that body does. Matter has an essence but it is not complex like the real essence of, say, the human body. The soul may be fire, *or* air, or *quinta essentia*, according to Cicero, and whichever it is constitutes the essential feature of the soul. In any case it is material. So something may be material and not have a structure or inner constitution.

I suggest that Locke sees an analogy between Cicero's view and his own because, although Cicero treats some Aristotelian elements as possible candidates for what stands under the essence of a person and Locke does not, yet they agree that the complex structure that constitutes this essence requires something unstructured to support it. This is so whether we are concerned with material bodies or immaterial spirits.

There is further evidence against the identification of substance and real essence to be gleaned from changes made by Locke in the *Essay* between the third and fourth editions.

1. In the 1st to 3rd editions, at II.xxiii.5, Locke wrote '*We have as clear a Notion of the Nature, or Substance of Spirit, as we have of Body*' but in the 4th and 5th editions he omitted the words 'Nature, or'.[8]
2. In the 1st to 3rd editions, also at II.xxiii.5, Locke wrote 'It being as rational to affirm, there is no Body, because we cannot know its Essence, as 'tis called, or have no *Idea* of the Substance of Matter; as to say there is no Spirit because we know not its Essence or have no *Idea* of a Spiritual Substance'. But in the 4th and 5th editions he wrote 'It being as rational to affirm, there is no Body, because we have no clear and distinct *Idea* of the *Substance* of Matter; as to say there is no Spirit, because we have no clear and distinct *Idea* of the *Substance* of a Spirit'.[9]

It is important to note that the letters to Stillingfleet were written between the 3rd and the 4th editions[10] and that in footnotes to II.xxiii in the 5th edition Locke refers to the arguments of that correspondence. This may indicate a change of mind but, in view of other evidence, it seems more likely that the changes were intended to remove any suggestion that he ever identified substance and real essence.

VI

I now come to my suggested interpretation of Locke's view of substance-in-general. This interpretation lies between the view that substance is an absolutely indeterminate and unknowable substratum in which all qualities, whether observable or not, and whether material or mental, must inhere and the view that it is simply identical with real essence and is a support merely for the observable qualities of bodies and so not *in principle* indeterminate and indescribable. One of the advantages of my interpretation is that it falls out easily from my interpretation of Locke's views on qualities, the distinction between primary and secondary qualities and the corpuscular philosophy, which I have previously argued for, and so makes Locke's view as a whole more coherent and more plausible than other interpretations.

I begin, as usual, with Boyle. He says

I agree with the generality of philosophers so far as to allow, that there is one catholick or universal matter common to all bodies, by which I mean a substance extended, divisible, and impenetrable. (*O.F.Q.* 15, *S.* 18)

He frequently talks of 'the universal matter' (*e.g., O.F.Q.* 16) and he says 'the matter of all natural bodies is the same; namely, a substance, extended and impenetrable' (*O.F.Q.* 35, *S.* 50), and that motion does not belong to the essence of matter. He also says 'bodies having but one common matter can be differenced but by accidents...' (*O.F.Q.* 35, *S.* 49) and 'bodies thus agreeing in the same common matter, their distinction is to be taken from those accidents that do diversify it' (*O.F.Q.* 35, *S.* 50).

Boyle argues (*O.F.Q.* 17, *S.* 21–2) against what he alleges to be a scholastic doctrine of substance as independent of qualities, in the sense of being capable of existing without qualities, which he finds 'either unintelligible or manifestly contradictious'. Substance, on that view, is a necessary something in which qualities inhere but it is separable from them. But, Boyle says, if substance is separable from qualities, qualities must be separable from substance. So, he alleges, qualities are treated by implication as substances by those he is attacking. I am not concerned to consider the validity of this argument; what matters here is that Boyle thought it was valid and that it throws light on his view of matter.

I suggest that, although Locke does not follow Boyle in every detail, he does take the natural philosophers' universal matter as a starting point and that this is crucial to his view of substance. This much is surely plausible given the obvious importance of the corpuscular philosophy to the argument of the *Essay*.

The main clue to what I take to be Locke's view of substance is contained in Book III. Up to this point Locke has been attacking certain conceptions of substance but also making it clear that some such conception is necessary. Now, in considering language, he can show how mistaken ideas about substance as an unknowable, undifferentiated but independently existing thing have arisen and, at the same time, indicate what sense there is in the idea of substance-in-general. The passage in question occurs where Locke is giving an instance in which '*names taken for Things* are apt to *mislead the Understanding*'. He says

How many intricate Disputes have there been about *Matter*, as if there were some such thing really in Nature, distinct from *Body*; as 'tis evident, the Word *Matter* stands for an *Idea* distinct from the *Idea* of Body? For if the *Ideas* these two Terms stood for, were precisely the same, they might indifferently in all places be put one for another. But we see, that tho' it be proper to say, There is *one Matter of all Bodies*, one cannot say, There is *one Body of all Matters*: ... though *Matter* and *Body*, be not really distinct, but where-ever there is the one, there is the other; Yet *Matter* and *Body*, stand for two different Conceptions, whereof the one is incomplete, and but a part of the other. For *Body* stands for a solid extended figured Substance, whereof *Matter* is but a partial and more confused Conception, it seeming to me to be used for the Substance and Solidity of Body, without taking in its Extension and Figure: And therefore it is that speaking of *Matter*, we speak of it always as one, because in truth, it expresly contains nothing but the *Idea* of a solid Substance, which is every where the same, every where uniform. This being our *Idea* of *Matter*, we no more conceive, or speak of different *Matters* in the World, than we do of different Solidities; though we both conceive, and speak of different Bodies, because Extension and Figure are capable of variation. But since Solidity cannot exist without Extension, and Figure, the taking *Matter* to be the name of something really existing under that Precision,[11] has no doubt produced those obscure and unintelligible Discourses and Disputes, which have filled the Heads and Books of Philosophers concerning *Materia prima*; which Imperfection or Abuse, how far it may concern a great many other general Terms, I leave to be considered. (III.x.15)

Inattention to words has misled people into thinking of matter as distinct from body in the sense of existing separately from it. 'Matter' and 'body' are distinct words standing for distinct ideas; the idea of matter is distinct from the idea of body, as our different uses of these words show, but if we think that this allows us to say that matter is, in reality, distinct from body, having a separate existence as a 'thing', then we are confused. Matter and body are not really distinct because the idea of matter is just part of the idea of body, though they are different ideas; the most we can say is that 'matter' refers to just part of what 'body' refers to. Something in reality corresponds to the idea of matter but it is not a separately existing thing; it exists wherever and whenever body exists. 'Body' stands for 'a solid, extended, figured, Substance' but 'Matter' is a partial and more confused conception used for 'the Substance and Solidity of Body, without taking in its Extension and Figure'. So we speak of matter 'always as one' because the idea of it is just 'the *Idea* of a solid Substance, which is every where the same, every where uniform'. It would be odd if Locke meant by this merely that our *idea* of matter is equally confused and lacking in content wherever it occurs. The word 'uniform' alone suggests more than this.

We 'no more conceive, or speak of different Matters in the World than we do of different Solidities; though we both conceive, and speak of different Bodies, because Extension and Figure are capable of variation'. There are not, at this level, different solidities because the solidity of corpuscles is absolute (Boyle's impenetrability) and admits of no variation. We speak of different hardnesses on the level of experience where hardness is something different, depending upon corpuscles *and* empty space between them, and admits of variation. Extension and figure, because they are capable of variation, may differentiate bodies and so are *qualities*, whereas solidity is not. The matter is the same in different corpuscles (bodies) and so is the solidity, whereas the extension and figure may not be. Matter, unlike body, is the same everywhere, that is, in all material things.

However, solidity cannot exist without extension and figure, *some* extension and *some* figure, so we must not take matter to exist independently of body; that would lead us into all the difficulties of the scholastic *materia prima*.

All this relates to Boyle's corpuscular hypothesis and the passages quoted earlier from Boyle. Impenetrability or solidity does not figure in Boyle's list of primary qualities or inseparable accidents that diversify bodies: it is that in respect of which universal matter is everywhere the same. Matter is a solid stuff which is what, in material bodies, is qualified by specific shapes, sizes and mobilities. My suggestion is that this is what Locke meant by substance-in-general for material things. It is not featureless because it is solid and solidity is its essential characteristic although it is not a quality; it does not exist independently of qualities since being solid entails having shape and size.

However, both Locke and Boyle were, I believe, dualists and matter is not the only substance-in-general; there is also spiritual substance-in-general. In the *Essay*, especially up to the chapter on substances and substance (II.xxiii), Locke is mainly concerned with the scientific account of the material world; the conception of qualities and the primary/secondary quality distinction are discussed in relation to this; with that in mind we can regard the passages in which he seems to regard substance-in-general as the same everywhere as meaning the same in all *material* things. However, in the chapter on substances and substance he suggests that there is (must be) a corresponding spiritual substance-in-general involved in all mental phenomena.

Already in Chapter xiii he has said something relevant to this. If he is asked whether empty space is substance or accident, he says, he will reply that he does not know until he is 'shown' a clear, distinct idea of substance (II.xiii.17). He then warns against 'taking words for things' and goes on

I desire those who lay so much stress on the sound of these two Syllables, *Substance*, to consider whether applying it, as they do, to the infinite incomprehensible GOD, to finite Spirit, and to Body, it be in the same sense; and whether it stands for the same *Idea*, when each of these three so different Beings are called *Substances*? If so, whether it will not thence follow, That God, Spirits, and Body, agreeing in the same common nature of *Substance*, differ not any otherwise than in a bare different modification of that *Substance*; as a Tree and a Pebble, being in the same sense Body, and agreeing in the common nature of Body, differ only in a bare modification of that common matter; which will be a very harsh Doctrine. (II.xiii.18)

Note, in the first place, that he appears at the end of this passage clearly to equate substance and matter. I regard him as taking it as clear that a tree and a pebble *do* differ only in a bare modification of (material) substance and saying that it would be a very harsh doctrine if God and spirits were said to differ from trees and pebbles only in being further modifications of the *same* substance.

He continues

If they say, That they apply it to God, finite Spirits, and Matter, in three different significations, and that it stands for one *Idea*, when GOD is said to be a *Substance*; for another, when the Soul is called *Substance*; and for a third when a Body is called so. If the name *Substance*, stands for three several distinct *Ideas*, they would do well to make known those distinct *Ideas*, or at least to give three distinct names to them, to prevent in so important a Notion, the Confusion and Errors, that will naturally follow from the promiscuous use of so doubtful a term... (II.xiii.18)

The next section contains the first passage about the poor Indian.

I believe that Locke here means just what he says: if we use different senses of 'substance' we must make it clear what those senses are. He does not say that this is either unacceptable or impossible. He may even think it unnecessary or incorrect to refer to God as a substance and *still* wish to talk about both material and spiritual substance-in-general. However, if we can say no more about substance than that 'without knowing what it is, it is that which supports *Accidents*' (II.xiii.19) then we make no distinction between substances and are forced into 'the very harsh Doctrine'. Much of what he says in the correspondence suggests that he is wrestling with the problem of characterising spiritual substance-in-general.

My view, then, is that Locke, as a dualist, wishes to assert the existence of material and spiritual substances as substrata for material things and mental activities, respectively, but that he sees the necessity of saying more about them than that they are something we know not what. He gives reasons for rejecting the Cartesian characterization of matter by extension and mind by thinking. Extension will not adequately characterize the essentially solid 'stuff', matter, since it does not distinguish it from empty space; I shall come shortly to his reasons for rejecting thinking as characterizing spiritual 'stuff'. He seems fairly confident that *solidity* is what characterizes material substance and, in the *Essay*, that the power of *thinking* is what characterizes spiritual substance, but in the correspondence he appears to reject thinking for this purpose.

The beginning of the two-substance view in the *Essay* appears to be at II.xxiii.3 where Locke says

> when we speak of any sort of Substance, we say it is a *thing* having such or such Qualities, as Body is a *thing* that is extended, figured, and capable of Motion; a Spirit a *thing* capable of thinking; ... the Substance is supposed always *something* besides the Extension, Figure, Solidity, Motion, Thinking, or other observable *Ideas*, though we know not what it is.

This, as I have said, is mainly critical of the idea that substance-in-general is an independent existent. It does not, however, rule out other ideas of substance or the idea of *two* substances-in-general. It appears to allow that spiritual substance underlies thinking just as material substance underlies extension. It may be objected to my view that solidity appears in the second list, as if material substance-in-general underlies *solidity* as well as extension but this clearly concerns *observable* solidity, or hardness, and not absolute solidity as is evident from the phrase 'or other observable Ideas'.

Two sections later, Locke refers to, and appears to contrast, the 'Substance of Spirit' and the 'Substance of Body'. He talks of *operations of the mind*

> which we concluding not to subsist of themselves, nor apprehending how they can belong to Body, or be produced by it, we are apt to think these the Actions of some other *Substance*, which we call *Spirit*... It being as rational to affirm, there is no Body, because we have no clear and distinct *Idea* of the *Substance* of Matter; as to say, there is no Spirit, because we have no clear and distinct *Idea* of the *Substance* of a Spirit. (II.xxiii.5)

However, the passage that perhaps best supports my view and most obviously leads on to the discussion in the correspondence is this

> *we have as much Reason to be satisfied with our Notion of immaterial Spirit, as with our Notion of Body; and the Existence of the one, as well as the other.* For it being no more a contradiction, that Thinking should exist, separate, and independent from Solidity; then it is a contradiction, that Solidity should exist, separate, and independent from Thinking, they being both but simple *Ideas*, independent one from another; and having as clear and distinct *Ideas* in us of Thinking, as of Solidity, I know not, why we may not as well allow a thinking thing without Solidity, *i.e. immaterial*, to exist; as a solid thing without Thinking, *i.e. Matter*, to exist. (II.xxiii.32)

In II.xiii.11 Locke has argued that body and extension are not the same and that those who say they are confuse solidity and extension. Solidity is necessary to body and implies extension but extension, *e.g.* of the space between corpuscles, does not imply solidity. Now, in chapter xxiii, he says

Our *Idea* of Body, as I think, is an extended solid Substance, capable of communicating Motion by impulse: and our *Idea* of our Soul, as an immaterial Spirit, is of a Substance, that thinks, and has a power of exciting Motion in Body, by Will, or Thought. These, I think, are *our complex* Ideas *of Soul and Body, as contra-distinguished*; and now let us examine which has most obscurity in it, and difficulty to be apprehended. I know that People, whose Thoughts are immersed in Matter, and have so subjected their Minds to their Senses, that they seldom reflect on any thing beyond them, are apt to say, they cannot comprehend a thinking thing, which, perhaps, is true: But I affirm, when they consider it well, they can no more comprehend an extended thing. (II.xxiii.22)

It appears from the correspondence that Locke may have changed his mind about what characterizes spiritual substance. He there seems much more certain that solidity characterizes material substance than that thinking characterizes spiritual substance.

Early in the first letter Locke mentions that the Bishop has said that upon his (Locke's) principles it 'cannot be proved that there is a spiritual substance in us' (*L.S.* 32) since he has admitted the possibility that God may superadd a power of thinking to matter. On that supposition, 'it may be a material substance that thinks in us' (*L.S.* 33). Locke now distinguishes between *spiritual* and *immaterial* substance; a substance is spiritual if it thinks, whether it is material or not, but is immaterial only if it is not solid. So if there is thinking matter then there is a spiritual substance but not necessarily an immaterial substance, although God would have to be immaterial if he is able to superadd thinking to matter. (See *Essay*, IV.x.16 which is referred to by Locke at this point.)

The relevant passage reads

your lordship will argue, that by what I have said of the possibility that God may, if he pleases, superadd to matter a faculty of thinking, it can never be proved that there is a spiritual substance in us, because upon that supposition it is possible it may be a material substance that thinks in us. I grant it; but add, that the general idea of substance being the same every where, the modification of thinking, or the power of thinking joined to it, makes it a spirit, without considering what other modifications it has, as whether it has the modification of solidity or no. As on the other side, substance that has the modification of solidity, is matter, whether it has the modification of thinking or no. And therefore, if your lordship means by a spiritual an immaterial substance, I grant I have not proved, nor upon my principles can it be proved, (your lordship meaning, as I think you do, demonstratively proved) that there is an immaterial substance in us that thinks. Though I presume, from what I have said about the supposition of a system of matter thinking [IV.x.16] (which there demonstrates that God is immaterial) will prove it in the highest degree probable, that the thinking substance in us is immaterial.[12] (*L.S.* 33)

This is perhaps the most difficult passage for my view that Locke accepts two substances-in-general because of his apparent acceptance that substance is the same everywhere, whether in a body or in a thinking thing. He appears to be saying firmly that there is an undifferentiated and unknowable substratum underlying everything, which may become material by the addition of solidity and spiritual by the addition of the power of thinking. However, this is a very early stage in the argument and it is noticeable that in the correspondence Locke is careful to deal with one thing at a time. He is considering the extreme view attributed to him by Stillingfleet that what thinks in us is, in fact, matter.

I believe that Locke is taking it that, *on that supposition*, there would be just one substance-in-general supporting both material and spiritual things and then solidity and thinking *would* be different modifications of the same substance. He is suggesting that even the most extreme view attributed to him does not have the consequences that Stillingfleet supposes. Moreover, he is beginning to make it clear that he does not hold this extreme view and that although it is logically possible that God could have superadded the power of thinking to matter it is highly improbable that he has done so. This interpretation is amply reinforced as the correspondence proceeds.

The course of Locke's argument is made even clearer shortly after this. He says

If by spiritual substance your lordship means an immaterial substance in us ... I grant what your lordship says is true, that it cannot, upon these principles, be demonstrated. But ... it can be proved, to the highest degree of probability. If by spiritual substance your lordship means a thinking substance, I must dissent ... and say, that we can have a certainty ... that there is a spiritual substance in us ... from the idea of thinking, we can have a certainty that there is a thinking substance in us; from hence we have a certainty that there is an eternal thinking substance. This thinking substance, which has been from eternity, I have proved to be immaterial. This eternal, immaterial, thinking substance, has put into us a thinking substance, which, whether it be a material or immaterial substance, cannot be infallibly demonstrated from our ideas; though from them it may be proved, that it is to the highest degree probable that it is immaterial. (*L.S.* 36–7)

I take it, then, that Locke, believing he has established with high probability that the spiritual substance in us is immaterial, goes on to consider what can be said about his two substances-in-general to contrast them. I shall therefore refer to these as 'material' and 'immaterial' substances to keep that point in view.

In his third letter, in the course of a longer discussion of 'thinking matter', Locke says much that may throw light on his view of substance-in-general. For example,

Your first argument I take to be this, that, according to me, the knowledge we have being by our ideas, and our idea of matter in general being a solid substance, and our idea of body a solid extended figured substance; if I admit matter to be capable of thinking, I confound the idea of matter with the idea of a spirit: to which I answer, No; no more than I confound the idea of matter with the idea of a horse, when I say that matter in general is a solid extended substance; and that a horse is a material animal, or an extended solid substance with sense and spontaneous motion. (*L.S.* 460)

That is, it is not contradictory to suppose that some complex organization of matter may be capable of thinking and if that is so there is no need to postulate an immaterial substance as well; to say that this is to 'confound the idea of matter with that of a spirit' is just to refuse to consider the possibility. God may ordain that matter thinks; he may, even if this is incomprehensible to us, have 'actually endued some parcels of matter, so disposed as he thinks fit, with a faculty of thinking' (*L.S.* 466). In fact, God may 'superadd' to matter any powers he pleases as long as no contradictions occur. Wherever there is an extended solid substance there is matter whatever other qualities not contained in the *essence* of matter it has. Locke is stressing conceptual distinctions. Thus

Hitherto it is not doubted but the power of God may go, and that the properties of a rose, a peach, or an elephant, superadded to matter, change not the properties of matter; but matter is in these things matter still. But if one venture to go on one step further, and say, God may give to matter thought, reason, and volition, as well as sense and spontaneous motion, there are men readily present to limit the power of the omnipotent Creator, and tell us he cannot do it; because it destroys the essence, 'changes the essential properties of matter.' To make good which assertion, they have no more to say, but that hought and reason are not included in the essence of matter. I grant it; but whatever excellency, not contained in its essence, be superadded to matter, it does not destroy the essence of matter, if it leaves it an extended solid substance; wherever that is, there is the essence of matter; (*L.S.* 460–1)

The essential characteristic of matter is solidity; whatever is *superadded*, that is not changed. It is important here that, given the way matter is defined, the essence of matter must be different from the real essence of particular substances such as gold or lead. Such real essences are constituted by patterns of corpuscles. A single corpuscle must involve the essence of matter but it can involve no such pattern; there are no gold or lead corpuscles. Just as the real essence of gold must be the same, within limits, in any piece of gold, so the essence of matter must be the same in any portion of matter. That can only be solidity.

Next comes a consideration that appears to lead Locke to doubt that thinking is to immaterial substance as solidity is to material substance. It first appears when he says

> God has created a substance; let it be, for example, a solid extended substance: is God bound to give it, besides being, a power of action? that, I think, nobody will say. He therefore may leave it in a state of inactivity, and it will be nevertheless a substance; for action is not necessary to the being of any substance, that God does create. God has likewise created and made to exist, *de novo*, an immaterial substance, which will not lose its being of a substance, though God should bestow on it nothing more but this bare being, without giving it any activity at all. Here are now two distinct substances, the one material, the other immaterial, both in a state of perfect inactivity. (*L.S.* 464)

Why does Locke not say that a *power* of thinking would serve, since a power is not an activity and a thing may have a power without exercising or exhibiting it *i.e.* be in a state of perfect inactivity? The reason, I think, is that for Locke a power is complex and permits of analysis or explanation; it makes sense to say that something has a power only if *either* it exhibits that power *or* a description of its structure makes it evident that it has that power. Locke is looking for features of substance-in-general that are absolutely basic, simple, unanalysable and intrinsic. Absolute solidity is one such feature; what is the corresponding feature of immaterial substance?

Locke then says that thinking is an activity. This thinking can be removed from immaterial substance without destroying its being as a substance. This point is made also at *Essay* II.xix.4. Moreover, given two bare substances, one solid, the other unsolid, what is to prevent an omnipotent God from giving the activity of thinking to either? What power can he give to one that he cannot give to the other, as long as the solidity or unsolidity remains undisturbed?

> Now I would ask, why Omnipotentcy cannot give to either of these substances, which are equally in a state of perfect inactivity, the same power that it can give to the other? Let it be, for example, that of spontaneous or self motion, which is a power that it is supposed God can give to an unsolid substance, but denied that he can give to a solid substance. (*L.S.* 464)

Both solid and unsolid substances are, then, without power. It follows that they are other than real essences since real essences necessarily have powers. In this part of the correspondence Locke more often talks of solid and unsolid substances, as if he were reasonably sure that material substance could be characterized by solidity but now unsure that immaterial substance could be characterized by thinking, or the power of thinking, and so must be provisionally characterized, merely negatively, as unsolid.

The idea of material substance is of central concern if we are attempting to understand the *Essay* so I concentrate on that. Further passages in this part of the correspondence support the view that substance-in-general, for bodies, is simply their common matter, characterized by solidity. In his first letter to Locke Stillingfleet says

We do not set bounds to *God's Omnipotency*: For he may if he please, change a body into an *Immaterial Substance*; but we say, that while he continues the Essential Properties of Things, it is as impossible for Matter to think, as for a Body by Transubstantiation to be present after the manner of a Spirit; and we are as certain of one as we are of the other... For if God doth not change the *Essential Properties of things*, their Nature remaining: then either it is impossible for a *Material Substance* to think, or it must be asserted, that a Power of thinking is within the *Essential Properties of Matter*, and so *thinking* will be such a Mode of Matter, as *Spinoza* hath made it...[13]

Locke reverts to this passage in his third letter. As we have seen, he denies that his view that God could give the power of thinking to matter entails that thinking would be an essential property of matter (*L.S.* 470–1). God's Omnipotence is limited only by the impossibility of giving *incompatible* properties to substance; God could not make a substance both solid and unsolid (*L.S.* 465) but the power of thinking entails neither solidity nor unsolidity.

What is important here is that Stillingfleet does not specifically mention solidity when, in these passages, he talks of the essential properties of matter, but when Locke tries to make sense of Stillingfleet's notion of changing a body into an immaterial substance he does so by supposing that this involves the removal of solidity leaving pure substance-in-general behind, as if this could be really the same in everything, whether material or spiritual. But Locke denies the possibility of that. When Stillingfleet says that God may 'change a body into an immaterial Substance' Locke glosses this by saying

i.e. take away from a substance the solidity which it had before, and which made it matter, and then give it a faculty of thinking, which it had not before, and which makes it a spirit, the same substance remaining. (*L.S.* 470)

Locke adds that if the same substance did not remain we should have the *substitution* of one substance for another rather than the *changing* of one substance *into* another. If I am right about Locke's conception of qualities, he may also have in mind here that Stillingfleet is treating solidity as if it were a quality, which it is not.

Locke comments that the taking away of one 'quality' from substance could not give it another; so if solidity were taken away we should be left with an immaterial substance 'without the faculty of thinking' or, indeed, any other, which conflicts with the Bishop's view. God could take away one without giving the other. (Locke presumably refers here to solidity as a quality, as he seldom does elsewhere, because that is how Stillingfleet is regarding it, not because he thinks that that is what it is.)

All this seems to me to support the view that Locke held that there were at least two substances-in-general, one material, one immaterial, each characterized by one characteristic which was neither a quality nor a power. The tone of these passages suggests that he thinks that if substance is not characterized by solidity or some spiritual feature or some other feature of a kind somehow related to experience then it is nothing; and unless that feature were somehow related to experience we could say nothing about it.

In his first letter to Stillingfleet, on the question of real essence, Locke says

> Here I must acknowledge to your lordship that my notion of these essences differs a little from your lordship's; for I do not take them to flow from the substance in any created being, but to be in everything that internal constitution, or frame, or modification of the substance, which God in his wisdom and good pleasure thinks fit to give to every particular creature, when he gives [it] a being: and such essences I grant there are in all things that exist. (*L.S.* 82)

This can be read in the spirit of my interpretation. Real essences do not 'flow from' substance-in-general because that is the same in all material things; real essence is rather superimposed upon, or superadded to, or a particular modification of substances. The expressions 'internal constitution', and 'frame' should *not* be read along with 'of the substance'; that phrase goes along with 'modification' alone. Every created thing has an internal constitution or frame which is its real essence: this is not pure substance-in-general but a modification of it, *i.e.* substance-in-general plus various particular qualities, solidity plus, among other things, specific shapes and specific sizes. Here, if anywhere, was an opportunity for Locke to identify substance-in-general with real essence if he thought they were identical or to express incurable ignorance about how it was to be described, but he does neither.

I am suggesting, then, that, like Descartes, he is accepting a two-substance theory but characterizing the two substances differently. Extension will not do for material substance because it is not the same everywhere; thinking will not do for immaterial substance because it is an activity. Extension is replaced by solidity but Locke is doubtful about the replacement for thinking. There are signs in the correspondence that he is moving towards replacing it by sensation which, since the mind is passive with respect to it, at least might be, for him, a better candidate than thinking.

There is perhaps some further evidence for this to be gleaned from changes made by Locke in the *Essay* II.xxi between the first and the fourth editions. In the second edition (1694) he replaced section 46 by section 72 and announced a change of opinion.[14] The significant difference is that in the later edition he talks of both active and passive powers and applies the distinction to material bodies and to minds. We have, he says, ideas of two sorts of action, namely, motion and thinking. But both may include ideas of active and passive powers so further distinctions are necessary. About thinking he says

a Power to receive Ideas, or Thoughts, from the operation of an external substance, is called a *Power* of thinking: But this is but a *Passive Power*, or Capacity.

Later in the same paragraph he says

this Proposition, I see the Moon, or a Star, or I feel the heat of the Sun, though expressed by a *Verb Active*, does not signify any *Action* in me whereby I operate on those Substances; but the reception of the *Ideas* of light, roundness, and heat, wherein I am not active but barely passive... (II.xxi.72)

In section 47 in the first edition (replaced by 73 in later editions) Locke sets out the 'original *Ideas*' from which all the rest are derived and mentions as those received from our minds by reflection

Thinking, and the
Power of Moving

In the fourth edition (1700) he changed this to

Perceptivity, or the Power of perception, or thinking; *Motivity*, or the Power of moving;

and added

I crave leave to make use of these two new Words, to avoid the danger of being mistaken in the use of these which are æquivocal. (II.xxi.73)

Between the third and fourth editions he had had his correspondence with Stillingfleet and, as I have shown, now stresses that thinking is an activity and so not suitable for characterizing mental substance-in-general. So he was looking for some other way of characterizing it than as 'unsolid'. If we read the new version of section 73 as dividing what Locke previously called 'thinking' into two parts, a passive part '*Perceptivity*, or the Power of perception' and an active part 'thinking' then we can see him as groping for a possible positive characterization. He may have been considering the possibility that perceptivity is to mental substance what solidity is to material substance. It might not be absurd to wonder whether Locke had got wind of some of Leibniz's ideas. Leibniz completed his *Discourse on Metaphysics* in 1686, although it was not published until later.

Independent support for the view that Locke held a two-substance theory may be found in the work of Anthony Collins (1676–1729) whom Locke regarded as an important disciple and as probably understanding the *Essay* better than anyone else.[15] However, Collins was not a slavish follower of Locke and he objected to Locke's view of substance-in-general on the grounds that it was a mistake to postulate an unknowable substance. This appears in the third of a series of attacks by Collins on the arguments of Samuel Clarke.[16] In the course of this Collins takes Locke to be asserting the existence of two substances, material and immaterial, and distinguishing them as solid and unsolid, respectively.

Collins says

Mr. *Locke*, to justify the Consideration of *Substance* as an unknown *Support*, in which Propertys inhere, says, *A Philosopher that says, Substance (or that which supports Accidents) is something he knows not what, and a Countryman that says the Foundation of the great Church at* Harlem *is supported by something he knows not what...; in this respect talk exactly alike [Letter to the Bp. of* W. p.16].[17] Now I humbly conceive, that they may not talk exactly alike in this respect for the Countryman ... may have a clear abstract Idea of solid Being or Matter, a Species whereof the Countryman may imagine the Church at *Harlem* ... supported by; whereas the Philosopher has no Idea at all: For let the Philosopher strip any part or piece of Matter of Solidity, and nothing conceivable remains, nothing on which Solidity can inhere... (pp.82–3)

It is not clear to me that Collins has quite understood Locke here for Locke continues the passage quoted by expressing some hope that the philosopher may in time achieve a clearer idea of this support for accidents. I have explained how I think he seeks to do this as he develops his arguments against Stillingfleet. However, in the next paragraph, Collins continues

But as far as I can judg, all this talk of the Essences of things being unknown, is a perfect mistake: and nothing seems clearer to me than that the Essence or Substance of Matter consists in Solidity, and that the Essence or Substance of a Being, distinct from Matter, must consist in want of Extension, and is truly defin'd an *unextended Being*... And therefore to make immaterial Being extended (as Mr. *Clark* does) is to make immaterial Being material. (p.83)

Collins is mainly attacking Clarke's arguments against thinking matter in the course of which Clarke holds that there are different kinds of substances and that 'we are ignorant of the Substance or Essence of all things' (p.77). Collins argues that if we claim that substances are unknowable then we cannot claim that there are two of them because there would be no means of distinguishing them. He seems then to go on to apply this criticism to Locke who, he says, claims that substances are unknowable.

However, I think it is unclear how much of this criticism is intended to apply to Locke, even though he refers to Locke explicitly. My main point is that he attributes to Locke the postulation of two substances. When he appears to suggest that Locke is mistaken in regarding material substance as unknowable and says that the 'Substance of Matter consists in Solidity' I think he is agreeing, without realizing it, with a view that Locke later holds, as I have argued. Locke appears to be criticizing Stillingfleet for not seeing that if the philosopher 'strip any part or piece of Matter of Solidity', then 'nothing conceivable remains' for solidity to inhere in, which is Collins' point at the end of my first quotation from him.

I do not wish to suggest that my interpretation leaves Locke with no serious problem. How *is* immaterial substance to be characterized and what is a characteristic which is not a quality and so does not have to inhere in anything? If solidity characterizes material substance and something else characterizes immaterial substance are these characteristics not, after all, being treated as qualities of a higher order which differentiate substances-in-general? What I do wish to claim is that this interpretation is an improvement on others in that it allows that something more may be said about substance-in-general than that it is an unknowable and undifferentiated somewhat that supports qualities. This makes Locke's view look a little less foolish than it is usually made to look.

12

LANGUAGE AND MEANING

I

It is often said that, for Locke, words stand for ideas and that ideas are private mental objects. It is an easy and obvious step to regarding him as a feeble and inconsistent precursor of Berkeley and his idealist phenomenalism. It suggests how sense can be made of Berkeley's alleged view that for clarity in philosophy we would do well to abandon words and stick to ideas.[1]

Hilary Putnam, in his essay 'Language and Philosophy'[2] dates the tradition attacked by Wittgenstein from Locke's 'way of ideas'. According to this tradition, Putnam says, 'whatever concepts and ideas were, they were clearly mental objects of some kind' and so 'a concern with introspective psychology was central' to it (p.7). This account 'suggests that finding out that someone has a concept is finding out that he has a particular mental presentation, and finding out that two people have the same concept is finding out that they have identical mental presentations' (p.8), where the mental presentation in question is usually an image. It is not clear whether Putnam is characterizing Locke's own view and that is probably not relevant to his purpose but Locke has perhaps usually been taken to have held a view correctly characterized in some such way.

He is clearly so taken by W. P. Alston in his *Philosophy of Language*,[3] an introductory book which is read by, and corrupts, many beginning students of philosophy. Under the heading 'The Ideational Theory', which is the theory 'that two expressions have the same use if and only if they are associated with the same idea(s)', Alston says

> The classic statement of the ideational theory was given by the 17th Century British philosopher, John Locke... 'The use, then, of words is to be sensible marks of ideas; and the ideas they stand for are their proper and immediate signification'... According to this theory, what gives a linguistic expression a certain meaning is the fact that it is regularly used in communication as the 'mark' of a certain idea; the ideas with which we do our thinking have an existence and a function that is independent of language. If each of us were content to keep all his thoughts to himself, language could have been dispensed with; it is only because we feel a need to convey our thoughts to each other that we have to make use of publicly observable indications of the purely private ideas that are coursing through our minds. (pp. 22–3)

For this theory to work, he continues, it would have to be true that 'for each distinguishable sense of a linguistic expression, there would have to be an idea such that when any expression is used in that sense, it is used as an indication of the presence of that "idea"'. So 'whenever an expression is used in that sense, 1. the idea must be present in the mind of the speaker, and 2. the speaker must be producing the expression in order to get his audience to realize that the idea in question is in his mind at that time. Finally 3. insofar as communication is successful, the expression would have to call up the same idea in the mind of the hearer...' (pp. 23–4).

Alston, without enquiring any more closely into what Locke meant by all this, subjects the view to well-known criticisms. It is just not true that we are aware of distinguishable ideas corresponding to words like 'when', 'in' and 'becomes' whenever we use them; the questions this raises, he says, are particularly disturbing because we do not even know how to set about answering them. Moreover, ideas cannot form a basis for the explanation of meaning unless the presence or absence of an idea is decidable independent of determining in what sense words are being used (pp. 24–5). Finally, the theory will not work even for words that have an obvious connection with mental imagery; different mental images, if they occur at all, may accompany my use of 'dog' in the same sense on different occasions.

I believe that Locke held a view that was more subtle and less obviously mistaken, and that separates him more clearly from Berkeley, than this interpretation suggests. The interpretation suffers by failing to note what Locke says about the relations between ideas, on the one hand, and qualities and things, on the other, by ignoring what he says about the learning of language and the conveying of ideas and by failing to consider what he may mean, or even says he means, by 'ideas' in various contexts. It appears to depend upon only *one* of the *eleven* chapters that Locke devotes directly to language and meaning.

II

Book III of Locke's *Essay* entirely concerns language and meaning. In the second of its eleven chapters Locke makes the statement that has usually been taken to epitomize his theory of meaning; much of the rest of this Book is often ignored and some abridged editions even omit some vital passages altogether.[4]

Locke begins the second chapter by saying that men, wishing to communicate their thoughts which were invisible and hidden from others in each of them, needed to find external sensible signs to stand for their ideas. They hit upon articulate sounds or *words* which could be made arbitrary signs of their ideas. They could use these to record their thoughts, to assist their memories and to 'lay them before the view of others'. So we come to the much-quoted and, I believe, much-misused, statement

> *Words, in their primary or immediate Signification, stand for nothing, but the* Ideas *in the Mind of him that uses them*... (III.ii.2)

This, taken by itself, gives rise to all the problems for which I referred to Putnam and Alston. It is true that Locke made things difficult for himself by putting this more generally than he intended and by using the expression 'nothing, but' misleadingly and he made it worse, apparently, by adding

> it is a perverting the use of Words, and brings unavoidable Obscurity and Confusion into their Signification, wherever we make them stand for any thing, but those *Ideas* we have in our own Minds. (iii.ii.5)

So, people have concluded, for Locke *all* words stand for ideas and for nothing else but the user's own ideas. The difficulties are insuperable if ideas are taken as private mental images, or private concepts or private sensations since this would ensure that men don't just happen to make their words stand for their own ideas and nothing else but that they logically could not make them stand for anything else. So when Locke later talks about words standing for qualities and substances he is taken to be contradicting his own theory which was no use anyway because it would make communication logically impossible. In consequence, Locke's view of language tends to be treated as an awful warning to students and nothing more.

I think that in this particular chapter Locke intended to do no more than to put, in the famous statement quoted, the rough basis of his theory to be developed later and to make the rather elementary point that in order to communicate a person must have something to communicate, namely an idea. How can one communicate an idea one does not have? Words are 'voluntary signs of *Ideas*' but a person cannot make a word stand for an idea he has never had; he may use words to stand for the ideas of others only if he has, or shares, those ideas. 'A Man cannot make his Words the Signs of Qualities in Things or of Conceptions in the Mind of another, whereof he has none in his own.' Given this, a man may suppose his ideas 'to correspond with the Conceptions of another Man' and 'consent to give them the same Names, that other Men do' (III.ii.2). It is necessary that 'all use the Words they speak (with any meaning) all alike' (III.ii.3).

Another puzzle concerns the apparent suggestion in this Chapter that *all* words stand for ideas. It has often been pointed out that this is implausible. Locke contradicts it, both earlier and later, as we shall see. One possible way out of this puzzle is that Locke begins Chapter ii by giving a genetic account of language. The general statement upon which so much weight has been put may have been intended to indicate the first insight of men searching for a way of communicating their thoughts: the first necessity is for words to stand for the ideas about which they want to talk and only later comes the realization that other sorts of words are necessary for indicating connections between those ideas.

Involved in all this there is a problem that has been much discussed in recent years. At III.ii.1 and 2 Locke appears to be saying that men first had thoughts but no language, then desired to record their own thoughts and to communicate them to others, so they created language. This seems to suggest two beliefs that are unacceptable to many, namely, that thinking can go on independently of language and that private languages are possible. However, I think it is not clear that Locke accepted either of these in any objectionable sense.

First, by language useful for communication he appears to mean articulate *sounds* or *marks on paper*. He says nothing to rule out the possibility that we think in an 'inner language' for which we have to invent sounds or marks if we wish to communicate and record our thoughts.

Second, since this private language is translatable into or representable in a public language it is not *necessarily* private.

At III.ix.2 in talking of recording our own thoughts he says

> *for the recording our own Thoughts* for the help of our own Memories, whereby, as it were, we talk to ourselves, any Words will serve the turn. For since Sounds are voluntary and indifferent signs of any *Ideas*, a Man may use what Words he pleases, to signify his own *Ideas*, to himself: and there will be no imperfection in them, if he constantly use the same sign for the same *Idea*: for then he cannot fail of having his meaning understood, wherein consists the right use and perfection of Language.

At IV.v. Locke distinguishes between *mental* and *verbal* propositions and signs (*Ideas* and Words). A mental proposition is

> nothing but a bare consideration of the *Ideas*, as they are in our Minds stripp'd of Names.

This idea of a 'language of thought' was fairly common at the time. It is to be found in the influential *Port Royal Logic* of which Locke thought highly and later in Berkeley. It does not, I think, conflict with Locke's arguments about innate ideas and fits in with what Leibniz[5] thought Locke really meant. We have been, most of us, brought up to think such an idea absurd but it is at least an idea that people nowadays think worth discussing.[6]

III

It is of the utmost importance that Locke says that it is '*in their primary or immediate* signification' that words stand for ideas and that he adds this qualification, often italicized, almost always when he talks of words standing for ideas. He also says 'Nor can anyone apply [words], as Marks, immediately to any thing else, but the *Ideas* that he himself hath' (III.ii.2). The qualification, and the contrast mediately/immediately, must be taken seriously, as Norman Kretzmann has shown in an important article.[7] Locke clearly allows that words may stand not only, immediately and primarily, for ideas but also, mediately and secondarily, for things and qualities. Later, in Chapter iv, he says

> The *Names of simple* Ideas *and Substances*, with the abstract *Ideas* in the Mind, which they immediately signify, *intimate* also *some real Existence*, from which was derived their original pattern. (III.iv.2)

The word 'intimate' is contrasted with 'immediately signify'. Words refer beyond, or through, ideas to real existences, that is, qualities or substances.

In Chapter ii he says that men 'in their Thoughts give [words] a secret reference to two other things' besides ideas in the mind of the speaker, namely, to ideas in the minds of other men and to the reality of things (III.ii.4 and 5). If this were not so, men believe, they would fail to communicate or would be talking only of their own imaginings. Some critics have taken the expression 'secret reference' to indicate irony; there certainly is irony hereabouts but that is not where it lies. Locke clearly thinks, as is amply confirmed by what comes later, that men are *right* in these beliefs; communication would be impossible or worthless were it not for these secret references and communication is neither of these. He never says that men are *mistaken* in these beliefs. Indeed, he says

> For though Men may make what complex *Ideas* they please, and give what Names to them they will; yet if they will be understood, when they speak of Things really existing, they must, in some degree, conform their *Ideas* to the Things they would speak of: Or else Men's Language will be like that of *Babel*; and every Man's Words, being intelligible only to himself, would no longer serve to Conversation, and the ordinary Affairs of Life, if the *Ideas* they stand for, be not some way answering the common appearances and agreement of Substances, as they really exist. (III.vi.28)

and, later,

if our Knowledge of our *Ideas* terminate in them, and reach no farther, where there is something farther intended, our most serious Thoughts will be of little more use, than the Reveries of a crazy Brain; and the Truths built thereon of no more weight, than the Discourses of a Man, who sees Things clearly in a Dream, and with great assurance utters them. But, I hope, before I have done, to make it evident, that this way of certainty, by the Knowledge of our own *Ideas*, goes a little farther than bare Imagination: and, I believe it will appear, that all the certainty of general Truths a Man has, lies in nothing else. (IV.iv.2)

Why does Locke talk of 'secret reference'? Because he thinks, I believe, that we are immediately aware of the correspondences between our words and our own ideas but not *immediately* aware of the correspondence between our words and the ideas of others or the reality of things. Some *work* is required to become aware of them. Moreover, some ideas are caused in us by things and qualities; although our ideas of things and qualities may not be, or be known by us to be, accurate we do at least know that *some* things or qualities, of which our ideas *may* be accurate, cause those ideas. 'Secret' may mean 'obscure and recondite' as well as 'not openly expressed'. There is some 'secrecy' or obscurity about just what precisely are the characters of the referents of our words beyond our ideas.

From the beginning to the end of Book III Locke relates ideas to things. In the introductory chapter, stressing the need for generality in language, he says

It is not enough for the perfection of Language, that Sounds can be made signs of *Ideas*, unless those *signs* can be so made use of, *as to comprehend several particular Things*. (III.i.3)

At III.ii.6 he says that words may, with constant use, come to excite ideas in people just as if the objects 'which are apt to produce' those ideas were present. At III.iii.6, as is well known, he says 'Words become general by being made the signs of general *Ideas*' and adds that ideas become general by abstraction. What is less often commented on is the next sentence, in which Locke says that this process makes ideas 'capable of representing more individuals than one'. At III.iii.11 he says that general ideas represent many particular things according to similarities we observe in them and that abstract ideas are 'as it were, the bonds between particular Things that exist and the Names they are to be ranked under'.

Two things become evident from these passages. First, words and ideas, whether particular or general, may represent things. Second, that any account of Locke's view that says merely that, for him, words stand for ideas tells only part of the story and, in so doing, makes his view look more ridiculous than it is.

Very well, then; Locke *wants* words to stand for things as well as ideas and for ideas in others' minds. Does his theory allow this? What can justify the belief in these secret references? How can I ever know that I use words for the same ideas as others without 'looking into their minds' and comparing the ideas I find with those I find by introspection? How can I ever know that words stand for things if all I ever have before my mind is ideas? If ideas were mental images the whole project would seem to be impossible. But Locke does not think it impossible although he sees the problems and spends much of the rest of the *Essay* trying to solve them. How does he do this?

Can Locke give any justification for thinking that words may stand for 'things as they really are' (III.ii.5)? It is necessary to go to Book IV to see that he was aware of the problems and how he attempted to solve them. A prior problem to that concerning words is, of course, how we can know even that our *ideas* stand for the reality of things. As he says

'Tis evident, the Mind knows not Things immediately, but only by the intervention of the *Ideas* it has of them. *Our knowledge* therefore is *real*, only so far as there is a conformity between our *Ideas* and the reality of Things. (IV.iv.3)

He already hints at the problems by using the word 'intervention'. He immediately goes on

But what shall be here the Criterion? How shall the Mind, when it perceives nothing but its own *Ideas*, know that they agree with Things themselves? This, though it seems not to want difficulty, yet, I think there be two sorts of *Ideas*, that, we may be assured, agree with Things. (IV.iv.3)

Locke now mentions *three* sorts of ideas: simple ideas, complex ideas other than those of substances and complex ideas of substances. Of these we may be assured that simple ideas and complex ideas other than those of substances agree with the reality of things. We may think that complex ideas of substances agree with the reality of things but be mistaken, unless we are very careful.

It must be remembered that underlying the whole of the *Essay* is the initially unargued hypothesis that there are external objects which act upon us. This is part of the grand hypothesis which includes also the corpuscular hypothesis and which is to be supported by the comprehensiveness and efficacy of the explanations it allows as shown by the *Essay* as a whole.

The justification for thinking that simple ideas agree with the reality of things is that the mind is passive in respect of them and they are just presented to it. So they 'must necessarily be the product of Things operating on the Mind in a natural way'. They

represent to us Things under those Appearances which they are fitted to produce in us: whereby we are enabled to distinguish the sorts of particular Substances, to discern the states they are in, and so to take them for our Necessities, and apply them to our Uses. Thus the *Idea* of Whiteness, or Bitterness, as it is in the Mind, exactly answering that Power which is in any Body to produce it there, has all the real conformity it can, or ought to have, with Things without us. And this conformity between our simple *Ideas*, and the existence of Things, is sufficient for real Knowledge. (IV.iv.4)

Different qualities in bodies 'naturally' produce different ideas in us; if one object appears blue to me that is just because it has such an inner constitution as is able to produce that appearance, or idea, in me and if another object, under the same conditions, appears red that is because it has a different inner constitution fitted to produce a different appearance. The difference in the appearances conforms to a difference in the objects appearing.

Complex ideas other than those of substances, that is ideas of modes and relations (see II.xii.3), are '*archetypes* of the Mind's own making, not intended to be the Copies of anything, nor referred to the existence of any thing, as to their Originals' and therefore 'cannot want any conformity necessary to real knowledge' (IV.iv.5). Such ideas cannot in themselves fail to agree with what they purport to represent because they do not purport to represent anything actual when they are made. We freely construct these complexes without reference to any corresponding complex in the world (or our experience). We apply them to anything in the world that *we find* to conform to them. Thus in advance of any discovery of Euclidean triangles in nature I form the idea of 'triangle' which is, say, 'plane figure bounded by three straight lines'. I then apply the word only to anything I find in the world that conforms exactly to that description. Similarly with 'justice' or 'adultery'. 'So that' Locke says, 'we cannot but be infallibly certain, that all the Knowledge we attain concerning these *Ideas* is real, and reaches Things themselves' (IV.iv.5).

Complex ideas of substances are referred to '*Archetypes* without us'. These ideas may differ from their archetypes and so our knowledge about them 'may come short of being real'. They are intended to be made up of simple ideas 'found united in things themselves' but if we are not careful we may include ideas that do not always occur together in the substance in question or exclude ideas that do. So they may fail to conform to things (IV.iv.11).

I say then, that to have *Ideas* of *Substance*, which, by being conformable to Things, may afford us *real* Knowledge, it is not enough, as in Modes, to put together such *Ideas* as have no inconsistence, though they did never before so exist ... *our Ideas of Substances* being supposed Copies, and referred to *Archetypes* without us, must still be taken from something that does or has existed ... Herein therefore is founded the *reality* of our Knowledge concerning *Substances*, that all our complex *Ideas* of them must be such, and such only, as are made up of such simple ones, as have been discovered to co-exist in Nature. (IV.iv.12)

Thus complex ideas of substances may be conformable to things if we are careful about the simple ideas we include in them. We can get agreement about simple ideas because the mind is passive in respect of them and, as I argued in Chapter 4, we can 'show' them to one another. The mind is passive in respect also of complex ideas of substances at least to the extent that we can't control which simple ideas they produce together in us; we can check this against the observations of others by 'showing' them the simple ideas occurring together. If I find brittleness in gold and no one else does then there is something wrong with my observation. I can also explain to others, in words, which simple ideas I am including in the complex idea of gold or, for that matter, of jealousy.

The central point that Locke is making in distinguishing between complex ideas of substance and other complex ideas is, I think, this. If by 'jealous' I mean 'exhibiting features *a*, *b* and *c*' then if I find someone clearly exhibiting these features I cannot be mistaken in applying 'jealous' to him. If someone disagrees while admitting the presence of features *a*, *b* and *c*, then we mean different things by 'jealous'. There is not something in the world which is indubitably jealousy of which we are seeking to discover the characteristics. It is a matter for decision which combination of characteristics it is convenient to call 'jealousy'; we have long ago reached agreement about 'Euclidean triangle' but we did not reach it by the close examination of things in the world. Gold is different: if I find brittleness in some samples of gold and then someone produces something that he and I and everybody else call 'gold' but is not brittle then if I continue to call it 'gold' I am mistaken and had better revise my list of simple ideas making up the complex idea of gold. On the other hand, if I define 'gold' as 'yellow', malleable, soluble in *aqua regia*' and if all samples of gold have also ductility, I may call 'gold' something that is not gold and I had better add 'ductile' to my list.

In sum, if I take 'jealous' to mean 'exhibiting features *a*, *b*, *c*' and someone indubitably exhibits those features, I cannot be mistaken in calling that person jealous; but if I take 'gold' to mean 'a body exhibiting features *x*, *y*, *z*' and a metal indubitably exhibits those features I may be wrong in calling it gold. There are things in the world already called 'gold', 'silver', etc. and the meaning of 'gold' is already determined by the general agreement about that; there is not something in the world already called 'jealousy' about which there is general agreement. I am free to define 'jealousy' as I wish and then attempt to get general agreement. I am not free to define 'gold' as I wish because I must preserve distinctions between different metals already agreed upon. Of course, if I discover a new metal I may invent what name I like for it but I am not free to adopt any definition that fails to distinguish it from already known, and named, metals.

At the very beginning of Book III Locke distinguishes names for 'common sensible *Ideas*' from names for 'Things that fall not under our Senses' (III.ii.5). Although these are different, the latter are derived somehow from sensible ideas. Chief among the 'Things that fall not under our Senses' are the operations of our own minds. The words '*Imagine, Apprehend, Comprehend, Adhere, Conceive, Instill, Disgust, Disturbance, Tranquillity*' are, he says, perhaps somewhat obscurely, 'Words taken from the Operations of sensible Things, and applied to certain Modes of Thinking'. Men, he says,

> to give Names, that might make known to others any Operations they felt in themselves, or any other *Ideas*, that came not under their Senses, ... were fain to borrow Words from ordinary known *Ideas* of sensation, by that means to make others the more easily to conceive those Operations they experimented [*sc.* experienced] in themselves, which made no outward sensible appearances... (III.i.5)

I am not concerned here with the plausibility of the details of this passage but with Locke's reasons for insisting on the point. Why is it so important for him?

Note that the passage suggests at once that Alston's 'ideational theory' is not Locke's. If it were, Locke should see no distinction in this respect between particular sensible ideas and others. There would be no need for, or possibility of, deriving words for the latter from words for the former; all ideas would have the same status as being private and yet providing the meaning of words.

I believe that the reasons why Locke insists on the distinction and the derivation are, first, that particular sensible ideas depend upon patterns in external things and thus give some hope of allowing public criteria for meanings and, second, that this is further facilitated by my being able to point to, in some sense, the appearances of things, which are particular sensible ideas. That is why Locke says, in the passage just quoted, that operations of the mind 'made no outward sensible appearances' in contrast to the patterns of external things. Words for things that make no outward sensible appearances are to be derived from words for things that do make such appearances.

The first clue to the correct interpretation of the view occurs, I think, at III.iii.10 in the account of general terms. There he says, casually and without much discussion because, no doubt, he will return to the point in detail,

> For Definition, being nothing but making another understand by Words, what *Idea*, the term defined stands for, a definition is best made by enumerating those simple *Ideas* that are combined in the signification of the term Defined...

and, shortly after, he says that definition is merely the explaining of one word by several others, 'so that the meaning, or *Idea* it stands for', may be made known *and*, he says, *certainly* known.

Later, he puts this more fully.

> *a Definition is* nothing else, but *the showing the meaning of one Word by several other not synonymous Terms*. The meaning of Words, being only the *Ideas* they are made to stand for by him that uses them; the meaning of any Term is then shewed, or the Word is defined when by other Words, the *Idea* it is made the Sign of, and annexed to in the Mind of the Speaker, is as it were represented, or set before the view of another, and thus its Signification ascertained. (III.iv.6)

Note, that the meaning of the word, that is, the idea it stands for, is 'shewed', or 'set before the view' of the person being given the definition and that it is made 'certainly known'.

Of course, not all words can be defined and Locke brings this out in contrasting his classes of names. He makes six contrasts, of which I immediately mention the first three. First, names of simple ideas and substances and the ideas in the mind signified by them 'intimate some real Existence, from which was derived their original pattern' whereas names of mixed modes do not (III.iv.2). Second, names of simple ideas and modes always signify both the real and the nominal essences of their species while names of substances rarely signify anything more than their nominal essences (III.ix.3). Third, names of simple ideas cannot be defined whereas names of complex ideas can (III.iv.4).

For the moment I concentrate on the third item. Locke says that he is not concerned with the general point that to say that all words are definable is to embrace an infinite regress; he has in mind the more specific point that simple ideas are *simple* and have 'no composition'. Definition is the analysis of words for ideas into words for the simple ideas that make them up and simple ideas have no constituent ideas. The ignoring of this he says has led, especially in the schools, to the putting forward of circular, empty and incomprehensible 'definitions'.

Words for simple ideas, then, are fundamental and indefinable and the defining of words for complex ideas depends upon them. So how can I ever know that the idea for which I take 'red' to stand is the same as the ideas in the minds of others when they use the word? Simple ideas, Locke says, '*are only* to be *got by* those *impressions* Objects themselves make on our Minds, by the proper Inlets appointed to each sort' (III.iv.11). No amount of explaining in words will give a person who has never tasted pineapple the idea of the taste of pineapple for which those words stand.[8] No definition of 'lightness' or 'redness' is better able to produce in us the relevant ideas than the words 'light' or 'red' themselves. The only way of making a person understand a name whose represented simple idea he has never had is 'by applying to his senses the proper Object', his senses being in proper working order.

The names of simple ideas are generally less doubtful and uncertain in their signification than those of mixed modes and substances because each stands only for one simple perception. So 'Men, for the most part, easily and perfectly agree in their signification' and there is little room for mistake (III.iv.15). Simple ideas are of particulars such as red, yellow or blue; the idea of *colour* under which they fall, is just the complex idea of simple ideas that are 'produced in the Mind only by the sight and have entrance only through the Eyes' (IV.iii.16).

It is worth noting that Locke also says that we cannot be given a simple idea by being given an account of its cause, even given a true causal account of our seeing light, in terms, say, of the corpuscular structure of a reflecting object and the sending by it of corpuscles to impinge on our eyes, and so on. He says

> yet the *Idea* of the cause of *Light*, if we had it never so exact, would no more give us the *Idea* of *Light* it self, as it is such a particular perception in us, than the *Idea* of the Figure and Motion of a sharp piece of Steel, would give us the *Idea* of that Pain, which it is able to cause in us. (III.iv.10)

In order to convey the meaning of, or the idea signified by, a word for a simple idea, then, I have to rely on ostension. Here the relevance of what I said earlier about ideas becomes obvious. If I am right, ostension is valuable because I can point to the red appearance of the book, the idea of its colour, at least to the extent of indicating its whereabouts. I can point at what I see, where I see it, and say 'I use "red" of *that*'. I can be sure, at least, that you look in the right place and I can hope that you get the same idea. Of course, there are ways, also through ostension, of ensuring that you don't think I am indicating shape rather than colour. But words can also help. Locke says

> *For the explaining* the signification of *the Names of Substances* as they stand for the *Ideas* we have of their distinct Species, both the fore-mentioned ways, *viz.* of *shewing and defining, are requisite*, in many cases, to be made use of. (III.xi.19)

But how can I ever be sure? How can I tell whether or not you have acquired the same idea as I have? Again, I believe, Locke has an answer about how I may get further evidence, first indicated in the passage about 'secret reference'. The paragraph in which he says that men suppose their words to be '*Marks of the* Ideas *in the Minds also of other Men*' ends thus

> But in this, Men stand not usually to examine, whether the *Idea* they, and those they discourse with have in their Minds, be the same: But think it enough, that they use the Word, as they imagine, in the common Acceptation of that Language; in which case they suppose, that the *Idea*, they make it a Sign of, is precisely the same, to which the Understanding Men of that Country apply that Name. (III.ii.4)

It is, as I have suggested, not in the words 'secret reference' that Locke's irony lies but in the word 'usually'. Locke knows that men *never* look into the minds of others to discover what ideas they have because that would be impossible. Note that Locke says that men think it *enough* to rely on the common acceptation; they take it that to use a word in exactly the same way as others do *is* to make it stand for the same idea. Since using words to stand for the same ideas as others is a necessity of communication and since communication is possible we must be able to make words stand for the same ideas as others do and to know when we are doing so. We could not do this if it involved looking into the minds of others; we must do it by learning the common acceptation of words.

Common acceptation presumably *includes* ostension, saying the same word when pointing at the same thing, but this alone may still leave us in some doubt. It must also include an awareness of the sentential contexts in which the word is generally used and those in which it is not. When we have conveyed to a person the whole of the common acceptation of the word so that he uses it exactly as most people do in all contexts including ostensive ones then he has come to know, with certainty, that he uses it to stand for the same ideas as most people or, at least, as the understanding men of his linguistic community. What further tests could there be? It does not then matter if there is a difference in qualitative content between the idea one has and the idea another has when they look at the same thing; such a difference is undetectable by any means so this does not count, in the present context, as their having different ideas.

If Locke did not think this we should have expected him to comment on the passage last quoted that men were *wrong* in supposing that their words are marks of ideas also in the minds of others and in *thinking it enough* that they use words in their common acceptation. He nowhere does this. He is clearly giving part of what he takes to be a correct account of language and meaning.

This brings me to the vexed question of parrots. The first necessity of language, Locke says, is the ability '*to frame articulate Sounds*, which we call Words; but this is not sufficient for language since parrots have this ability but they do not have language' (III.i.1). What is further necessary is that the speaker '*be able to use these Sounds, as Signs of internal Conceptions*; and to make them stand as Marks for the *Ideas* within his own Mind' (III.i.2).

The question is, of course, how we can ever know that the parrot that talks with apparent rationality, like that instanced by Locke at II.xxvii.8,[9] is *not* using words as signs of internal conceptions, marks for the ideas within its own mind, and Locke has often been criticized for neglecting this problem. However, in at least two places he indicates what his answer would be. He says

Therefore some, not only Children, but Men, speak several Words, no otherwise than Parrots do, only because they have learn'd them, and have been accustomed to those Sounds. But so far as Words are of Use and Signification, so far is there a constant connexion between the Sound and the *Idea*; and a Designation, that the one stand for the other: without which application of them, they are nothing but so much insignificant Noise. (III.ii.7)

and later

Before a Man makes any Proposition, he is supposed to understand the terms he uses in it, or else he talks like a Parrot, only making a noise by imitation, and framing certain Sounds, which he has learnt of others; but not, as a rational Creature, using them for signs of *Ideas*, which he has in his Mind. (IV.viii.7)

But still, how do we tell that a parrot is just repeating what he has learnt, making a noise by imitation? If a parrot who claims to have looked after the chickens (II.xxvii.8) cannot say what chickens are, or points sometimes at a giraffe and sometimes at a fish while saying 'That's a chicken', or cannot give sensible answers to questions such as 'Is looking after chickens like looking after rabbits or like looking after computers or like looking after a man going down the road,...?' or cannot say, using even the words he can use, anything that his teacher knows he has not been taught, and so on, then he does not use them for internal conceptions or to stand for ideas he has in his own mind; at least we can gradually acquire a suspicion approaching certainty that he does not. That Locke thinks it makes sense to talk of parrots as he does supports the view that he attaches considerable importance to common acceptation.

If we now consider again what Locke says about definition we can see that there are parallel theses for simple and complex ideas. The meaning of a word for a simple idea may be certainly conveyed by teaching the common acceptation of that word; the meaning of a word for a complex idea may be certainly conveyed by giving a definition, which need not be part of the common acceptation.

Among words for complex ideas Locke considers the word 'triangle'. He says that the name 'triangle' is the name of the abstract idea 'Figure including a Space between Three lines' (III.iii.18). The idea a word stands for, he says, is '*shewn, set before the view*,' by its definition (III.iv.6); the idea for which 'triangle' stands *is* 'Figure including a space between three Lines' and not a picture, a mental image or anything essentially private.[10] That is why the definition can make the idea certainly known to anyone who knows the meanings of the words used in the definition. The parallel feature of words for simple ideas is that what is pointed to in ostension *is* the idea for which the word stands. Thus we have criteria with some sort of publicity for words for both simple and complex ideas.

The general view I have been putting is supported also by what Locke says about words for 'mixed modes', such as '*Murther* or *Sacrilege*'. Ideas of mixed modes are complex ideas consisting of simple ideas of different kinds, which combinations, unlike ideas of substances, do not refer to real Beings that have a steady existence' (II.xxii.1). They are mainly associated with human actions and relate closely to morality. Among the examples Locke gives are '*Obligation*, *Drunkenness*, a *Lye*'. Words for mixed modes are those about which it is most difficult to discover whether people are using them in the same way. This is because the ideas they stand for are their own archetypes and refer to no external standards; they are purely human constructions made for our own convenience. One of the main reasons for the difficulty of establishing sameness of meaning of say, 'murder' or 'sacrilege' used by different speakers is that

> There be many of the parts of those complex *Ideas*, which are not visible in the Action it self, the intention of the Mind, or the Relation of holy Things, which make a part of *Murther*, or *Sacrilege*, have no necessary connexion with the outward and visible Action of him that commits either: and the pulling the Trigger of the Gun, with which the Murther is committed, and is all the Action, that, perhaps, is visible, has no natural connexion with those other *Ideas*, that make up the complex one, named *Murther*. (III.ix.7)

Ideas of mixed modes are, thus, distinguished from both ideas of substances and simple ideas. Note the view that some parts of these complex ideas are 'visible in the Action' e.g. the pulling of the trigger. One idea which is part of the complex idea of (one sort of) murder is the idea of pulling the trigger which represents the action of pulling the trigger; the idea is the appearance of the trigger being pulled which is presented when we see the action. This, however, is insufficient for the idea of murder; it requires also an intention, which need not 'appear' in the action. I may not be able to just *see* that it was murder rather than accidental killing. When I *think* about the murder later I am operating with the concept or thought or memory of *inter alia* the pulling of the trigger, which *is* private, in contrast to the original appearance.

In talking of difficulties attached to settling the meaning of words for mixed modes Locke says, in his chapter 'Of the Imperfections of Words',

> 'Tis true, *common Use*, that is the Rule of Propriety, may be supposed here to afford some aid, to settle the signification of Language; and it cannot be denied, but that in some measure it does. Common Use *regulates the meaning of Words* pretty well for common Conversation; but no body having an Authority to establish the precise signification of Words, nor determine to what *Ideas* any one shall annex them, common Use is not sufficient to adjust them to philosophical Discourses; there being scarce any Name, of any very complex *Idea*, (to say nothing of others,) which, in common Use, has not a great latitude, and which keeping within the bounds of Propriety, may not be made the sign of far different *Ideas* ... it is often matter of dispute, whether this or that way of using a Word, be propriety of Speech, or no. From all which, it is evident, that the Names of such kind of very complex *Ideas*, are naturally liable to this imperfection ... (III.ix.8)

The instances he gives of 'such kind' of ideas are '*Glory*' and '*Gratitude*'.

This passage suggests that, as I have said, common use is of great importance for Locke in establishing meaning and even for words of mixed modes it has a place. For such words, however, it is insufficient for reasons already mentioned, namely, that some of the simple ideas making up the complex ideas are not 'visible in the Actions'. Because of this a use may not have been commonly established. Common use is adequate for everyday conversation until we get into disputes that depend on different uses, which can be settled only by closer attention to the meanings of these words such as we engage in in philosophy. But this introduces no difference in principle.

Moreover, philosophers use technical terms and cannot expect to find common uses for these already established; it is important that they establish among themselves common uses for these terms (III.xi.10). It is sometimes desirable also that philosophers tidy up the common uses of everyday terms when these are not firmly enough established (III.xi.12).

At III.xi.25 Locke says

> Nor is it a shame for a Man not to have a certain Knowledge of any thing, but by the necessary ways of attaining it; and so it is no discredit not to know, what precise *Idea* any Sound stands for in another Man's Mind, without he declare it to me, by some other way than barely using that Sound, there being no other way, without such a Declaration, certainly to know it.

In other words, I cannot know what ideas a word stands for in another man's mind unless he tells me but as Locke frequently stressed, he *can* tell me.

Anthony Collins, to whom I referred in the last chapter as much respected by Locke for his understanding of the *Essay*, is regarded by Berman[11] as being a champion of Locke's theory of meaning. Collins takes words to stand for ideas that are known and then continues[12]

> The words must stand for those Ideas which the use of Language has appropriated them to, or for what the Author or Relator says he understands by them, or for what the visible design of the Author or *Relator* determins them to signify. These are all the ways of knowing the meaning of Words: and if we cannot find out the meaning of a Man's words by comparing him with himself, by the use of the Language in which he writes, or by his particular Definitions, it is all one to the Reader or Hearer, as if the words stood for Ideas he knew nothing of...

Note that what the author *says* and his *visible* design are what count. This clearly supports what I have said about the importance of common acceptation, regular use and definition in Locke's account of language and there is no suggestion that either Locke or Collins thought that we could, or needed to, discover a man's ideas independently of his words.

IV

The account I have given so far bears mainly on Alston's second and third criticisms of Locke; his first criticism gives me a little more trouble but not in the way that Alston intends. The criticism is that we are not aware of distinguishable ideas in our minds corresponding to each of the words in a sentence that we understand. He mentions particularly such words as 'when', 'in', 'course' and 'becomes' in the sentence beginning 'When in the course of human events, it becomes necessary for one people to...' I think my interpretation avoids this criticism, at least in the spirit in which it was meant, but it does point to another problem.

At the very beginning of his discussion Locke says

> Besides these Names which stand for *Ideas*, there be other words which Men make use of, not to signify any *Idea*, but the want or absence of some *Ideas* simple or complex, or all *Ideas* together; such as are *Nihil* in Latin, and in English, *Ignorance* and *Barrenness*. (III.i.4)

Again, in his much-neglected chapter 'Of Particles' he says that a great many words are used 'to signify the connexion that the Mind gives to *Ideas*, or *Propositions, one with another*' (III.vii.1) and that these words, called 'particles', are *not* names of ideas. Yet in between, especially in Chapter ii, he has repeatedly said, apparently meaning it quite generally, that words stand for ideas.[13] So he appears to contradict himself.

Since, in Chapter ii where he first says this, he appears to be giving his general account of the signification of words it is not surprising that the usual criticisms of his theory are directed against this. However, attention to the chapter on particles leads one to see that he held a more satisfactory theory and a slight re-wording of his general statements would have removed the contradiction without having unwanted consequences for him or conflicting with other things he said.

I am inclined to think that the worse crime of which he can be convicted in this connection is carelessness of expression. He sees the difficulties attached to the view that *all* words stand for ideas. First, if ideas are, in some respects private to each of us it is difficult to explain how communication can occur; this problem he tries to meet by means of common acceptation and by rejecting a purely referential theory. Second, some words appear not to stand for determinate ideas; this problem he tries to meet by giving a different account of such words. These attempts appear to me to be reasonably successful; they are at least discussable and not to be rejected out of hand as ridiculous. What is unsatisfactory is that he failed to bring all these things together and to modify his general statements in Chapter ii in the light of them. Even so, read in their context, these general statements should not have misled commentators as much as they have.

Locke begins the chapter on particles by making a distinction, thus

> The Mind, in communicating its thoughts to others, does not only need Signs of the *Ideas* it has then before it, but others also, to shew or intimate some particular action of its own, at that time, relating to those *Ideas*. This it does in several ways; as, *Is*, and *Is not*, are the general marks of the Mind, affirming or denying. But besides affirmation, or negation, without which, there is in Words no Truth or Falsehood, the Mind does, in declaring its Sentiments to others, connect, not only the parts of Propositions, but whole Sentences one to another, with their several Relations and Dependencies, to make a coherent Discourse. (III.vii.1)

He mentions words showing '*Connexion*, *Restriction*, *Distinction*, *Opposition*, *Emphasis*'; words indicating '*Cases* and *Genders*, *Moods* and *Tenses*, *Gerunds* and *Supines*'; and '*Prepositions* and *Conjunctions*'. These cover many words of the kinds mentioned by Alston.

Of these words Locke says that they 'are not truly, by themselves, the names of any *Ideas*' (III.vii.2). What he appears to have in mind here no doubt relates to what he has to say about definition, namely, that these words cannot be defined independently of context, or, in his terms, said to stand for particular ideas. We might think that he could have said that these words stand for general ideas e.g., of connection, restriction, and so on, but he brings out problems about even this.

Suppose we say that 'but' stands for the idea of disjunction or, as Locke has it, of 'discretive Conjunction'. Even this, he thinks, is doubtful because the word seems to 'intimate several relations, the Mind gives to the several Propositions or Parts of them, which it joins by this monosyllable' (III.vii.5). The examples he gives include '*BUT to say no more*' where it 'intimates a stop of the Mind', '*I saw BUT two Planets*' where it shows 'that the Mind limits the sense to what is expressed, with a negation of all other' and '*All Animals have sense*; *BUT a Dog is an Animal*' where it shows simply 'that the latter Proposition is joined to the former, as the *Minor* of a Syllogism' (III.vii.5). There is no one general idea for which 'but' stands such that all its occurrences stand for instances of that general idea. Its various meanings can be explained only with reference to these and other contexts.

Locke concludes the chapter by saying

> I intend not here a full explanation of this sort of Signs. The instances I have given in this one, may give occasion to reflect upon their use and force in Language, and lead us into the contemplation of several Actions of our Minds in discoursing, which it has found a way to intimate to others by these Particles, some whereof constantly, and others in certain constructions, have the sense of a whole Sentence contain'd in them. (III.vii.6)

One of the mistakes that underlie many criticisms of Locke relates to confusion about his use of 'idea'; I think he was less confused about this, and his critics more confused, than they suppose. That he used the word in various senses does not alone imply that he confused them; we can usually, by close attention to various passages, and to what we know from experience, see which sense he is using it in. A favourite device of critics is to seize on the fact that he has a technical sense, as when he talks of e.g., seeing as having ideas, to regard ideas in this sense as mental images, to stress the implausibility of this and then to attack, in the light of this, any passage in which ideas figure importantly. However, having allowed that Locke used the word in various senses, it would be only charitable to try interpreting occurrences of it in these different ways to see whether good sense can be made of them.

If one reads carefully the whole of Book III, including the chapters that deal with mistakes arising out of the imperfection and misuse of language, one can find a case for supposing that in relating words and ideas Locke is sometimes using 'ideas' in a familiar informal, non-technical sense. He says

> Men would often see what a small pittance of Reason and Truth, or possibly none at all, is mixed with those huffing Opinions they are swell'd with; if they would but look beyond fashionable Sounds, and observe what *Ideas* are, or are not comprehended under those Words, with which they are so armed at all points, and with which they so confidently lay about them. (III.v.16)

Do we not frequently say of a politician or a trades unionist or a poet or even a philosopher such things as 'He has great fluency but not an idea in his head' or 'He is not good at expressing his ideas' and that the ideas behind such expressions as 'at this moment in time', 'hopefully we will reduce inflation' or 'at the end of the day' are unclear, confused or non-existent and wish that those who use them would think before they speak and consider what, if any, ideas lie behind them? Do we not often ask such questions as 'What is your idea of freedom?' and expect an answer in words or 'What is your idea of beige?' and expect to be shown a colour? And do we suppose that those questions could only really be answered by looking into other people's minds? I don't think that Locke was any more foolish in this respect than we are. The most mysterious aspect of all this is why, given the interpretations people put on his work, he was ever regarded as the philosopher of common-sense.

V

An important and controversial part of Locke's discussion of language is his account of general terms and abstract general ideas. This is yet another topic on which I think he was seriously misinterpreted by Berkeley and has been misinterpreted ever since. I conclude this chapter by trying to set the record straight.

A language must have general terms and, indeed, most of the words in any language are general. Locke gives three reasons for this. First, the human mind just could not cope with a language in which every particular thing had a 'distinct peculiar Name' (III.iii.2). Second, even if this were possible language would be useless for communicating our thoughts; if words stand for ideas in my mind, nobody would be able to understand my words unless he had in his mind just those particular ideas that I have, that is, unless he had experienced just those particular things that I have experienced and use those words to stand for (III.iii.3). Third, even if this were somehow possible, having a peculiar name for every particular thing would not be of much use in improving our knowledge since the most important kind of knowledge is general knowledge (III.iii.4).

Thus Locke thinks it necessary to consider '*how general Words come to be made*'. Since all existing things are particulars, although it is easy to see how a name may be attached to an existing thing as a label, it is not immediately obvious how general terms come about or what they stand for. Locke's answer is that we make them by abstraction: we can abstract from a number of different particular things those features they have in common and give a name to the collection of them. Such a name is not a name of a *thing*. He says

Words become general, by being made the signs of general *Ideas*: and *Ideas* become general, by separating from them the circumstances of Time, and Place, and any other *Ideas*, that may determine them to this or that particular Existence. By this way of abstraction they are made capable of representing more Individuals than one; each of which, having in it a conformity to that abstract *Idea*, is (as we call it) of that sort. (III.iii.6)

For example, a child comes to recognize its mother and father and uses those words to stand for particulars, its own mother and father. Then it notices that there are other things in the world that resemble in shape, and certain other qualities, its mother and father, the idea of anything having those qualities in which they resemble is formed and given a general name 'man'.

And *thus they come to have a general Name*, and a general *idea*. Wherein they make nothing new, but only leave out of the complex *Idea* they had of *Peter* and *James*, *Mary* and *Jane*, that which is peculiar to each, and retain only what is common to them all. (III.iii.7)

A further step of abstraction leads to the idea of anything with 'a Body, with Life, Sense and spontaneous Motion', but with no specified shape, and the name 'animal' to stand for it. And so on, through ideas for which the words 'body', 'substance', 'being' and 'thing' stand. Thus we arrive at the ideas of *genera* and *species*, each term standing for a species of what the name next above it stands for, which is its genus (III.iii.9).

It is important that Locke immediately proceeds to talk, in the next section, of *definitions*, in which '*we make use of the Genus*, or next general Word that comprehends it'. He says

Definition being nothing but making another understand by Words, what *Idea*, the term defined stands for, a definition is best made by enumerating those simple *Ideas* that are combined in the signification of the term Defined... (III.iii.10)

However, instead of listing those simple ideas we may use the next more general term, the name of the genus. If I am defining 'man' I can define it as 'rational animal' instead of listing all the simple ideas combined in the meaning of 'animal'.

I stress this not for what Locke says about genus and species but for the fact that he talks of definition at this point. Its importance relates to a way in which I believe he was misunderstood by Berkeley. I shall return to this in a moment.

Locke sums up the discussion so far by saying that 'General' and 'Universal' belong 'not to the real existence of things'. They are '*Inventions and Creatures of the Understanding*' and they '*concern only Signs*, whether Words, or *Ideas*'. They belong only to signs because each thing is particular; there are no general or universal *things*.

Words are general, as has been said, when used, for Signs of general *Ideas*; and so are applicable indifferently to many particular Things; and *Ideas* are general, when they are set up, as the Representatives of many particular Things: but universality belongs not to things themselves, which are all of them particular in their Existence, even those Words, and *Ideas*, which in their signification, are general. (III.iii.11)

Later in the *Essay*, Locke considers the general idea of a triangle; it would be as well to discuss it now because it is relevant to the interpretation of what he says here. In a famous passage he says of 'the general *Idea* of a *Triangle*' that

it must be neither Oblique, nor Rectangle, neither Equilateral, Equicrural, nor Scalenon; but all and none of these at once. In effect, it is something imperfect, that cannot exist; an *Idea* wherein some parts of several different and inconsistent *Ideas* are put together. (IV.vii.9)

Berkeley, in objecting to Locke's view, not that there are general ideas but that there are *abstract* general ideas, quotes this passage at length.[14] He then gives an account of the way in which general conclusions are reached in geometrical proofs even though particular diagrams are used. He next comments on the passage from Locke, as follows

it must be acknowledged that a man may consider a figure merely as triangular, without attending to the particular qualities of the angles, or relations of the sides. So far he may abstract: but this will never prove, that he can frame an abstract general inconsistent idea of a triangle. In like manner we may consider Peter so far forth as man, or so far forth as animal, without framing the forementioned abstract idea, either of man or of animal, in as much as all that is perceived is not considered.[15]

The first point I have to make is that this contains at least one flagrantly obvious misinterpretation of Locke's words. I refer to the expression 'abstract general inconsistent idea of a triangle'. Locke, in the passage referred to does not call the abstract idea 'inconsistent' nor, as far as I know, does he do so anywhere else. He refers to the abstract ideas as one 'wherein some parts of several different and inconsistent *Ideas* are put together' (IV.vii.9). What Locke surely means is that the idea of a right-angled triangle, for example, is inconsistent with the idea of an equilateral triangle, which ideas are in turn inconsistent with the idea of a scalene triangle. No triangle can be both right-angled and equilateral or both isosceles and scalene. The general idea of a triangle is abstract because it is the idea of what all triangles have in common, namely, something that can be expressed by 'plane figure bounded by three straight lines' or, as Locke has it, 'Figure including a space between three Lines' (III.iii.18). That is, the idea is given by the definition of 'triangle'; indeed, since Locke says that the idea a word stands for is *shewn, set before the view*, by its definition I think we may say that the abstract general idea of 'triangle' *is* 'Figure including a space between three Lines'. Idea and definition are identical.

Now there is no inconsistency in these words so when Berkeley talks of framing 'an abstract general *inconsistent* idea of a triangle' he is clearly misinterpreting Locke. Parts of 'several different and inconsistent ideas' need not themselves be inconsistent and Locke never thought he was talking about general ideas as inconsistent.

When Locke says that 'it is something imperfect that cannot exist' what he must mean is 'such a triangle' i.e. there cannot be an actual triangle with *only* the properties mentioned in the definition; nor can there be a picture of one. He does not mean that we cannot have a general idea of a triangle *i.e.* that the idea cannot exist. He is perhaps warning us about the ambiguity of the expression 'general idea of a triangle'; 'a triangle' may stand for an individual existent figure or a concept under which such figures fall. When we define 'a triangle' we are talking about the concept.

It is true that Locke confused the issue by saying of the general idea of a triangle that 'it must be neither Oblique, nor Rectangle, neither Equilateral, Equicrural, nor Scalenon; *but all and none of these at once*'. The first confusing feature is the reference of the pronoun 'it'; we might think that it refers to the triangle being defined but that is awkward because what we are defining is not an actual thing that can be referred to but rather 'triangularity'. It is better to regard 'it' as referring to the idea and to regard the sentence as elliptical, saying that the idea being defined must not contain the ideas equicrural, equilateral, and so on.

The phrase 'all and none of these at once' clearly floored Berkeley and it must floor anyone who thinks that Locke is here talking about individual triangles or images or pictures of them. Consistently with what I have said so far, I think it is to be interpreted thus: the concept 'triangle' is 'all and none of these at once' because it *applies* to all these particular sorts of triangles without containing the ideas that make them the particular sorts of triangles they are and distinguish them from one another. It applies to all of them just because it does not specify any one of them; that is why anything allegedly described by it alone would be 'something imperfect that cannot exist'. It is interesting that if this account of Locke is correct then he may have got closer to what Berkeley intended for general words than Berkeley did himself.

Berkeley says that the problem is how a geometrical proof using a particular diagram, say, of a triangle can reach a conclusion about all triangles. As he puts it

For because a property may be demonstrated to agree to some one particular triangle, it will not thence follow that it equally belongs to any other triangle, which in all respects is not the same with it.

According to him, Locke's challenge is 'how we can know any proposition to be true of all particular triangles, except we have first seen it demonstrated of the abstract idea of a triangle which equally agrees to all'. Berkeley answers thus

though the idea I have in view whilst I make the demonstration [that the angles of a triangle equal two right angles], be, for instance, that of an isosceles rectangular triangle whose sides are of a determinate length, I may nevertheless be certain it extends to all other rectilinear triangles, of what sort or bigness soever. And that, because neither the right angle, nor the equality, nor determinate length of the sides, are at all concerned in the demonstration.[16]

This, he says, avoids the problems of the abstract idea of a triangle.

Berkeley sums up his view by saying what he conceives a general idea, rather than an *abstract* general idea, to be. He says

an idea, which considered in itself is particular, becomes general, by being made to represent or stand for all other particular ideas *of the same sort.*[17]

The weakness of this lies in the phrase 'of the same sort' and Berkeley lets it pass without remark. When are the ideas of a scalene triangle, an isosceles triangle and a right-angled triangle, to which the conclusion of a proof applies, *all of the same sort*? That is the very question at issue. No doubt what Berkeley has in mind is what he says shortly after this about geometrical proofs: that in proving with the help of a particular triangle a conclusion that applies to all triangles we are in the end ignoring any feature of that particular triangle that is not used in the proof, that is, in the example Berkeley takes, the equality of angles, the sizes of angles and the lengths of sides.

The question then arises as to how this differs from Locke's view. Locke, in saying that the abstract idea of a triangle is 'Figure including a Space between three Lines' or 'plane figure bounded by three straight lines' is saying just what 'of the same sort' amounts to for triangles. Moreover, in calling the idea 'abstract' he is saying that we reach it by ignoring those properties that differentiate particular triangles from one another. This, I suggest, is just what Berkeley tells us to do when he tells us to ignore those features of particular triangles that do not figure in the proof.

There appears, if I am right, to be considerable but sometimes unrecognized agreement between the views of Locke and Berkeley. However, there does appear to be a difference between their understanding of mathematical proofs. Berkeley seems to take it that we set out to prove various things about triangles each time with the idea of a particular triangle (scalene, isosceles, etc.) in mind, represented by a diagram or constituted by an image. When we reach the end of the proof we see that *the proof is such* that we have in fact proved something more general than we intended, perhaps about all possible triangles or about all triangles of a certain sort.

Locke, by contrast, seems to take it that we begin such proofs not with the idea of a particular triangle in mind, even if we use a diagram, but with the general idea of *triangularity* in mind, in the form of a general definition. That is, we *set out* to prove something generally about the triangles and we succeed if we take for granted *only* what is specified in the definition.

It might be thought that this difference is trivial. The reason I am inclined to favour Locke's view is that it seems to me to show an awareness of the nature of mathematical systems and proofs, in particular, the system of Euclidean geometry. In this system proofs begin from definitions and axioms and ultimately rely entirely upon what they contain. We need not use diagrams. There is, of course, the remaining problem about how our conclusions apply to the world and I suspect that Locke is also tentatively grappling with that problem here.

Given that I have found a considerable agreement between the views of Berkeley and Locke, where does the disagreement lie? We can begin to see the answer to this when Berkeley says

> it will be objected, that every name that has a definition is thereby restrained to one certain signification. For example, a *triangle* is defined to be a *plane surface comprehended by three right lines*; by which the name is limited to denote one certain idea and no other. To which I answer that in the definition it is not said whether the surface be great or small, black or white, nor whether the sides are long or short, equal or unequal, nor with what angles they are inclined to each other, in all which there may be great variety, and consequently there is no one settled idea which limits the signification of the word *triangle*.[18]

This seems to be just false and to suggest that Berkeley does not understand definition as well as Locke does. The definition gives the necessary and sufficient conditions of being a triangle and so it does 'limit the signification of the word *triangle*' by ruling out some figures that cannot be called 'triangles' i.e. those that do not satisfy the necessary conditions. Berkeley may also be saying that a definition does not ensure that the word can be used only in that meaning. A definition, indeed, does not ensure that there is only '*one* settled idea which limits the signification of the word'; some words have more than one meaning and defining one of them does not abolish the others and it does not preclude metaphorical uses of the sense defined. Defining 'well' as 'hole in the ground used for obtaining natural liquids' does not prevent our using 'well' to mean 'not ill' and 'The Well of Loneliness' does not refer to a hole in the ground although it is related to that sense. It does, however, limit the signification of 'well' in the sense defined because it does not allow us to refer to a bomb crater, for example, as a well. Definitions limit the significations of words for particular contexts; this is clearest for words such as 'triangle'. These difficulties do not touch Locke's concern in the present context.

However, the next and final sentence of that paragraph suggests that something else is worrying Berkeley. It reads

> 'Tis one thing for to keep a name constantly to the same definition, and another to make it stand everywhere for the same idea; the one is necessary, the other useless and impracticable.

The argument is really about the term 'idea'. Because Berkeley thinks of ideas as mental images, a sort of pictures, he refuses to call what is given in the definition of 'triangle' an 'idea'; an idea of a triangle must be, for him, an idea of a particular triangle having particular lengths of sides, sizes of angles and even a particular colour. As I have argued, Locke seldom means 'mental image' by 'idea', particularly in this context, and is willing to call 'plane figure bounded by three straight lines' an 'idea' and to refuse to make the distinction in Berkeley's last sentence.

This interpretation of Berkeley is supported by the fact that Berkeley appears to think that in conducting geometrical proofs we must necessarily use diagrams (pictures); diagrams must necessarily be particulars and so have features that are not used in the proofs. Locke could entertain the idea that a geometrical proof could be based on a definition but it appears that Berkeley could not. Berkeley appears to think also that an idea must share the particularity of a diagram. In *An Essay Towards a New Theory of Vision*[19] he says

> Now I do not find that I can perceive, imagine, or any wise frame in my mind such an abstract idea as is here spoken of. A line or surface which is neither black, nor white, nor blue, nor yellow, etc., nor long, nor short, nor rough, nor smooth, nor square, nor round, etc. is perfectly incomprehensible.

And in the *Principles*

> What more easy than for any one to look a little into his own thoughts, and there try whether he has, or can attain to have, an idea that shall correspond with the

description that is here given of the general idea of a triangle, which is, *neither oblique, nor rectangle, equilateral, equicrural, nor scalenon, but all and none of these at once?*[20]

Of course, we can form no *image* of a figure having these properties or of a figure having *only* the properties mentioned in the definitions. However, if we can understand these words, and I have suggested a sense, or the words of the definition of 'triangle', we can have a Lockean abstract general idea.[21] Berkeley admits that we can understand the definition but denies that this gives us an idea as he must be taking idea as a mental image or picture. But in admitting understanding he gives Locke all he requires.

13

ESSENCES, SPECIES AND KINDS

Locke attaches considerable importance to the distinction between *real* and *nominal* essences.[1] He first introduces it formally when he says that the ideas we have of substances

> have in the Mind a double reference: 1. Sometimes they are referred to a supposed real Essence of each Species of Things. 2. Sometimes they are only design'd to be Pictures and Representations in the Mind, of Things that do exist, by *Ideas* of those qualities that are discoverable in them. (II.xxxi.6)

The second of these Locke later (III.iii.15) calls 'nominal essences' and he then relates nominal essences to species. Here he says that these ideas are imperfect and inadequate, whichever way we take their references. He now summarizes the view he is to expand in Book III.

Men normally think that words they use for substances, such as 'gold' or 'iron', refer to real essences that make particular pieces of matter gold or iron respectively but when questioned they cannot tell us what these real essences are. This refers to *scholastic* real essences or substantial forms, inherited from the medieval Aristotelians. Locke explicitly mentions substantial forms at the end of this section. He says

> when I am told, that something besides the Figure, Size, and Posture of the solid Parts of that body is its Essence, something called *substantial form*, of that, I confess, I have no *Idea* at all, but only of the sound Form; which is far enough from an *Idea* of its real Essence, or constitution. (II.xxxi.6)

This may remind us of Boyle and I believe that Locke is here using the same strategy. Scholastic essences or substantial forms are not only unknown but also we are given no clues about the kinds of thing they are, what would be the form of descriptions of them, if indeed a request for that would even make sense. Locke's proposal is to keep the expression 'real essences' but to give it a different meaning from the scholastic one; he means by it the inner corpuscular constitutions of things. Although we may not know the inner corpuscular constitution of any familiar substance, at least the corpuscular theory tells us what sort of thing it is and how it operates, in general, to produce observable interactions between things. It gives us the form of the analysis of inner constitutions or real essences that at least makes it sensible to talk about them.

The man in the street operates perfectly well with the ideas of genus and species: he can, for example, recognize gold, lead and iron. For Locke, 'gold' stands for a complex idea made up of simple ideas we have when we observe pieces of gold, that is, ideas of such things as a particular *'Colour, Weight, Hardness, Fusibility, Fixedness'* together with ideas of changes wrought in the corresponding qualities by physical interaction with other substances (II.xxxi.6). Those ideas constitute the abstract idea of the *nominal* essence of gold. However, the man in the street does not think this; he thinks that we identify gold by reference to its scholastic real essence. If that were so we ought to be able to *deduce* from this real essence all the properties we discover in gold and the necessary connections between them; but how could we do that since scholastic real essences are admittedly unknown and, one suspects, unknowable?

The real essence of a substance, for Locke, is that inner corpuscular structure common to different samples of it, 'the Figure, Size and Connexion of its solid Parts', which is responsible for the observable qualities of that substance (III.vi.6). The complex idea we have of a substance that constitutes its nominal essence

> cannot be the real Essence of any Substance; for then the Properties we discover in that Body, would depend on that complex *Idea*, and be deducible from it, and their necessary connexion with it be known; as all Properties of a Triangle depend on, and as far as they are discoverable, are deducible from the complex *Idea* of three Lines, including a Space. But it is plain, that in our complex *Ideas* of Substances, are not contained such *Ideas*, on which all the other Qualities, that are to be found in them, do depend. (II.xxxi.6)

> The complex *Ideas* we have of substances, are, ... certain Collections of simple *Ideas*, that have been observed or supposed constantly to exist together. (II.xxxi.6)

It remains true, however, that we have no knowledge of the particular inner constitutions of substances, so if we suppose that a word for a substance, e.g. 'gold', refers to the real essence of gold, in Locke's sense, we are supposing that the word stands for our idea of that real essence but we have no such idea. These ideas are therefore inadequate.

In this chapter Locke is discussing generally what he calls 'adequate' and 'inadequate' ideas. He begins by saying that adequate ideas are those 'which perfectly represent those Archetypes, which the Mind supposes them taken from, and to which it refers them' while inadequate ideas 'are such, which are but a partial, or incomplete representation of those Archetypes to which they are referred' (II.xxxi.1). Clearly our words for substances, conceived in the way Locke is rejecting, stand for inadequate ideas.

If we take the more sensible course of supposing that our names for substances refer to their nominal essences we are supposing them to stand for complex ideas made up of simple ideas obtained by observation of samples of the substances in question. However, these ideas are also inadequate, though presumably not *so* inadequate, because 'those Qualities and Powers of Substances, whereof we make their complex *Ideas*, are so many and various, that no Man's complex *Idea* contains them all' (II.xxx.8). There are two reasons. First, for clarity and economy we include as few simple ideas in our complex idea of any substance as is possible. Second, most of these simple ideas are of powers, and observation reveals powers to us only when we allow different substances to act upon one another and observe the effects. We can never be sure that we have not missed an important power unless we have tried the action of one substance upon all other substances and that is impossible.

Geometrical figures are importantly different from substances in this respect. If our ideas of geometrical figures had to be collected by observing them, in themselves and in relation to other figures, our ideas of them would be as imperfect and inadequate as are our ideas of substances. As it is, our ideas of them contain their whole essences because, Locke holds, they can be completely determined by definitions. If we knew the inner constitutions of substances, our ideas of them would be as perfect and as adequate as our ideas of circles and triangles (II.xxxi.11).

There are three sorts of 'abstract *Ideas*, or nominal Essences' in the mind. First, simple ideas are 'ectypes' or copies and are adequate. They are intended to express nothing but the powers in things to produce just such sensations. If the idea were not adequate to the power, a different idea would be produced. The word 'copy' must not be taken too literally; here it means not 'picture' or 'resembling image' but rather 'corresponding effect'. This is shown by the fact that Locke takes as his example the idea of whiteness produced by a piece of paper and he has made it abundantly clear earlier that ideas of colours do not resemble the qualities in objects that cause them. As he says

> the Sensation of White, in my Mind, being the Effect of that Power, which is in the Paper to produce it, is perfectly *adequate* to that Power; or else, that Power would produce a different *Idea*. (II.xxxi.12)

Second, complex ideas of substances are 'Ectypes, Copies, too' but not adequate ones because the mind cannot be sure that the collection of simple ideas making up the complex idea of a substance 'answer all that are in that Substance'. Most importantly, we may omit ideas that are required to distinguish the substance from others because our observation is never complete. Moreover, observation can never give us ideas of real essences of substances which would ensure the adequacy of our ideas of them (II.xxxi.13).

Third, complex ideas of modes and relations are not copies but 'Originals and Archetypes'. They are not 'made after the Patterns of any real Existence, to which the Mind intends them to be conformable, and exactly to answer'. These ideas must therefore be adequate. I have dealt with this more fully in Chapter 12.

Locke also says that most of our ideas of substances are *false* if we take them to refer to their real essences, on which their properties depend; as we do not know those real essences whatever content these ideas have is not of real essences (II.xxxii.5 and 18). A false idea is one of which the mind makes a tacit supposition of the conformity of that idea to an external thing where there is no such conformity (II.xxxii.4).

Locke returns to essences in the course of his discussion of general words. The generality of words is not strictly a feature of what they stand for but a feature of the way in which the understanding uses them. They are used to stand for general ideas and ideas are general when they are made to stand for many particular things. He says

> universality belongs not to things themselves, which are all of them particular in their Existence, even those Words, and *Ideas*, which in their signification, are general.

and

> the signification they have, is nothing but a relation, that by the mind of Man is added to them. (III.iii.11)

What general words signify is *sorts* of things and each of them does that by being a sign of an abstract idea in the mind. These abstract ideas are *essences* of sorts or species of things. For, Locke says,

> the having the Essence of any Species, being that which makes any thing to be of that Species, and the conformity to the *Idea*, to which the name is annexed, being that which gives a right to that name, the having the Essence, and the having that Conformity, must needs be the same thing: Since to be of any Species, and to have a right to the name of that Species, is all one. (III.iii.12)

The sorting of things, and the essences of sorts of things, are 'the Workmanship of the Understanding' since it is the understanding that abstracts.

Nature makes many things alike, notably animals and things propagated by seeds; we notice their similarities, sort things into classes or species on that basis and give the classes names. The principles of classification, the class-concepts, are abstract ideas and the essences of classes or species. These abstract ideas are 'the bonds between particular Things that exist and the Names they are to be ranked under' (III.iii.13). But these essences cannot be the real essences of substances if the real essences are different from our abstract ideas and they *are* always different because the abstract ideas are derived from simple ideas of sensation, not ideas of inner constitutions. This makes it clear that there are different senses of 'essence' and Locke now makes the distinction between real and nominal essences more explicit.

The original sense of the word 'essence' is 'the very being of any thing, whereby it is, what it is'. Thus it may be used for the internal constitutions of a thing on which its discoverable qualities depend. In substances this constitution is 'generally unknown'. The word is used in this sense when we speak of the essences of particular things 'without giving them any Name'. The significance of that reservation will emerge shortly.

However, the word 'essence' has almost lost its original sense through its use in scholastic discussions of *genus* and *species*; instead of being applied to the inner constitution of things it is almost always applied to 'the artificial Constitution of *Genus* and *Species*'. Things are ranked into sorts or species under names 'only as they agree to certain abstract *Ideas*' and the names refer to those abstract ideas. So the word 'essence' applied to a genus or sort has come to stand for the abstract idea determining the genus or sort and thus for a collection of simple ideas of observable qualities of things (III.iii.15). So if I talk of the essence of a particular piece of a substance, without any specification, I am using 'essence' in its original sense to refer to its inner constitution as long as I do not call it, say, 'gold' since 'gold' refers not to that constitution but to an abstract idea based upon sensation. Locke calls essence in the first sense 'real essence' and in the second sense 'nominal essence'.

There are two extant views about the real essences of substance, according to Locke. The first, scholastic, view relates species directly to real essences: there is a limited number of real essences in the world that determine things as being of this or that species; every thing has, exactly, one or other of these essences and so is of a particular species. These real essences are unknown. This is the scholastic idea of substantial forms and it has considerably hampered the advance of knowledge and has created problems about the apparent overlapping of species and borderline cases between them, as we shall see.

The second, and more rational, view is that the real essences of things are the inner unknown constitutions of insensible parts which determine the observable qualities by means of which we sort things into classes. This is the more sensible view of real essences because it confines them to common inner constitutions of individuals, which inner constitutions we can conceive of in terms of corpuscular structures. It does not see what we call species as natural kinds but as conventional kinds that we have adopted. However, pending knowledge of inner constitutions, even this idea of real essences is of no use to us in connection with our classification of things. We now sort things, that is, define genera and species, according to our ideas of observable qualities, which are known, rather than according to ideas of inner constitutions, which are unknown. Genera and species are of considerable use to us, even though they do not exist in nature as we know them, as long as we content ourselves with their basis in nominal essences and do not suppose the basis we use to be the inner constitutions of particular things. The idea of the real essences of substances does, I think Locke believes, make sense and is of use to us in constructing our general picture of the world and we may even hope that one day we may be able to describe in detail the inner constitutions of things and so the real essences of substances. In the meantime, we may regard them as causally responsible for the observed similarities in things upon which we base our ideas of the nominal essences of substances (III.iii.17).

In species of simple ideas and modes, real and nominal essences are identical; in species of substances they are always different. What Locke means by this is clearer in relation to modes and substances than in relation to simple ideas. His example of a mode is, once again, a triangle. He says

> Thus a Figure including a Space between three Lines, is the real, as well as the nominal *Essence* of a Triangle; it being not only the abstract *Idea* to which the general Name is annexed, but the very Essentia, or Being, of the thing it self, that Foundation from which all its Properties flow, and to which they are all inseparably annexed. (III.iii.18)

There is nothing to a triangle except what is contained in its definition, as given at the beginning of this passage, so that is its real as well as its nominal essence. Substances are different because, as we have seen, we do not base their nominal essences upon their unknown real essences.

Locke, unfortunately, does not immediately explain what he says here about simple ideas and, at first sight, there appear to be problems about it. The idea of the shape of a thing resembles its real shape, is an accurate idea of it; but the idea of its colour does not resemble any quality in the thing that gives rise to it. So how can Locke say that for *all* simple ideas nominal and real essences are identical? The only thing he can mean, I think, can be seen by comparison of simple ideas and ideas of substances. Our idea of gold is the idea of a species and its nominal essence. Corresponding to the nominal essence is a similarity of structure in actual samples of gold; real essences depend on common inner constitutions of individuals. Each sample of gold has a structure which has features in common with every other sample of gold but also may have other features in which it differs from other samples. Simple ideas, even of colours, differ in that, because they are simple, each corresponds to just one specific structure, or structural feature. A word for a specific shade of yellow, for instance, names a specific simple idea corresponding to a specific structure, so different samples of that shade have just that structure common to all of them. Thus, even though we cannot describe the structure, our name for the colour has a reference to a real essence that is unambiguous as well as to the nominal essence.

Essences of species of mixed modes differ from essences of species of simple ideas in the following ways: they are made by the understanding, they are made arbitrarily for the convenience of communication and their names 'lead our thoughts' to the mind and no further, that is, not to the external world. They differ from the essences of species of substances especially in that they are identical with real essences. What has been said of modes is also applicable to relations (III.v).

Essences are generally supposed to be ingenerable and incorruptible and Locke says that this supports his view about the importance of nominal essences. This is because real essences are subject to change, a consequence of their depending on inner constitutions of individuals. 'All things that exist, besides their Author, are liable to change; especially those Things we are acquainted with . . .' (III.iii.19). An individual animal dies and decays and that is a series of changes in inner constitution. Nominal essences, however, are not subject to decay and change since they are abstract ideas of sorts; even if horses went out of existence the species horse is defined by certain ideas, unless we change our minds, for all time and the name 'horse' would be available for use again if an animal having the relevant qualities were to appear. 'The Doctrine of the immutability of *Essences*, proves them to be only abstract *Ideas*.' Locke says

> whatever becomes of *Alexander* and *Bucephalus*, the *Ideas* to which *Man* and *Horse* are annexed, are supposed nevertheless to remain the same; and so the *Essences* of those Species are preserved whole and undestroy'd, whatever Changes happen to any, or all of the Individuals of those *Species*. (III.iii.19)

He holds that when I am considered as an individual none of my present characteristics is essential: I might be a different shape or colour, lack rationality or sense, or any combination of these. The idea that any of these is essential to me occurs only when I am classified as, and called, a member of the species man, given that these characteristics are included in the nominal essence of man. There might appear to be problems attached to this way of putting it; we might think that anything that lacked my present characteristics, or some of them, would, for that reason, not be *me*. However, that would be to consider me, covertly, as a member of the species man or person. I think that what Locke has in mind is, for example, that this collection of material corpuscles that is me might remain numerically the same but become differently organized so that my shape and colour changed, my sense and rationality disappeared, so the collection could not be regarded as the same man or as a man at all. More radically, the collection could break up into a few or many parts. If we now ask 'Is it the same?' the sensible reply would be 'The same *what*?' and that gives Locke his point. He is not strictly making the point about identity here, although that is connected with the point he is making. To say that real essences are unchanging would be to say that individuals cannot change; but we know that individuals *can* change. A species or kind cannot change unless *we* change it. If we do not, the collection of individuals included in the kind may change because a change in an individual may be sufficient to exclude it from that kind. As long as an individual continues to be included in the kind we take it that it has not changed in respect of the qualities included in the nominal essence of the kind; it does not make sense to ask whether an individual is the same as it was unless we can answer the question 'The same what?' and that *is* to consider it as a member of a kind; only nominal essences can be unchanging. Locke says 'It would be absurd to ask, Whether a thing really existing, wanted any thing essential to it' and

> particular Beings, considered barely in themselves, will be found to have all their Qualities equally *essential*; and every thing, in each Individual, will be *essential* to it, or, which is more true, nothing at all. (III.vi.5)

He sums up much of a long discussion by saying

> *Nature makes many particular Things, which do agree* one with another, in many sensible Qualities, and probably too, in their internal frame and Constitution: but 'tis not this real Essence that distinguishes them into *Species*; 'tis *Men*, who, taking occasion from the Qualities they find united in them, and wherein, they observe often several individuals to agree, *range them into Sorts, in order to their naming*, for the convenience of comprehensive signs; under which individuals, according to their conformity to this or that abstract *Idea*, come to be ranked as under Ensigns: so that this is of the Blue, that the Red Regiment; this is a Man, that a Drill: And in this, I think, consists the whole business of *Genus* and *Species*. (III.vi.36)

There is a mistake people make about the essences of substances, as Locke explains at the end of this chapter. It is important for communication that the names of substances stand for the same ideas for different people. However, a problem arises. If we call 'gold' something that is yellow, fixed and malleable but then discover further properties of samples we call 'gold' what is to prevent our adding these to the list? What reason have we for stopping at any point in the list? If we cannot find such a reason the properties discovered may be indefinitely many and some may be observed by some people and not by others so the list may begin to vary from person to person. Thus

if every distinct Quality, that were discovered in any Matter by anyone, were supposed to make a necessary part of the complex *Idea*, signified by the common Name given it, it must follow, that Men must suppose the same Word to signify different Things in different Men: since they cannot doubt, but different Men may have discovered several Qualities in Substances of the same Denomination, which others know nothing of. (III.vi.48)

Men have tried to overcome this problem in a misleading way.

To avoid this, therefore, they have *supposed a real Essence belonging to every Species*, from which these Properties all flow, and would have their name of the Species stand for that. But they not having any *Ideas* of that real Essence in Substances, and their Words signifying nothing but the *Ideas* they have, that which is done by this Attempt, is only to put the name or sound, in the place and stead of the thing having that real Essence, without knowing what the real Essence is; and this is that which Men do, when they speak of Species of Things, as supposing them made by Nature, and distinguished by real Essences. (III.vi.49)

This would mean that we should never know what gold is, as long as this real essence remained unknown, and so we should not know to what to apply the properties we wish to attribute to gold. So even if the proposition 'Gold is fixed' were true of this real essence we should be unable to apply it. However, that proposition *is* true and we do know how to apply it to what it is true of, that is, to anything having the nominal essence of gold.

Locke returns to this matter in his chapter on the abuse of words. His fifth variety of the 'abuse of words' is '*the setting them in the place of Things, which they do or can by no means signify*' (III.x.17). When we use the names of substances, of which we know only the nominal essences, in propositions, we commonly suppose or intend, tacitly that they stand for their real essences. When we say, for example, 'Gold is malleable' we intend not merely 'What I call gold is malleable' but 'What has the real essence of gold is malleable.' But in fact we have no right to mean this last. Locke says

'Tis true, the names of Substances would be much more useful, and Propositions made in them much more certain, were the real Essences of Substances the *Ideas* in our minds, which those words signified. (III.x.18)

We try to improve matters by a 'secret Supposition' that those words stand for real essences but when we do so we simply mislead ourselves; we rob the word 'gold' of its signification because we make it stand for no idea we have i.e. for nothing (III.x.19). The assumption that we make when we do this is that 'nature works regularly' and 'sets the Boundaries of each of those Species [*sc.* of substances] by giving exactly the same internal Constitution to each individual which we rank under one general name' (III.x.20). This assumption is especially misleading when it leads to the theory of substantial forms.

This, incidentally, is worth comparing with what Locke says about the secret suppositions we make about words standing for things and for ideas in the minds of others (see Chapter 12 above). The point of the comparison is to note the different way in which Locke treats a supposition that he takes to be misleading and one that he takes to be not misleading. We might think that Locke would say that, in the present state of our knowledge and abilities we are not equipped to discover whether our divisions of substances into species corresponds to divisions in nature depending on real essences. However, he puts the matter more strongly than that and says that to assume this is to make a false assumption. He says that one of the false assumptions we make when we suppose names of substances to stand for real essence is

> That there are certain precise Essences, according to which Nature makes all particular Things, and by which they are distinguished into *Species*. That every Thing has a real Constitution, whereby it is what it is, and on which its sensible Qualities depend, is past doubt: But I think it has been proved, that this makes not the distinction of *Species*, as we rank them; nor the boundaries of their names. (III.x.21)

This clearly concerns the existence of natural kinds, although I don't think Locke uses that expression anywhere. The question is: what conception of natural kinds is he rejecting? I believe that his view can be put in the following way. We classify things in the world into species on the basis of observed similarities between the qualities we find them to have. The qualities, of course, have their basis in the inner constitutions of things: having a different quality implies having a different inner constitution or real essence. Every individual in nature has a real essence but these real essences are not so arranged that there are sharp divisions between classes or kinds of things. The inner constitutions are so complex and varied that there are gradual transitions between things, some perhaps so minute that the resulting differences between them are undetectable by us in observation. When we classify things into species we use collections of observable qualities as a basis. To each such collection, that is, to each species there corresponds in the world a particular constitution but when we choose a certain collection as defining a species we abstract. That is, we do not mention every quality that corresponds to a specific feature of the inner constitution and we are incapable of taking account of every feature of the inner constitution. Not only is this impossible but it would be useless if it were possible. There would be no sharp divisions between species because there are no sharp divisions between types of inner constitution in nature. Every individual we regard as a member of a particular species may differ in some respect from every other and resemble in some respect individuals we regard as members of other species. When we classify things we abstract from a continuum of inner constitutions in nature.

Let *a, b, c,* etc. stand for features of inner constitutions and *A, B, C* etc. for corresponding observable qualities for which they are responsible. Suppose that we have classified as *metals* just those individuals that have qualities *A, B, C, D* and thus the inner constitutional features *a, b, c, d.* Now samples of metal have other inner constitutional features in which they may differ from one another but which we regard as irrelevant for the purpose of classification. So they may have such inner constitutions as

a b c d e/f g...
a b c d e/g h...
a b c d e/h i...
a b c d e/i j...

and so on. In classifying these as metals we do not consider features to the right of the oblique stroke and they may differ from or resemble one another in respect of these.

Now suppose that we find it useful to classify as another species individuals having the qualities *P, Q, R, S* and so the inner constitutional features *p, q, r, s.* Then if we were to discover a substance with qualities corresponding to the constitution

p q r s/a b c d e...

we should be inclined to classify it in both ways and regard it as a borderline case. Borderline cases are our creation and not nature's. There could, in nature, be individuals having the features

p q r s/a b c d f
p q r t/a b c d e
p q r s/a b c d g
p q r t/a b c d f

and so on. There could be many other constitutions varying by one or more features from each of those listed. That is, there can be a continuum of inner constitutions lying between any of those by which we have defined a species.

So everything in nature has its particular inner constitution or essence and there will be greater or lesser resemblances between them but no discontinuities sufficient to warrant the view that there are natural *kinds*. What is natural is enormous numbers of individuals with their inner constitutions to some of which, or some parts of which, our species will correspond, but not groups of these constitutions clearly distinct from one another. The classifications we make are, and no doubt will continue with advancing knowledge to be, *useful* to us for communication and manipulation but the more we come to know about inner constitutions the less it will appear that our species correspond to groupings in nature.

In an early journal entry[2] Locke says that species of animate things may be relatively clearly differentiated in nature and then continues

> But though there be such a measure to distinguish the species of things where there is propagation there are scarce any in inanimate which being but as it were severall collections of matter without any organicall constitutions or parts adapted to nutrition or generation may be capable of *infinite variety within the same species*, there being a latitude for a great variety of mixtures where we are able to make noe distinction whereas the principall difference of animate beings seems to depend on that internall principle that organizes the parts and contributes to and is the principall cause of generating the like. (My italics)

J. L. Mackie has argued that 'Locke sometimes goes too far in his denial of natural kinds and other considerations force him to recognize them' and, he adds, 'the doctrine of real essences that do not coincide with nominal essences is implicitly a doctrine of natural kinds'.[3] Mackie refers us in particular to two passages (III.iii.13 and III.vi.36–7) where, he says, Locke admits that there are natural kinds.

However, I am not convinced that Locke is admitting natural kinds in the sense in which he elsewhere argues against them. Mackie quotes Locke as saying

> *Nature makes many particular Things, which do agree* one with another, in many sensible Qualities, and probably too, in their internal frame and Constitution. (III.vi.36)

and adds 'that is, there are natural kinds'.[4] However, I do not think that this conflicts with Locke's rejection of natural kinds because I think that he is intent precisely on distinguishing between 'there are particular things that agree with one another in inner constitution' and 'there are natural kinds'. As I have argued, Locke thinks that nature does 'make many particular things, which do agree one with another in many [sic] sensible qualities and probably too in their internal frame and constitution', but this agreement is only in certain respects and nature does not make these things fall into groups with sharp boundaries. Perhaps his clearest statement of this is

> That which, I think, very much disposes Men to substitute their names for the real Essences of *Species*, is the supposition before mentioned, that Nature works regularly in the Production of Things, and sets the Boundaries to each of those *Species*, by giving exactly the same real internal Constitution to each individual, which we rank under one general name. Whereas any one who observes their different Qualities can hardly doubt, that many of the Individuals, called by the same name, are, in their internal Constitution, as different one from another, as several of those which are ranked under different specifick Names. (III.x.20)

In delineating species we arbitrarily, for our own purposes, cut off segments of the continuum of inner constitutions and give them names. We might, for different purposes, make the cuts in different places. This does no violence to nature unless we think that places where cuts must be made are determined by nature and allow no alternatives.

However, I believe that Locke's rejection of the idea that the things we classify into genera and species fall into natural kinds does not imply a rejection of natural kinds altogether. I shall return to this shortly. In the meantime, further light is thrown on this question by important passages concerning the distinction of species by genesis and the classification of changelings, i.e. mentally defective people.

First, Locke discusses the method of distinguishing species of plants and animals by their breeding true and their not interbreeding. He says

> Nor let anyone say that the power of propagation in animals by the mixture of Male and Female, and in Plants by Seeds, keeps the supposed real *Species* distinct and entire. (III.vi.23)

Having pointed out that this leaves out of account all substances except living ones he goes on to say that it won't do even for plants and animals. He uses without distinction mythical and highly dubious examples as well as genuine ones but they are all of a kind for which real parallels can be found. Women are reported to have conceived by 'drills' (a species of baboon); mules (progeny of an ass and a mare) and 'gimars' (jumarts: mythical progeny of a bull and a mare) are, he says frequent occurrences. He claims to have seen an offspring of a cat and a rat which 'had the plain marks of both about it'. All who admit these examples will find it hard, he says, even in the race of Animals to determine by the Pedigree of what *Species* every Animal's Issue is; and be at a loss about the real Essence, which he thinks certainly conveyed by Generation, and has alone a right to the specifick name. (III.vi.23)

If there were in nature a limited number of essences of one or other of which every individual thing partook then there would be no overlapping of species or borderline cases between them. But there *are* overlappings and borderline cases. Locke mentions freaks of various sorts and changelings at III.vi.22 and IV.iv.14–16. Each of these has its real essence but throws doubt on the view that there are distinct species in nature.

He says that it would be thought a paradox by some if he were to say that a changeling who had lived for 'some forty years' without any appearance of reason is 'something between a man and a beast' (IV.iv.13). It would be thought paradoxical because of the false supposition

that these two Names, *Man* and *Beast*, stand for distinct Species so set out by real Essences, that there can come no other Species between them.

There are presumably two mistakes here: that of supposing that we know the real essences for which 'man' and 'beast' stand and that of supposing that sharp divisions of things into a fixed number of species is a feature of nature. Then he says

if we should not fansy, that there were a certain number of these Essences, wherein all Things, as in Molds, were cast and formed, we should find that the *Idea* of the Shape, Motion and Life of a Man without Reason, is as much a distinct *Idea*, and makes as much a distinct *sort* of Things from Man and Beast, as the *Idea* of the Shape of an *Ass* with Reason, would be different from either that of Man or Beast, and be a Species of an Animal between, or distinct from both. (IV.iv.13)

Therefore, if Locke were asked 'If changelings are things between men and beasts, what are they?' he would answer 'Changelings'. There is a perfectly good abstract idea for which this word stands, as is indicated in the last quotation (IV.iv.14). There is nothing in nature to prevent our making as many species as we will find useful as long as we do not base a species upon ideas that do not occur together.

Religious objections will be raised to this. If what Locke says is correct, what will become of changelings in the next world? He replies that this is a matter not for him but for the Creator; whatever *we* decide about the nature of changelings will not affect their fate in the hereafter. We should not pretend to knowledge of such things when we are in considerable ignorance even about the world we experience. Besides, there are two false assumptions upon one or other of which such questions are based: that whatever has the outward appearance of a man and whatever is of human birth must necessarily have eternal life after death. To suppose that immortality has anything to do with outward bodily shape, rather than with qualities of soul and spirit, is to trivialize the idea of immortality; it is as absurd as to suppose that a man might achieve everlasting life because of the cut of his beard or the fashion of his coat. It may be replied that shape is a sign of a rational soul but Locke says there is no evidence for that and some against it. A newly-dead person may have a perfect outward shape but is incapable of actions showing the marks of reason. There may be nothing wrong with the shape of a changeling while his actions show fewer signs of rationality than those of many a beast (IV.iv.15).

It is equally questionable that whatever is the offspring of rational parents has a rational soul. Men nowhere, he says, accept this for they destroy ill-shaped monsters and not well-shaped changelings, presumably both the offspring of rational parents. Why should a bodily defect make a monster and not a mental defect, that is, a defect in the nobler and more essential part of men? On what grounds is a changeling, in spite of appearances, thought to have a rational soul and a monster not? Besides, do not those who take this view classify a monster as something between a man and a beast and refuse to do the same for a changeling (IV.iv.16)?

The view of natural kinds that Locke is attacking would involve there being a necessary connection between, for example, bodily deformity and the absence of a rational soul or between bodily perfection and the presence of a rational soul. We have no knowledge of real essences that would enable us to discover whatever necessary connections there are. So, in our present state, at least, we must rely on observation and induction; we are able to determine whether or not a thing is rational only by observing its behaviour. A changeling who constantly behaves irrationally gives us only evidence against the presence of a rational soul; a misshaped human being, however, may behave rationally and so give evidence of the presence of a rational soul. The origin of either in rational parents cannot count as evidence one way or the other.

I have argued that Locke is rejecting a vew of natural kinds according to which they correspond to our classification of things into species or could correspond, if not to our present classification, to some classification that we could devise. Locke would, I think, be unmoved by Mackie's statement

> Surely there really are natural kinds, for example chemical elements and compounds (as opposed to mixtures) such as gold, water, and common salt, and the various species of plants and animals. There are natural kinds because properties are not randomly and independently distributed among things, but tend to cluster.[5]

Locke's rejection of this, however, does not imply the rejection of natural kinds altogether. Indeed, given his realism, I think he is committed to a belief in natural kinds. He says

> The Things that, as far as our Observation reaches, we constantly find to proceed regularly, we may conclude, do act by a Law set them; but yet by a Law, that we know not: whereby, though Causes work steadily, and Effects constantly flow from them, yet their *Connexions* and *Dependancies* being not discoverable in our *Ideas*, we can have but an experimental Knowledge of them. (IV.iii.29)

That is, there are laws, involving as we have seen, necessary connections, responsible for the regularities we observe in phenomena, those laws being entirely independent of us. Laws must hold between classes of things and what can these be but natural kinds?

On the corpuscular theory, necessary connections arise ultimately from the primary qualities of the corpuscles and the arrangement of those corpuscles. Since there is a finite variety of individual corpuscles it seems that Locke is at least committed to natural kinds among corpuscles. This makes possible an indefinitely large or even infinite variety of patterns or organizations of corpuscles and thus justifies Locke's rejection of natural kinds on the level of stable groups of corpuscles.

The most basic laws would be laws of the behaviour of individual corpuscles such as: any corpuscle with its largest overall dimensions smaller than *n* units will pass through a space between corpuscles having dimensions of *n* units, whatever the shape of the corpuscle. Even at this level a law may involve abstraction since we can here ignore shape. More precise laws may, of course, involve all the primary qualities but depend upon there being corpuscles of the same shape and size. This allows laws at a higher level: several collections of corpuscles may resemble one another in certain structural features and a law may state that any collection with these features, whatever other features they have, behave thus and so. At this level, all laws may involve abstraction since it is possible that no two collections have all their structural features in common. This is a familiar characteristic of what we normally regard as scientific laws. It makes natural kinds, on the level of collections of corpuscles, unnecessary in order that there shall be laws, at least in the sense in which Locke is rejecting natural kinds on the level of substances. Laws may be possible even if there is no division of the world into natural kinds corresponding exactly to any classification of substances into species that we may ever make.[6]

Thus I think that the reply to Mackie is that Locke is committed to natural kinds and does not deny this but only at the 'deepest' level and not at the level of gold, water and common salt and the various species of plants and animals. It does not seem to me that he contradicts himself in arguing as he does against natural kinds and he does make it possible to make sense of laws at the level of the real essences of substances and regularities among our species of substances. He can also explain how it is possible for us to find apparent overlapping and borderline cases between species.

14

KNOWLEDGE

I

The last book of the *Essay* is in some ways a disappointment compared with the two preceding books; it has an air of scepticism and pessimism for which the reader has not been entirely prepared. It appears that Locke was uneasy about the degree of scepticism concerning scientific knowledge of the world that he seems to regard himself as forced into by what has gone before; there are passages which have a hopeful tone for future developments and, at one point, there is a direct attack on extreme scepticism. He has a firm belief in necessary connections in nature, as we have seen, but now he is sceptical about our ability to discover them. I begin by mentioning some of the more hopeful passages, some of which I shall deal with in more detail later.

In the first place there is the famous passage, already quoted, in which he refers to the corpuscular hypothesis as the best available for 'an intelligible Explication of the Qualities of Bodies' and continues

> I fear the Weakness of humane Understanding is scarce able to substitute another, which will afford us a fuller and clearer discovery of the necessary Connexion, and *Co-existence*, of the Powers, which are to be observed united in several sorts of them. This at least is certain, that which ever Hypothesis be clearest and truest, (for of that it is not my business to determine,) our Knowledge concerning corporeal Substances, will be very little advanced by any of them, till we are made see, what Qualities and Powers of Bodies have a *necessary Connexion or Repugnancy* one with another; which in the present state of Philosophy, I think, we know but to a very small degree: And, I doubt, whether with those Faculties we have, we shall ever be able to carry our general knowledge (I say not particular Experience) in this part much farther. Experience is that, which in this part we must depend on. And it is to be wish'd, that it were more improved. (IV.iii.16)

The beginning of this passage suggests that the corpuscular hypothesis has some chance of leading us to knowledge of necessary connections and that is in the hands of the natural philosophers. We already have a little knowledge of this sort so it is not in principle unattainable. However, until the corpuscular hypothesis is more fully developed we can do no more than rely upon experience. There is some doubt about the possibility of acquiring *general* knowledge, which is knowledge of necessary connections, by experience alone but there may be a possibility of improving our experience. It may be that much of the pessimism to be found elsewhere in this Book is pessimism on behalf of the metaphysical philosopher or the common man unaided by the work of the natural philosopher.

He then suggests that the progress of natural philosophy has been greatly impeded by the careless observation and dishonest reports of the 'Philosophers by fire', that is, the alchemists. Were it not for that 'our acquaintance with the bodies here about us, and our insight into their Powers and Operations had been yet much greater' (IV.iii.16).

Earlier in the same Chapter he has said, about knowledge generally,

> I do not question, but that Humane Knowledge, under the present Circumstances of our Beings and Constitutions may be carried much farther, than it hitherto has been, if Men would sincerely, and with freedom of Mind, employ all that Industry and Labour of Thought, in improving the means of discovering Truth, which they do for the colouring or support of Falshood, to maintain a System, Interest, or Party, they are once engaged in. (IV.iii.6)

It is true, as Locke immediately says, that we may never know all we would wish to know, solve all problems and surmount all the difficulties involved in the ideas we have but still there is room for such extensions of our knowledge.

It is perhaps important for the interpretation of Book IV as a whole that shortly after this Locke recommends to philosophers a becoming modesty in not pronouncing 'Magisterially, where we want that Evidence that can produce Knowledge' (IV.iii.6). He is here particularly discussing matters about which we may never get evidence but presumably the message is a general one.

Rather more striking than any of these passages is that in which Locke talks of our ignorance of 'the primary Qualities of the insensible Parts of Bodies' and then says that there is 'another and more incurable part of Ignorance' (IV.iii.12). It surely follows that a kind of ignorance compared with which another kind is more incurable cannot itself be entirely and in principle incurable. So Locke leaves open the possibility that the natural philosophers may one day acquire knowledge of the primary qualities of the insensible parts and so of the textures of bodies. The more incurable part of ignorance concerns the relations between those primary qualities and textures and our ideas produced by them and depends upon the insolubility of the mind/body problem. This in turn affects our knowledge of secondary qualities. Locke is greatly exercised, throughout Book IV, by this problem and I am inclined to think that it is this that is the ground of much of his scepticism. He certainly returns to it again and again even, sometimes, when it does not seem to be strictly relevant to the matter under discussion. I shall return to this.

II

Locke begins Book IV by defining 'knowledge' as 'nothing but *the perception of the connexion and agreement, or disagreement and repugnancy of any of our Ideas*' (IV.i.2). As the book goes on it becomes clear that nothing counts for Locke as knowledge, strictly, unless it is certain. This is made clearer in his second letter to Stillingfleet where he says 'What reaches to knowledge, I think may be called certainty; and what comes short of certainty, I think cannot be called knowledge...' (*L.S.* 145).

Immediately a puzzle is created by Locke's definition since it appears to leave open the possibility that knowledge may be the perception of an agreement or disagreement between an idea and something that is not an idea and yet the heading to this section reads '*Knowledge is the Perception of the Agreement or Disagreement of two* Ideas'. The first interpretation seems sometimes to be favoured by the text but further light is thrown upon this by a passage in Locke's third letter to Stillingfleet. He says

> your Lordship argues, that because I say, that the idea in the mind proves not the existence of that thing whereof it is an idea, therefore we cannot know the actual existence of any thing by our senses: because we know nothing, but by the perceived agreement of ideas ... you mistake one thing for another, viz. the idea that has by a former sensation been lodged in the mind, for actually receiving any idea, *i.e.* actual sensation; which, I think, I need not go about to prove are two distinct things ... Now the two ideas, that in this case are perceived to agree, and do thereby produce knowledge, are the idea of actual sensation (which is an action whereof I have a clear and distinct idea) and the idea of actual existence of something without me that causes that sensation. (*L.S.* 360)

It seems that ideas sometimes come associated with the idea of an external existent causing them and sometimes not. It is not easy to see how Locke thinks this occurs or how we distinguish the two and he is peculiarly reticent about this. It is particularly difficult to know what to say about his example 'God is' as being 'of real Existence' (IV.i.7).

Locke gives two preliminary examples to explain his meaning in general. When we know that 'White is not black' we perceive that these two *ideas* do not agree and when we know that 'The three angles of a triangle are equal to two right ones' we perceive that the idea of equality to two right angles agrees with or is inseparable from, the idea of the three angles of a triangle. However, he at once proceeds to say that there are four sorts of agreement or disagreement. They are

1. *Identity*, or *Diversity*;
2. *Relation*;
3. *Co-existence*, or *necessary connexion*;
4. *Real Existence*.

The first is involved in 'the first Act of the Mind' which is, when it has ideas, to 'perceive each *Idea* to agree with it self, and to be what it is; and all distinct *Ideas* to disagree, *i.e.* the one not to be the other' (IV.i.4). This it does immediately 'without pains, labour, or deduction'. It concerns particular ideas although we may later make general rules governing all cases, such as '*What is, is; and it is impossible for the same thing to be, and not to be.*' Identity and diversity are 'infallibly known' at first sight.

The second concerns relations between any two ideas of whatever sort, that is, similarities and differences in any respect, such as, lighter and darker (for colours), louder and softer (for sounds) and larger and smaller (for sizes) (IV.i.5).

The third concerns '*Co-existence* and *Non-co-existence*' in the same subject and relates particularly to substances. For example, in perceiving gold the ideas of yellowness, weight, solubility in *aqua regia*, and so on, are always accompanied by fixedness or stability and these all go together to make up our complex idea of gold (IV.i.6).

The fourth kind of agreement is 'that of *actual real Existence* agreeing to any *Idea*'. Here the idea 'has a real existence without the Mind' or, as Locke should say, if he were being strict is *of* something having a real existence without the mind. This is one of the places at which Locke seems to say that the agreement is between an idea and something that is not an idea (IV.i.7), but we have seen that his expression is elliptical.

Locke next classifies knowledge in two ways; first, according to the way in which the mind is 'possessed of Truth' (IV.i.8–9) and second, according to its degree of certainty, evidence or clarity (IV.ii.). There is a possibility of confusion here since he appears to use two different bases of classification in the two chapters but I think this is only apparent. I shall try to say clearly what Locke has in mind.

The first classification is into 'Actual' and 'Habitual' knowledge; when Locke calls these ways in which the mind 'is possessed of Truth' I think he does not mean 'acquires Truth' but something like 'has Truth before it'. The difference between actual and habitual knowledge is that the first does not involve memory while the second does. *Actual* knowledge is 'the present view the Mind has of the Agreement, or Disagreement of any of its *Ideas*, or of the Relation they have to one another' (IV.i.8), that is, the agreement or disagreement perceived between ideas when they are first directly presented, as distinct from being remembered. An example is my perception of the difference in colour between a red and a green book when the books are before me and I am seeing them. It appears that other examples are: examining geometrical diagrams or even thinking, for the first time, of, say, a triangle and a circle, or considering a definition or seeing each step in a mathematical proof.

Habitual knowledge arises from memory rather than present perceptions. A man is said to know any proposition

> which having once been laid before his Thoughts, he evidently perceived the Agreement, or Disagreement of the *Ideas* whereof it consists; and so lodg'd it in his Memory, that whenever that Proposition comes again to be reflected on, he, without doubt or hesitation, embraces the right side, assents to, and is certain of the Truth of it. (IV.i.8)

I will return to this after I have explained the classification of the degrees of knowledge contained in Chapter ii. Here Locke distinguishes three degrees: 1. *intuitive* knowledge; 2. *demonstrative* knowledge and 3. *sensitive* knowledge, the first two being superior to the third. Whatever comes short of intuitive and demonstrative knowledge is faith or opinion 'at least in all general Truths' but sensitive knowledge deserves to be called 'knowledge' although it concerns only particular truths, namely, the '*particular existence of finite Beings*'.

In *intuitive* knowledge we perceive immediately the agreement or disagreement of two ideas without the intervention of other ideas. It involves neither proving nor the close examination of ideas but is immediately seen; thus we perceive that white and black or a circle and a triangle differ. No doubt is possible. This is the clearest and most certain kind of knowledge and it is on this that all our knowledge ultimately depends. Thus

> a Man cannot conceive himself capable of a greater Certainty, than to know that any *Idea* in his Mind is such, as he perceives it to be; and that two *Ideas* wherein he perceives a difference, are different, and not precisely the same. (IV.ii.1)

In *demonstrative* knowledge we cannot compare two ideas directly but have to connect them by using intermediate ideas, as in mathematical proofs. For example, we cannot see by direct comparison of the three angles of a triangle and two right angles that they are equal. In this case

> the Mind is fain to find out some other Angles, to which the three Angles of a Triangle have an Equality; and finding these equal to two right ones, comes to know their Equality to two right ones. (IV.ii.2)

That is, reasoning and proof are needed. This, however, still depends upon intuitive knowledge because the following of one step in the proof from another is seen by direct comparison. Thus demonstrative knowledge is a series of pieces of intuitive knowledge and so an infinite regress is avoided. That is why intuitive knowledge is basic.

Locke holds that demonstrative knowledge, though certain, is less reliable than intuitive knowledge. He has in mind three points. First, the assent to demonstrative knowledge depends upon 'pains and attention'; the truths are not, as with intuitive knowledge, immediately obvious. Second, before the proof is constructed there may be a period of doubt. Third, when we come to the end of a long proof our assent depends upon remembering the intuition of each step; we may make mistakes by misremembering or failing to remember. If a proof has been fully and correctly constructed certainty is justified but there is always room for doubt that the proof has been fully and correctly constructed. It would have been better if Locke had said that demonstrative knowledge is as certain as intuitive knowledge but that we may be mistaken in thinking that we *have* demonstrative knowledge but not in thinking that we have intuitive knowledge. That, I believe, is what he means (IV.ii.4,5,6).

Now I return to the classification of his previous chapter. It seems that actual knowledge, in so far as the ideas involved are general, is entirely intuitive. When we are seeing two colours we are directly comparing two ideas. When we are working through a proof, presumably only one step is directly before our minds at any given moment so that is the only actual knowledge we have at that moment and it is intuitive.

Habitual knowledge, depending upon memory, has, Locke says, two varieties. We may remember two ideas that we were previously able to compare directly so clearly that we are able again to see their agreement or disagreement immediately; that, he says, is intuitive knowledge just as the original was. The other variety involves remembering a connection between ideas which originally had to be proved, that is remembering demonstrative knowledge. One might have expected Locke to mention two sub-varieties under this sub-heading: one in which we remember the proof in all its details as well as that the connection between the two ideas was proved; the other in which we remember *that* the connection was proved but do not remember the proof. He in fact appears to ignore the first of these and to go directly from the remembering of intuitive knowledge to

> *such Truths, whereof the Mind having been convinced, it retains the Memory of the Conviction, without the Proofs.* Thus a Man that remembers certainly, that he once perceived the Demonstration, that the three Angles of a Triangle are equal to two right ones, is certain that he knows it, because he cannot doubt of the truth of it. (IV.i.9)

This 'comes not short of perfect certainty, and is in effect true Knowledge'. It is still demonstrative knowledge.

When a man has forgotten the proof but remembers the proving he knows the connection between the ideas 'in a different way'. Their connection is perceived but by the intervention of different ideas from those used as intermediate in the proof. The new intermediate idea is

The immutability of the same relations between the same immutable things, ... that shews him, that if the three Angles of a Triangle were once equal to two right ones, they will always be equal to two right ones. And hence he comes to be certain, that what was once true in the case is always true; what *Ideas* once agreed will always agree; and consequently what he once knew to be true he will always know to be true, as long as he can remember that he once knew it.

It is because of this that proofs in mathematics give general knowledge (IV.i.9).

It seems likely that Locke doesn't mention under habitual knowledge the situation in which a proof is remembered in all its details because he has doubts about the possibility of this. A proof may contain many intermediate ideas and going through the proof involves seeing immediately the connection between each idea and its successor. The mind can have directly before it, and so intuitively grasp, only one of these connections at a time. Thus in constructing a proof we are relying at each step on memory of the former steps which can only be the memory *that* the step was intuitively seen; if we could not rely on memory and had to go back and intuit each step again we should never move forward. Thus constructing a proof involves habitual knowledge as well as intuitive knowledge. By the same token, we could not remember a proof in all its details; to do that would be to *re*construct the proof, that is, prove the connection all over again and that in turn would involve remembering *that* earlier steps had been seen to follow. Those familiar with Descartes will recognize this as a problem that exercised him.

I return now to Chapter ii and knowledge of the third degree, *sensitive* knowledge. This concerns 'the particular existence of finite beings without us' (IV.ii.14). When we perceive something we have intuitive certainty that particular ideas are in our minds. Some people doubt that there is anything more to be said such as, for instance, that we can assert or infer the existence of an external object causing those ideas because we know from experience that ideas that come into our minds in the presence of an object may also come into our minds when there is no such object. Locke here depends upon his earlier discussion of the 'reality' of ideas (II.xxx and see Chapter 4) and now he asks us to compare any sensible idea with the memory of it. He says

I ask anyone, Whether he be not invincibly conscious to himself of a different Perception, when he looks on the Sun by day, and thinks on it by night; when he actually tastes Wormwood, or smells a Rose, or only thinks on that Savour, or Odour? We as plainly find the difference there is between an *Idea* revived in our

Minds by our own Memory, and actually coming into our Minds by our Senses, as we do between any two distinct *Ideas*. (IV.ii.14)

Locke does not go into details about this particular passage but he no doubt has in mind various features in which the two experiences differ. In the ordinary way, barring hallucinations, most of us find a difference between seeing the sun and remembering seeing the sun; the sun does not look to be *there*, in the sky, when we are remembering; there is a difference in both vividness and immediacy; we normally don't confuse seeing and remembering. Moreover, even if we are adept at remembering seeing the sun, most of us are not adept at remembering, at the same time, all the sensations accompanying seeing the sun such as feeling its warmth, seeing it cause steam to rise from a wet pavement, seeing other things illuminated by it and seeing plants drooping under its heat. But even if we can remember all these things at once we do not mistake the memory-ideas for the thing remembered and suppose that that is how things are now. If we remember the smell of a rose it does not usually seem to come from a rose; what makes me talk of remembering the taste of wormwood rather than tasting wormwood is that there is no wormwood in my mouth. Of course, we may make mistakes but in the most usual circumstances we know that we are not making mistakes of the sort that would be relevant to the argument.

Locke now tries to meet the sceptic who says that it may all be a dream; it is perfectly possible that we should dream the differences we find between a sensible idea and that idea remembered and dream that an object appears to be present in one case and not in the other.

He has two replies. The first is that 'where all is but a Dream, Reasoning and Arguments are of no use, Truth and Knowledge nothing.' He does not explain this as fully as one would wish but he may have in mind the argument that to say that everything is a dream is to say something unintelligible because 'is a dream' only gets its meaning by contrast with 'is real' and on the hypothesis there is no means of drawing the contrast. He may also have in mind the idea that there is no basis for any argument with the extreme sceptic because there can be no agreed premisses so no argument can get started. If what the sceptic says is correct then the sceptic is not in a position to argue for it or even to say it intelligibly.

The second reply is rather more complex and difficult to assess. There is a difference, Locke says, between being in a fire and dreaming that one is. If the sceptic says that what Locke calls being in a fire is nothing but a dream and has no tendency to show that fire exists, Locke answers

That we certainly finding, that Pleasure or Pain follows upon the application of certain Objects to us, whose Existence we perceive, or dream that we perceive, by our Senses, this certainty is as great as our Happiness, or Misery, beyond which, we have no concernment to know, or to be. (IV.ii.14)

A similar argument occurs at IV.xi.8. The sense of it seems to be that, to us, nothing could be more real than our own happiness or misery, pleasure or pain, whether they are dreamt or not. To dream that one is miserable is to be miserable. Moreover, pain may warn us of harm to come and if we can distinguish between waking experience and dreams by the fact that one involves pain for us and the other does not or by the fact that pain in one is followed by actual harm while in the other it is not then that is all we need for the purposes of our life; we can learn from it to avoid objects which produce pain and harm in one sort of situation and that we need not trouble ourselves to do so in the other. This still, of course, leaves problems about how we are to know in time which sort of situation we are in.

Perhaps the more detailed thought is that if all life is a dream (call it $dream_1$) then we have found in experience that within it there are dreams (call them $dreams_2$) *from which we awake*. In $dream_2$ we may feel pain and even find that our fingers have been burnt by a fire but when we wake the pain and the scars are not there; that is, from $dream_1$ we have not suffered harm. If we never wake from $dream_1$ then any harm done to us in that dream matters; it is permanent and so, for all practical purposes, real. It matters that we believe in the existence of external objects causing our ideas in $dream_1$, but not in $dream_2$. Belief in external objects depends upon our faith in waking or not waking from the dream we are in. The trouble with this answer is that it merely pushes the problem farther back. It does not enable us to distinguish between a dream from which we will awaken from one from which we will not, whether we are in $dream_1$, $dream_2$ or $dream_n$, *while we are in that dream*. Thus we never know when our ideas of sensation are evidence for external objects, which conflicts with what Locke intends.

However, perhaps the emphasis is intended to be rather different. Determined adherence to the sceptic's argument drives us into an infinite regress in which case there is no basis for discussing those matters. We can stop the regress only by saying either that what we ordinarily regard as waking experience is really waking experience or that it is a dream from which we will never wake. In either case, we have as much reason for believing that there are external objects as we could possibly have or require. We may, of course, make mistakes on any given occasion but we cannot admit that we are always mistaken in general and in principle. Apart from anything else, that would make it impossible to say what 'being mistaken' meant.

Locke, at any rate, thinks, in the face of the extreme sceptic, that 'we are provided with an Evidence, that puts us past doubting' and that we may accept sensitive knowledge of the existence of particular external objects. The arguments that he has put previously will, he thinks, convince all but the extreme sceptic and there is no possibility of convincing him by *any* argument.

III

The most important chapter in Book IV is probably Chapter iii, 'Of the Extent of Humane Knowledge', in which Locke considers the sorts of things that are within and without the grasp of our knowledge. He here sets out to examine the consequences of his definition of knowledge and his allegedly exhaustive classification of it into intuitive knowledge (by intuition), demonstrative knowledge (by reason) and sensitive knowledge (by sensation).

Since knowledge is the perception of the agreement or disagreement of ideas we can have no knowledge without ideas and without the *perception* of their agreement or disagreement, by intuition, reason or sensation. 'Perception' must include intellectual perception as well as sense-perception. We cannot have intuitive knowledge of the relations between every pair of ideas we have because their agreement or disagreement is not in every case obvious in direct comparison; we cannot have knowledge by reason covering all our ideas because we cannot always find the intermediate ideas, or propositions, required by deduction; sensitive knowledge is narrower than either of these because it concerns only things actually present to our senses and so does not give *general* knowledge. It seems likely that the first and third of these express logical impossibilities while the second expresses only a practical impossibility. It follows that 'the extent of our knowledge comes not only short of the reality of Things, but even of the extent of our own *Ideas*' (IV.iii.6). I take this to mean that there are truths about the reality and even about the relations between our ideas that we shall never discover. Nevertheless, Locke does think that human knowledge can be considerably, perhaps indefinitely, improved if we will only take pains and exercise caution.

He gives a number of examples of different kinds of thing that it is probably impossible that we should know. We may never be able to find a circle equal in area to a given square and know that it is so; or to know whether or not it is possible for matter to think or whether God could bring that about; or to know how mind and matter interact; or to know the nature of the soul (IV.iii.6). It is noticeable that Locke does not here mention as probably impossible the knowledge of the inner constitutions of things, which suggests that he is not as sceptical about this as he has sometimes been thought to be; and that in this section he discusses the difficulties posed by the mind/body relation.

In considering the extent of knowledge Locke returns to his four sorts of agreement or disagreement of our ideas. Our knowledge of *identity and diversity*, in so far as it is intuitive, extends to all our ideas and appears to him to be unproblematic; simply to have an idea is to know it as it is and to distinguish it from others. This is the familiar empiricist view of the period; it makes no sense, ideas being what they are, to distinguish between how an idea seems to us and how it really is. Locke is assisted in this by his view, mentioned in Chapter 12 above, of the possibility of both mental and verbal propositions since no problems need arise for him about inaccurate *descriptions* of our ideas.

Our knowledge of *co-existence* is more limited because it depends upon a kind of experience which involves the possibility of mistakes: we know that the idea of gold involves the ideas of heaviness, yellowness, malleability and fusibility because we have 'found these ideas together' in gold but we cannot know by experience either that this is true of all samples of gold or what other ideas would be found along with these on closer observation. This is because, as far as observation can tell us, simple ideas are independent of one another so they 'carry with them, in their own Nature, no visible necessary connexion, or inconsistency' with one another (IV.iii.10). However regular the connections found in experience it does not reveal to us necessary connections.

I am inclined to think that Locke's scepticism about scientific knowledge in Book IV springs very largely from his concern about the mind/body problem and there now follows an important, and difficult passage about this. At first sight this passage appears to withdraw much of what has gone before and suggest a greater pessimism than might have been expected. I shall try to give an interpretation which I think it will bear and which to some extent mitigates that scepticism.

Our complex ideas of substances depend most upon secondary qualities; our ideas of sensation, even of primary qualities, depend upon ideas of secondary qualities because, for example, we would not be able to see shapes without differences in colour. However, secondary qualities depend upon the primary qualities of 'the minute insensible parts' of bodies because they are analysable into relations between them. We could only know which secondary qualities are necessarily connected or inconsistent with one another by knowing the texture of bodies, that is, the combinations of primary qualities involved. (IV.iii.11) Thus

For not knowing the Root they spring from, not knowing what size, figure, and texture of Parts they are, on which depend and from which result those Qualities which make our complex *Idea* of Gold, 'tis impossible we should know what other Qualities result from, or are incompatible with the same Constitution of the insensible parts of *Gold*; and so consequently must always *co-exist* with that complex *Idea* we have of it, or else are *inconsistent* with it. (IV.iii.11)

Locke says that unless we have knowledge of the primary qualities of the corpuscles of bodies and their textures it is impossible that we should know the necessary connections and inconsistencies between secondary qualities. Some commentators have taken him to be saying also that it is impossible that we should have knowledge of textures. I believe that, here at any rate, he is not ruling out such knowledge as impossible but is hopeful that the natural philosophers will one day be able to provide some knowledge of them.

My main reason for saying this is that in the next paragraph Locke refers to a 'more incurable part of Ignorance'; the less incurable part must surely be at least partially curable. He says

Besides this Ignorance of the Primary Qualities of the insensible Parts of Bodies, on which depend all their secondary Qualities, there is yet another and more incurable part of Ignorance, which sets us more remote from a certain Knowledge of the *Co-existence* or *Inco-existence* (if I may so say) of different *Ideas* in the same Subject; and that is, that there is no discoverable connection between any *secondary Quality, and those primary Qualities* that it depends on. (IV.iii.12)

It looks almost as if Locke were here withdrawing what he said in the previous paragraph[1] if it were not for his reference to 'a more incurable part of Ignorance'. This suggests another interpretation to which I think the next paragraph gives a clue; it appears that Locke is concerned about the mind/body problem.

First he says

That the size, figure, and motion of one Body should cause a change in the size, figure, and motion of another Body, is not beyond our Conception; the separation of the Parts of one Body, upon the intrusion of another; and the change from rest to motion upon impulse; these and the like, seem to us to have some *connexion* one with another. And if we knew these primary Qualities of Bodies, we might have reason to hope, we might be able to know a great deal more of these Operations of them one upon another... (IV.iii.13)

If we knew the primary qualities of the corpuscles of a body in detail we should also know their arrangement i.e. the texture of the body, and we should in turn know how one texture would affect another in interactions between bodies. So since, as I have argued, secondary qualities *are* textures we would have some knowledge of the secondary qualities of bodies, in one sense at least. It looks as if Locke is saying that such knowledge is not logically impossible since all this is not 'beyond our conception'. However, there is a difference between a complete knowledge of the texture of a body and a knowledge of which features of that texture are secondary qualities, rather than, e.g. qualities of the third sort, that is, of which features produce ideas of colours, sounds, tastes and smells in us. Secondary qualities are just those features of textures that produce such ideas in us.

Thus Locke immediately goes on to say

> But our Minds not being able to discover any *connexion* betwixt these primary qualities of Bodies, and the sensations that are produced in us by them, we can never be able to establish certain and undoubted Rules, of the Consequence or *Co-existence* of any secondary Qualities, though we could discover the size, figure, or motion of those invisible Parts, which immediately produce them. We are so far from knowing what figure, size, or motion of parts produce a yellow Colour, a sweet Taste, or a sharp Sound, that we can by no means conceive how any *size, figure, or motion* of any Particles, can possibly produce in us the *Idea* of any *Colour, Taste,* or *Sound* whatsoever; there is no conceivable *connexion* betwixt the one and the other. (IV.iii.13)

That is, we cannot conceive how material qualities of things can produce *ideas* of colours, and so on, in our minds, and that is the mind/body problem.

Since I have insisted on the distinction between secondary qualities and ideas of them I must show why Locke thinks that the mind/body problem infects the relation between primary and secondary qualities. There is another relevant passage in a later chapter, where Locke says

> Nor if it were revealed to us, what sort of Figure, Bulk, and Motion of Corpuscles, would produce in us the Sensation of a *yellow* Colour, and what sort of Figure, Bulk, and Texture of Parts in the superficies of any Body, were fit to give such Corpuscles their due motion to produce that Colour, Would that be enough to make *universal* Propositions with *certainty*, concerning the several sorts of them, unless we had Faculties acute enough to perceive the precise Bulk, Figure, Texture, and Motion of Bodies in those minute Parts, by which they operate on our Senses, that so we might by those frame our abstract *Ideas* of them. (IV.vi.14)

The whole argument, I think, goes in this way. If we knew the inner constitutions of bodies we should know their textures, which include their secondary qualities. However, there would still be much that we did not know about secondary qualities: we should not know which textures were secondary qualities and which of the third sort since to know that we should have to know which textures, or features of textures, were directly responsible for ideas and that is something we could never know because of the mind/body problem. It might be thought that we could discover this by the elimination of the third sort of qualities since their operation is purely mechanical but this is not so because there may be textures, or features of them, that are neither of the third sort nor produce ideas in us. Suppose we have a body that looks yellow and tastes sour; even if we knew its texture completely we should not know which feature of its texture produced the idea of yellow and which produced the idea of sour because we cannot comprehend how physical textures can have mental effects. We might, of course, be able to establish correlations between texture and ideas by examining other bodies that were yellow but not sour or sour but not yellow but such correlations would not tell us whether the yellow colour and the sour taste were necessary concomitants. Correlations might just indicate empirical regularities unless we could discover the 'mechanics' by which the ideas were produced. This may account for Locke's statement (IV.iii.12) about there being 'no discoverable connection between any *secondary Quality, and those primary Qualities* that it depends on'. Thus knowing textures would amount to knowing qualities but in a general and unspecific way; not knowing which particular textures corresponded with particular ideas we should not be able to identify those textures which were secondary qualities.

I am here reading 'secondary quality' as 'texture, or textural feature, responsible for a specific simple idea of colour, sound, taste or odour' which is consistent with the account I have given earlier. To say that secondary qualities are textures is not to say that every texture, or textural feature, is a secondary quality.

I am suggesting then that, for Locke, the 'more incurable part of ignorance' is absolutely incurable because the mind/body problem is insoluble by philosophical, scientific, or any other means. The less incurable part I take to be the ignorance of textures; it is not logically incurable because there is no inconceivability involved in the relations between the primary qualities of the minute parts of bodies and their textures. But it would be possible to know every detail of these textures without knowing which were secondary qualities.

Locke mentions some few qualities, mainly primary, that have 'a necessary dependence and visible connexion' with one another, for example, shape implies size, but this is true of very few of them, so we can

> by Intuition or Demonstration, discover the co-existence of very few of the Qualities are to be found united in Substances: and we are left only to the assistance of our Senses, to make known to us, what Qualities they contain. For of all the Qualities that are *co-existent* in any Subject, without this dependence and evident connexion of their *Ideas* one with another, we cannot know certainly any two to *co-exist* any farther, than Experience, by our Senses, informs us. (IV.iii.14)

It is at this point that Locke mentions the corpuscular hypothesis

I have here instanced in the corpuscularian Hypothesis, as that which is thought to go farthest in an intelligible Explication of the Qualities of Bodies; and I fear the Weakness of humane Understanding is scarce able to substitute another, which will afford us a fuller and clearer discovery of the necessary Connexion, and *Co-existence,* of the Powers, which are to be observed united in several sorts of them. (IV.iii.16)

It may well be that Locke is expressing some doubts about our ability to apply even the most satisfactory available hypothesis or to construct a better. He may be saying that the corpuscular hypothesis would afford us a clear account ('discovery') of necessary connections but at the same time be doubting the possibility of establishing it in detail. He may, however, be leaving open the possibility that the natural philosophers may devise means of establishing it which we laymen cannot, in our present state, conceive. Until they do, we must rely on observation and induction. It is perhaps worth remembering that Locke appears to have thought action at a distance unintelligible and, at the same time, that we must take Newton seriously when he postulates it. (See Chapter 7 above.) There may be some parallel here.

Locke's third sort of knowledge, the agreement or disagreement of our ideas in other relations, that is, relations other than identity and diversity and co-existence, is the largest field of our knowledge and its extent is difficult to determine. The reason for this is that relations between general ideas that are not co-existent have to be discovered by discovering 'intermediate ideas' that connect them. This is one point at which Locke puts his view that it is not only ideas of quantity that allow of demonstration; even morality may allow of demonstration. What makes people think that this is not possible is the complexity of moral ideas compared with mathematical ones but mathematics itself suggests a way of dealing with this; we can make clear our moral concepts by definition. Locke appears to regard morality as forming a system in which principles are deducible from the definitions of moral concepts. This is clearly connected with his discussion of mixed modes. (See Chapter 12 above.)

Finally, Locke's fourth kind of knowledge, of the '*real, actual, Existence* of Things' is of three kinds. We have intuitive knowledge of our own existence, demonstrative knowledge of God's existence and merely sensitive knowledge of the existence of other things. Sensitive knowledge, however, 'extends not beyond the Objects present to our Senses' (IV.iii.21); it is knowledge of particulars, not of general truths.

IV

Locke now considers the causes of our ignorance. There are three main causes: the want of ideas, the want of a discoverable connection between our ideas and the failure to examine our ideas (IV.iii.22).

First, we may lack ideas which we would have if our senses were more or better. Our five senses may not be the only possible ones so we should not suppose that everything that *could* be learnt by sense-experience is available to *us*. There may be creatures with different senses which can detect things inaccessible to our senses but, of course, lacking their senses, we cannot even conceive what sorts of idea they might have (IV.iii.23). However, we may also lack ideas that we are capable of having and which we can conceive of as capable of being known by us. We have ideas of size, shape and mobility but not of the particular combinations of size, shape and mobility of the corpuscles which produce the effects we observe. We may have the conception of corpuscles with shape, size and mobility and of their operations depending upon these but we do not have ideas of them by sensation. Some things are hidden from us because they are too remote, or too small to cause sensations in us. It is the minuteness of the corpuscles and their consequent inaccessibility to the senses that prevents our arriving at general truths about their necessary connections. Thus he says

> I doubt not but if we could discover the Figure, Size, Texture, and Motion of the minute Constituent parts of any two Bodies, we should know without Trial several of their Operations one upon another, as we do now the Properties of a Square, or a Triangle. Did we know the Mechanical affections of the Particles of *Rhubarb*, *Hemlock*, *Opium*, and a *Man*, as a Watchmaker does those of a Watch, whereby it performs its Operations, and of a File which by rubbing on them will alter the Figure of any of the Wheels, we should be able to tell before Hand, that *Rhubarb* will purge, *Hemlock* kill, and *Opium* make a Man sleep; as well as a Watch-maker can, that a little piece of Paper laid on the Balance, will keep the Watch from going, till it be removed; or that some small part of it, being rubb'd by a File, the Machin would quite lose its Motion, and the Watch go no more. (IV.iii.25)

Locke continues by applying this use of the corpuscular hypothesis to the dissolving of gold in *aqua regia* and silver in *aqua fortis*. This passage appears to be strongly influenced by Boyle's famous passage about the lock and the key (*O.F.Q.* 18–19) and it is interesting that all the examples here used by Locke are also used by Boyle in that passage. Boyle certainly thought that further scientific investigation would eventually lead to detailed accounts of these matters.[2] If Boyle did not think that project hopeless I think it unlikely that Locke did, so great was his respect for Boyle.

It is true that Locke says, in this same section, 'our want of precise distinct *Ideas* of their [*sc.* the corpuscles'] primary Qualities, keeps us in an uncurable Ignorance of what we desire to know about' the natural operations of bodies, but this is still consistent with the hope that we might one day acquire those precise distinct ideas and the ignorance become curable. This possibility is again left open at the end of the section when he says

> whilst we are destitute of Senses acute enough, to discover the minute Particles of Bodies, and to give us *Ideas* of their mechanical Affections, we must be content to be ignorant of their properties and ways of Operation; nor can we be assured about them any farther than some few Trials we make, are able to reach (IV.iii.25)

If we could artificially improve our senses e.g. by the improvement of microscopes, we might possibly cure this sort of ignorance. With hindsight and our knowledge of electron microscopes we might say that Locke would not have been foolish to have thought so. However Locke has made it difficult for himself to accept this if he indeed regards the corpuscles as in principle unobservable. He would presumably point out that the use of electron microscopes involves large inferences, cannot count as direct seeing of structures and so cannot give certainty about them.

The next section appears to sound a more pessimistic note still about our ever achieving 'general, instructive, unquestionable Truths' about these things. However, what he says is

> I am apt to doubt that, how far soever humane Industry may advance useful and *experimental* Philosophy *in physical Things, scientifical* will still be out of our reach: because we want perfect and adequate *Ideas* of those very Bodies, which are nearest to us... (IV.iii.26)

and it is clear that the bodies nearest to us are *observable* bodies. This may be interpretable in a way that is consistent with what I have just been saying. Locke uses the word 'experiment' for 'experience', i.e. sense-experience, in a number of places. Experimental philosophy may thus be philosophy that relies on direct sense-experience, that is, what I have been referring to as observation and induction. Now he may be saying that out of this, out of our observational 'knowledge', alone, of the bodies around us can never come 'scientifical' knowledge, that is, certain knowledge of inner constitutions and necessary connections. But that does not imply that no method of investigation different from ordinary observation could give us 'scientifical' knowledge. This interpretation is supported to some extent by the rest of the section since he immediately points out that scientific knowledge demands adequate ideas of bodies but we never get *adequate* ideas from the observation of bodies 'that fall under the Examination of our Senses'. He then continues

> *Certainty* and *Demonstration*, are Things we must not, in these Matters, pretend to. By the Colour, Figure, Taste, and Smell, and other sensible qualities, we have as clear, and distinct *Ideas* of Sage and Hemlock, as we have of a Circle and a

Triangle: But having no *Ideas* of the particular primary Qualities of the minute parts of either of these Plants, nor of other Bodies which we would apply them to, we cannot tell what effects they will produce; Nor when we see those Effects, can we so much as guess, much less know, their manner of production. (IV.iii.26)

If observation gave us *adequate* ideas we should have not only the idea which is the taste of sage but also the idea of the textures responsible for that taste.

The second cause of our ignorance is the want of a discoverable connection between ideas we actually have and here again the mind/body problem looms large. Locke says

As the *Ideas* of sensible secondary Qualities, which we have in our Minds, can, by us, be no way deduced from bodily Causes, nor any correspondence or connexion be found between them and those primary Qualities which (Experience shews us) produce them in us; so on the other side, the Operation of our Minds upon our Bodies is as unconceivable. (IV.iii.28)

There are some ideas between which necessary connections can be shown and Locke instances, once again, the idea of a triangle which is necessarily connected with the idea of the equality of its angles to two right angles. It is only in such cases, where certain connections are 'visibly included' in the nature of the ideas themselves, that we are capable of certain and universal knowledge. In physical matters, as well as in the interaction of mind and body, where the laws of communication of motion have no natural connection with the ideas involved we can have no more than an experimental knowledge and we must ascribe the connections to 'the arbitrary Will and good Pleasure of the Wise Architect'.

What exactly is being ascribed to the wise architect? This is not occasionalism; I believe that what Locke means is that our faith in the intelligibility of the universe, depending upon necessary connections in nature, as long as we have to rely for information about the universe gained by observation, must rest on a belief in the wisdom and goodness of a God who set the whole machine going in the first place. What is arbitrary is the character of the original creation and not day-to-day intervention, which would not be likely to provide the regularities we find and rely on. God *knows* the connections which to him are not arbitrary although they appear so to us.

By sensitive knowledge we have access to *particular* things and events but no general knowledge of connections between them. So

as to a perfect *Science* of natural Bodies (not to mention spiritual Beings,) we are, I think, so far from being capable of any such thing, that I conclude it lost labour to seek after it. (IV.iii.29)

I take this to be less general and so less sceptical than it appears. Still the mind/body problem is lurking in the background and still Locke is concerned about the possibility of building a *science* of nature on direct observation alone. He says

> The Things that, as far as our Observation reaches, we constantly find to proceed regularly, we may conclude, do act by the Law set them; but yet by a Law, that we know not: whereby, though causes work steadily, and Effects constantly flow from them, yet their *Connexions* and *Dependancies* being not discoverable in our *Ideas*, we can have but an experimental Knowledge of them. (IV.iii.29)

And again

> Several effects come every day within the notice of our Senses, of which we have so far *sensitive Knowledge*: but the causes, manner, and certainty of their production ... we must be content to be ignorant of. (IV.iii.29)

This section is headed '*Instances*' and what it gives instances of is what is discussed in section 28 '*Want of a discoverable connexion between* Ideas *we have*'. As I have pointed out in discussing that section, Locke clearly has in mind ideas of sensation and he mentions the connection between the 'mechanical Affections' of bodies and the sensations they produce in us. The other main burden of that section concerns the consequences of having to rely, in the present state of natural philosophy, on observation and experiment and the impossibility of their revealing necessary connections to us. This I take to be what the sceptical passage above refers to as 'lost labour to seek after'. The scepticism concerns the power of observation and does not extend to the possibility of new methods in the natural sciences for discovering necessary connections through the discovery of the textures of things.

The third cause of ignorance arises from a mere failure to acquire, examine and compare ideas which we could or do have. This is often contributed to, Locke says, by 'the ill use of words', especially the use of words of 'undetermined and uncertain signification' (IV.iii.30). If we do not take the trouble to use our words in clear and precise ways we shall always be unclear about the precise ideas they stand for and our statements and descriptions will be vague and uncertain even about the things that are accessible to observation.

V

In Chapter iv Locke tries to make the situation look a little more promising. He deals with an anticipated criticism: if knowledge is the perception of the agreement or disagreement of our own ideas does this not allow that knowledge may be merely a matter of men's imaginations? Or, at any rate, would there be a distinction between the perception of agreement between the products of the imagination and the perception of agreement between the products of the reasoning of a sober man?

That an Harpy is not a Centaur, is by this way as certain Knowledge, and as much a Truth, as that a Square is not a Circle.

But *of what use is all this* fine *Knowledge of Men's own Imaginations*, to a Man that enquires after the reality of Things? It matters not what Men's Fancies are, 'tis the Knowledge of things that is only to be prized: 'tis this alone gives a value to our Reasonings, and preference to one Man's Knowledge over another's, that it is of Things as they really are, and not of Dreams and Fancies. (IV.iv.1)

Locke replies by saying what kinds of *real* knowledge we can have, referring back to his discussion of the reality of ideas. Knowledge is real 'only so far as there is a conformity between our *Ideas* and the reality of Things' and, as we have seen, there are two kinds of ideas that we can be sure 'agree with Things': simple ideas and complex ideas other than those of substances. There is also real knowledge of mathematical truths (IV.iv.6) and there is the possibility of real knowledge of moral truths (IV.iv.7–10). We may have real knowledge of substances also but here there is more room for error (IV.iv.11–12).

About substances, Locke says

I say then, that to have *Ideas* of *Substances*, which, by being conformable to Things, may afford us *real* Knowledge, it is not enough, as in Modes, to put together such *Ideas* as have no inconsistence, though they did never before so exist. *V.g.* the *Ideas* of *Sacrilege* or *Perjury, etc.* were as real and true *Ideas* before, as after the existence of any such fact. But *our Ideas of Substances* being supposed Copies, and referred to *Archetypes* without us, must still be taken from something that does or has existed; they must not consist of *Ideas* put together at the pleasure of our Thoughts, without any real pattern they were taken from, though we can perceive no inconsistence in such a Combination. (IV.iv.12)

If we *find* certain simple ideas going together in nature *e.g.* those that make up our idea of gold, then there must be corresponding features of the inner constitution of gold that produce those ideas. Different collections of simple ideas making up complex ideas of different substances reflect differences of inner constitution. Locke says

Herein therefore is founded the *reality* of our Knowledge concerning *Substances*, that all our complex *Ideas* of them must be such, and such only, as are made up of such simple ones, as have been discovered to co-exist in Nature.... Whatever *Ideas* we have, the Agreement we find they have with others, will still be knowledge. (IV.iv.12)

Locke then adds, rather mysteriously,

If those *Ideas* be abstract, it will be general Knowledge.... Whatever simple *Ideas* have been found to co-exist in any Substance, these we may with confidence join together again, and so make abstract *Ideas* of Substances. For whatever have once had an union in Nature, may be united again. (IV.iv.12)

I say this is mysterious because it seems that Locke is ignoring the problem of induction, of which he has elsewhere shown himself to be very well aware. It looks as if he is saying: we find particular samples of gold to be yellow, malleable, fixed and ductile and we may generalize this into 'Gold is yellow, malleable, fixed and ductile' and this is as certain as the original singular statements. However, how can we be sure that samples of gold as yet unexamined will have these properties? This is clearly to misunderstand his meaning. What he must mean is this: having found that samples of what we call gold have these properties we will turn this into a definition, an abstract idea of gold, so that in future nothing that does not have these properties will be counted as gold. Since these properties have been found together in nature we are very likely to find them together again. What is certain is that if we don't, we won't find gold. This, of course, is real knowledge of nominal essences and is real because nominal essences have a 'foundation in nature'. This passage is followed by Locke's arguments about natural kinds which I have already discussed. (See Chapter 13 above.)

The chapter ends thus

Where-ever we perceive the Agreement or Disagreement of any of our *Ideas* there is certain Knowledge: and where-ever we are sure those *Ideas* agree with the reality of Things, there is certain real Knowledge. Of which Agreement of our *Ideas* with the reality of Things, having here given the marks, I think I have shewn wherein it is, that *Certainty, real Certainty*, consists. (IV.iv.18)

VI

Chapter xi of Book IV, 'Of our Knowledge of the Existence of Other Things' may appear to contradict some of the things I have said in this chapter and in earlier ones. In it Locke seems, at least at times, to be saying that we infer the existence of external objects from our ideas. For example, he says,

'Tis therefore the actual receiving of *Ideas* from without, that gives us notice of the *Existence* of other Things [*sc.* other than ourselves and God], and makes us know, that something doth exist at that time without us, which causes that *Idea* in us... (IV.xi.2)

However, I have claimed that he assumes from the beginning, as part of the grand hypothesis, that there are external objects that cause our ideas. Must I now withdraw that?

I suggest that in this chapter he is doing at least two things. I do not think he doubts the existence of external objects or wishes to reject part of the hypothesis with which he began, or that he no longer sees this hypothesis as being confirmable by the success of the explanations that can be based upon it. There may, however, still be those who are sceptical about the efficacy of this method and for their benefit he is adding further, direct, arguments for a belief in external objects. That may be one of his aims here. The other, and perhaps more important, task is to consider not how we can know that external objects exist and cause our ideas but how we can know that this or that particular object exists here and now. This is important since we may have ideas that are not caused by existent objects but are constructed by us out of simpler ideas by our imagination. Perhaps we can make inferences about this from the nature of the ideas themselves. This is suggested by the penultimate section of the chapter where, in the course of comparing particular propositions with general propositions and considering our knowledge of them, he says

> There is one sort of Propositions *concerning* the *Existence* of any thing answerable to such an *Idea:* as having the *Idea* of an *Elephant, Phoenix, Motion*, or an *Angel*, in my Mind, the first and natural enquiry is, Whether such a thing does any where exist? And this Knowledge is only of *Particulars*. No existence of any thing without us, but only of GOD, can certainly be known farther than our Senses inform us. (IV.xi.13)

The argument begins in Chapter ix of this book. He is moving from a consideration of abstract ideas to a consideration of 'particular Existences' and he discusses in turn our own existence, that of God and that of other things. He says that

> all *particular Affirmations or Negations*, that would not be certain if they were made general, are only concerning *Existence*; they declaring only the accidental Union or Separation of *Ideas* in Things existing, which in their abstract Natures, have no known necessary Union or Repugnancy. (IV.ix.1)

He argues, in this and the next chapter, 'that we have Knowledge of *our own Existence* by Intuition; of the *Existence of* GOD by Demonstration,' and then in Chapter xi he proceeds to argue that we have the knowledge of other things, that is external objects, by sensation. He begins by saying

> no particular Man can know the *Existence* of any other Being, but only when by actual operating upon him, it makes it self perceived by him. For the having the *Idea* of any thing in our Mind, no more proves the Existence of that Thing, than the picture of a Man evidences his being in the World... (IV.xi.1)

and

'Tis therefore the actual receiving of *Ideas* from without, that gives us notice of the *Existence* of other Things, and makes us know, that something doth exist at that time without us, which causes that *Idea* in us, though perhaps we neither know nor consider how it does it... (IV.xi.2)

Note the particularity of this. The question is about the existence of a particular thing 'at that time' which causes 'that *Idea*' in us, and not about the general matter of the existence of external objects that cause our ideas.

The example he uses immediately is *white*. As I write this, he says

I have, by the Paper affecting my Eyes, that *Idea* produced in my Mind, which whatever Object causes, I call *White*; by which I know, that that Quality or Accident (*i.e.* whose appearance before my Eyes, always causes that *Idea*) doth really exist, and hath a Being without me. (IV.xi.2)

He is talking about the existence *now*, in *this* situation, of an object of a particular sort that I call white. We can be almost certain of such things. It is true that we have less certainty of them than we have in intuitive or demonstrative knowledge but we gain thereby 'an assurance that *deserves the name of Knowledge*'. When ideas are externally caused I can be almost certain that objects that regularly cause ideas of whiteness, for example, have in common a particular quality, that is, a particular texture.

This depends, I think, on two considerations, both of them relating to my interpretation of Locke's view of ideas. On that view, to see something is just for it to appear in some way to me; the appearance is the idea. To have an idea of this kind is to see the object. Support for this occurs here, when Locke says 'For I think no body can, in earnest, be so sceptical, as to be uncertain of the Existence of those Things which he sees and feels' (IV.xi.3). What we see and feel, in having ideas, are objects and their qualities; *they* appear in some way to us.[3]

My interpretation of 'ideas' also involved saying that Locke saw, and relied upon, a difference between, on the one hand, ideas of sensation and, on the other, ideas of memory and imagination. This too is confirmed here. He says 'there is no body who doth not perceive the difference in himself, between contemplating the Sun, as he hath the *Idea* of it in his Memory, and actually looking upon it...' (IV.xi.5). I can remember an idea that involved physical pain without its now involving pain; I can remember the pain of hunger and thirst without now experiencing them. In certain favourable cases, at least, we are able to tell the difference between ideas that are now being caused by objects without us and those that are not; we usually know which ideas are being caused by external objects. So we can know that there are such things as elephants and motion although we may never know whether there are such things as phoenixes or angels.

Those who are still sceptical, Locke says, should consider that certain ideas are inaccessible to people lacking a sense-organ, that some ideas are forced upon us while we can banish others at will, that the different senses can confirm one another's findings and, finally, that it is vain to expect the kind of certainty given by demonstration about things not capable of demonstration.

One might say, at the risk of sounding anachronistic, that although Locke himself regards the sceptic's doubts as unreasonable and even unreal, even the sceptic may be encouraged to make from the phenomena an 'inference to the best explanation'.

VII

I shall now try to summarize Locke's position. We have knowledge of identity and diversity but this is knowledge only of individual ideas; it is *real* when it concerns simple ideas of sensation because when we discern differences between them there must be differences in their causes, that is, in the qualities of external things. Knowledge of co-existence covers both substances and non-substances. We have knowledge of a substance when the idea we have of it contains only simple ideas that have been found together; we have knowledge of non-substance, such as jealousy, when we apply the word 'jealous' to something in which we find together those ideas that we have used in defining it. Both these are real knowledge; because we *find* simple ideas together in sensation we can be sure that this happens because their causes occur together.

Knowledge of the agreement or disagreement of ideas of relations other than identity, diversity and co-existence, depends upon 'intermediate' ideas and is certain only if deductive steps connect two ideas *via* these intermediate ideas. This concerns primarily mathematical ideas although there is a possibility of extending it to ethical ideas. If we apply the word 'triangle' only to those things in the world that clearly fit the definition of the word then real knowledge results: the relations we can prove to exist between the ideas involved must hold also in the real things to which we thus apply the word.

Finally, knowledge of the actual real existence of things is certain and real but it is only of individuals and is not universal.

This, of course, has important consequences for scientific explanation. There are necessary connections in nature, according to the corpuscular hypothesis, because the real essences of things, their inner corpuscular constitutions, determine their properties and their possible or actual interactions with other possible or actual inner constitutions. If we knew these real essences we should be able to predict with certainty and without experiment how they would react with one another. The possibility of such explanations is at least intelligible to us and it is not in principle impossible that we should one day achieve some knowledge of inner constitutions; we should then have general scientific truths about the natural world comparable to those we now have in mathematics. Unless and until we reach this situation we have no alternative but to rely upon observation and induction. Observation can give us certainty about individual occurrences but induction cannot give us certain general truths; the most we can hope for is probable conjectures and predictions that have a greater or lesser probability. We cannot claim that this is knowledge, however high the probability less than certainty, according to Locke, because he takes a narrow and rigid view of knowledge. In the end, therefore, Locke's view of the procedures and powers of science, in the present state of the game, is not much different from that of more extreme empiricists such as Hume or Mach.

There is one area in which our ignorance is permanent and incurable. Because of the insolubility of the mind/body problem we shall never understand how external bodies produce ideas in us, although it is certain that they do. Only God can understand this. It is interesting that Boyle appears to take the same view but adds that if it were possible for us to talk to God he might be able to explain it to us even though it is not possible for us to explain it on our own. He says that some atomists claim that any phenomenon that cannot be explained by their methods cannot be explained at all. But, he continues,

> who has demonstrated to us, that men must be able to explicate all nature's phænomena, especially since diverse of them are so abstruse, that even the learnedest atomists scruple not to acknowledge their being unable to give an account of them. And how will it be proved, that the omniscient God, or that admirable contriver, Nature, can exhibit phænomena by no ways, but such as are explicable by the dim reasons of man? I say explicable, rather than intelligible; because there may be things, which though we might understand well enough, if God, or some more intelligent being than our own, did make it his work to inform us of them, yet we should never of ourselves find out these truths.[4]

We might wonder what is the justification for holding a metaphysical view of the universe which, because of our limitations, we may never be able to establish by any means open to us for investigating nature. One of Locke's defences would be, I think, one that conforms to a pattern adopted by other, more thoroughgoing metaphysicians, such as Leibniz, namely, that the kind of world that emerges from scientific investigation is the kind of world we would expect if reality is as postulated. He might go farther than this and say that the increasing success of the explanatory methods of science, based on the corpuscular account of this reality, gives some evidence, though never conclusive evidence, that the metaphysical account is correct; and that it may be possible, eventually, to investigate inner constitutions sufficiently closely to give us some knowledge of necessary connections between things, even if we cannot now imagine what the method could be.

Another advantage that I think Locke sees in the metaphysical account is that it justifies a limited faith in the power of observation and induction; if there are necessary connections in nature then it is sensible to look for regularities which, even if we cannot establish their necessity, depend upon and reflect those connections.

However, it is likely that Locke would take an even more rationalistic line and say simply that the corpuscular hypothesis, with its consequent necessary connections, is simply the most reasonable and plausible kind of hypothesis about the structure of the natural world. It is perhaps Locke's willingness to adopt such attitudes and to suppose that the hypothesis might one day be established that accounts for his wavering, in Book IV, between pessimism and optimism about knowledge and which prevents his being plunged into the sceptical gloom that drove Hume for relief to backgammon and coffee.

We may wonder whether even the extent of scepticism Locke allows himself is justified. His deep respect for the work of Boyle, who was far from sceptical of the power of science, and his own use of the hypothetico-deductive method do give him some hope that the improvement of scientific techniques and apparatus will give us more and more insight into the inner constitutions of things. In his important passage on probability and analogy he says

> observing that the bare rubbing of two Bodies violently one upon another, produces heat, and very often fire it self, we have reason to think that what we call Heat and Fire, consists in a violent agitation of the imperceptible minute parts of the burning matter: Observing likewise that the different refractions of pellucid Bodies produce in our Eyes the different appearances of several Colours; and also that the different ranging and laying the superficial parts of several Bodies, as of Velvet, watered Silk, *etc.* does the like, we think it probable that the Colour and shining of Bodies, is in them nothing but the different Arrangement and Refraction of their minute and insensible parts. (IV.xvi.12)

Later in the section he says that 'a wary Reasoning from Analogy leads us often into the discovery of Truths, and useful Productions, which would otherwise lie concealed'.

The examples Locke takes here are closely related to many of Boyle's experimental investigations of heat and colour and Boyle clearly thought that such indirect methods would gradually lead to the knowledge of inner constitutions. Such methods, aided by the indefinite improvement of optical and other equipment which Locke does not think impossible, would give information converging on hypotheses approaching certainty about inner constitutions. Of course, these methods remain inductive because their conclusions concern things beyond our senses. Nevertheless they are in principle capable of giving us highly probable statements that justify rational assent. The approach to them can be methodical and, in a modern sense, scientific. They remove the justification for an all-embracing scepticism about knowledge of inner constitutions with which Locke has often been credited and no doubt account for his allowing a less incurable ignorance.

I believe that, in fact, Locke would not have been surprised by the enormous developments in science, very much along the lines proposed by Boyle, since his day. He would not have called it *science* or allowed that it yielded *knowledge* because he regarded both those terms as involving certainty but he did attach much importance to probability and analogy in the investigation of nature. That I take to be the whole point of his doubts about the possibility of a 'scientifical' knowledge of inner structures (IV.iii.26) but he would never have said that what *we* call scientific investigation was useless or that it should be abandoned.

He appears more sceptical about this than he really is, I suggest, because he was opposing an *a prioristic* approach to nature infected by an overwhelming faith in the certainty of conclusions based on Aristotelian and alchemical methods and in the face of this the appropriate things to recommend were caution, close attention to clarity in language, close investigation of the phenomena and the refusal to theorize without reliance on evidence. A statement he made primarily with reference to problems about the immateriality of the soul has for him, I think, a more general reference. It is:

> I think not only, that it becomes the Modesty of Philosophy, not to pronounce Magisterially, where we want that Evidence that can produce Knowledge; but also that it is of use to us, to discern how far our Knowledge does reach; for the state we are at present in, not being that of Vision, we must, in many Things, content our selves with Faith and Probability... (IV.iii.6)

That is perhaps a suitable last word.

COLLATION OF EDITIONS FOR QUOTATIONS FROM BOYLE

Pagination in 1772 edition and 1st edition
(Page numbers are given only for displayed quotations)

The Sceptical Chymist

1772	*1st*	*1772*	*1st*
468	16	500–1	131–3
469	19–20	506	152
470–1	21–4	516	187–8
476	44	525	214–16
479	55	528	226–7
481	61	550	307
438–4	68–71	551	309
487	82	556	328–9
488	85–6	557	331–4
489–90	90–1	562	350
491–2	99–100	569	376

The Origin of Forms and Qualities

1772	*1st*	*1772*	*1st*
1–14	Not paginated	29	71
15	3–6	32	81
16	8–9	35	95–7
18	17–19	37	144
22	36–8	40	157
23	41	41	159–60
24	47	45	178
25	49–52	46	185
26	56–7	47	186
27	60		

NOTES

Introduction

1 See the editor's introduction to John Locke *An Essay Concerning Human Understanding*, ed. P. H. Nidditch. In what follows references will be to and quotations from this edition. References will be given in the usual form: thus IV.iii.16, without a code letter, refers to Book IV, Chapter iii, section 16 of the *Essay*.

2 See *Locke and Berkeley*, eds. C. B. Martin and D. M. Armstrong. Martin, in his section of the Introduction to this collection of essays, lists some of the mistakes current up to the 1960s.

3 G. A. J. Rogers, 'Boyle, Locke and reason', *J.Hist.Ideas*, XXVII (1966), 205–16.

4 eg. H. M. Bracken, *The Early Reception of Berkeley's Immaterialism*. David Berman, 'On missing the wrong target', *Hermathena*, CXIII (1972), 54–67.

5 John Locke, *An Essay Concerning Human Understanding* : *An Abridgment*, ed. John W. Yolton. Yolton's complete Everyman edition of 1961 was much more satisfactory.

6 He had over 3000 items in his personal library, including works by contemporary and ancient scientists as well as philosophers. See *The Library of John Locke*, ed. John Harrison and Peter Laslett.

7 Pierre Gassendi is an interesting figure about whom surprisingly little has been written. For a general introduction to his work and influence see G. S. Brett, *The Philosophy of Gassendi* and M. H. Carré, 'Gassendi and the new philosophy', *Phil.*, XXXIII (1958), 112–20.

8 A variety of views about influences on Locke may be found in the following books and articles: Ira O. Wade, *The Intellectual Origins of the French Enlightenment*; R. I. Aaron, *John Locke*; L. Laudan, 'The clock metaphor and probabilism', *Annals of Sci.*, XXII (1966), 73–104; G. A. J. Rogers, 'Descartes and the method of English science', *Annals of Sci.*, XXIX (1972), 237–55; H. A. S. Schankula, 'Locke, Descartes and the science of nature', *J.Hist.Ideas*, XLI (1980), 459–77; G. A. J. Rogers, 'The empiricism of Locke and Newton', in *Philosophers of the Enlightenment*, ed. S. C. Brown (R.I.P. Lectures, vol. 12), Sussex, Harvester Press, 1979, pp. 1–30; R. Specht, 'Uber empiricistische ansätze Lockes', *Allgemeine Z. Phil.*, III (1977), 1–35; R. Brandt, 'Historical observations of the genesis of the three-dimensional optical picture (Gassendi, Locke, Berkeley)', *Ratio*, XVII (1975), 176–90; G. Coirault, 'Gassendi et non Locke créatur de la doctrine sensualiste moderne

sur la génération des idées', *Actes du Congrès du Tricentenaire de Pierre Gassendi*, Digne, 1955, pp. 71–94; M. J. Osler, 'Providence and the divine will: the theological background to Gassendi's views on scientific knowledge', *J.Hist.Ideas*, XLIV (1983), 549–60; 'John Locke and the changing ideal of scientific knowledge', *J.Hist.Ideas*, XXXI (1970), 3–16; E. A. Driscoll, 'The influence of Gassendi on Locke's hedonism', *Int.Phil.Q.*, XII (1972), 87–110.

9 Maurice Cranston, *John Locke, A Biography*, pp. 337–8.

10 See *Essay* II.viii.11 and 12 and Nidditch's notes thereto, and Chapter 7 below.

11 G. A. J. Rogers, 'Locke's *Essay* and Newton's *Principia*', *J.Hist.Ideas*, XXXVII (1978), 217–32; 'The systems of Locke and Newton' in *Contemporary Newtonian Research*, ed. Z. Beckler, pp. 215–38; Marie Boas Hall, *Robert Boyle on Natural Philosophy*, p. 110.

12 *Essay*, pp. 9–10.

13 Boas Hall, *Robert Boyle*, p. 107; Boyle, *Works*, Vol. V, pp. 655–83.

14 Cranston, *John Locke*, pp. 75–6, 88, 353, 361; R. I. Aaron, *John Locke*, pp. 12–14; M. A. Stewart, 'Locke's professional contacts with Robert Boyle', *Locke Newsletter* 12 (1981), 19–44.

15 R. H. Kargon, *Atomism in England from Hariot to Newton*, chapter X and p. 129.

16 John W. Yolton, *Locke and the Compass of Human Understanding*, pp. 58 and 63, note 3; Maurice Mandelbaum, *Philosophy, Science and Sense Perception*, pp. 7–8, takes a different view.

17 L. Laudan, 'The nature and sources of Locke's view on hypotheses', *J.Hist.Ideas*, XXVIII (1967), 211–23.

18 Robert Boyle, *Experiments, Notes, etc. About the Mechanical Origin or Production of Divers Particular Qualities* (1675), *Works* vol. IV, pp. 230–354. This quotation from p. 232. Quoted by Boas Hall, *Robert Boyle*, p. 234.

19 M. A. Stewart (ed.), *Selected Philosophical Papers of Robert Boyle*, p. viii.

1 Boyle on Empirical Investigation

1 A. E. Waite (ed.), *The Hermetic and Alchemical Writings of Paracelsus*, vol. II, p. 380; J. R. Partington, *A History of Chemistry*, vol. II, p. 134.

2 For further background see:

Partington, *History*; R. P. Multhauf, *The Origins of Chemistry*; A. G. Debus, *The English Paracelsians*; A. G. Debus, *The Chemical Philosophy*; Marie Boas, *Robert Boyle and Seventeenth Century Chemistry*.

3 Debus, *English Paracelsians*, Chap. 1.

4 Boas, *Seventeenth Century Chemistry*, Chaps. 1 and 2.

5 Robert Boyle, *Experiments and Considerations Touching Colours* (The *Experimental History of Colours*) (1664) in Boyle, *Works*, vol. I. Quotations are from this edition, indicated by *E.C.T.C.* followed by a page number. References are also given to a facsimile of the first edition ed. Marie Boas Hall, indicated by *F* followed by a page number. Mercury was used as an agent for the production of saliva in medical treatment.

6 Boas, *Seventeenth Century Chemistry*, Chaps. 1 and 2 and especially pp. 63–74.
7 Boas, *Seventeenth Century Chemistry*, pp. 95–7.
8 Quotations from *The Sceptical Chymist* are taken from Boyle, *Works*, vol. I, indicated by *S.C.* followed by a page number. References are also given to the edition by M. M. Pattison Muir (Everyman Library) indicated by *E.* followed by a page number.
9 See, e.g., K. R. Popper, *Conjectures and Refutations*, p. 14. Popper sees Bacon as rejecting 'the method of conjecture or hypothesis'. It is also of interest in this connection that Bacon said 'in establishing any true axiom the negative instance is the most powerful'. Bacon, *Novum Organum* (1600), Book I, Aphorism XLVI, p. 392.
10 Debus, *English Paracelsians*, p. 27.
11 Debus, *English Paracelsians*, p. 29.
12 Quoted by Debus, *English Paracelsians*, p. 28. See also Partington, *History*, vol. II, p. 142.
13 It is not entirely clear what 'alcali' meant. It appears to have been used for substances, thought to be sulphureous, which were slippery to the touch. Phlegm: a watery, tasteless and odourless product of distillation.
14 Boyle misreports van Helmont who also omitted the weight of the leaves because he clearly says that he did not take them into account. The relevant passage is in *Ortus Medicina* (Amsterdam, 1748) but a translation by Naphtali Lewis is to be found in *Great Experiments in Biology*, ed. M. L. Gabriel and S. Fogel, p. 155.
15 Aristotle, *de Generatione et Corruptione*, 329b.
16 Thomas Erastus, *Disputationes de Medicina nova Paracelsi* (1572); see Debus, *English Paracelsians*, p. 38. Erastus was a German-Swiss theologian who wrote on medicine, astrology, and alchemy.

2 Boyle and the Peripatetics

1 Robert Boyle, *The Origin of Forms and Qualities*, in *Works*, vol. III, pp. 1–112. Quotations are from this edition, indicated by *O.F.Q.* followed by a page number. The passages quoted will also be found in a modernized version in Stewart (ed.) *Philosophical Papers*. References are also given to this edition, indicated by *S.* followed by a page number.
2 Sennert (1572–1637) is often referred to by Boyle as one worthy of respect but sometimes misguided. This passage is quoted from Sennert, *De Chymicorum cum Aristotelicis et Galenicis Consensu ac Dissensu Liber* (1619) at *S.C.* 551, *E.* 166–7, and I am indebted to Mr S. J. Tester for the translation of Boyle's version of it.
3 Lucretius, *De Rerum Natura*, I.823–9 and II.688–99.
4 See, e.g. Multhauf, *Origins*, p. 122; W. D. Ross, *Aristotle*, pp. 105–8 and 167–76.

5 This idea may have been derived from Aristotle *de Gen. et Corrup.*, I.3.
6 Bacon, *Novum Organum*, Book II, Aphorism V, pp. 451–2.
7 Boas, *Seventeenth Century Chemistry*, pp. 82–3; Boas Hall, *Robert Boyle*, p. 59.
8 Boas, *Seventeenth Century Chemistry*, p. 77.
9 See, for example: W. Charlton (ed.), *Aristotle's Physics*, I, II, pp. 129–45; C. J. F. Williams (ed.), *Aristotle's de Generatione et Corruptione*, pp. 102–3, 154–5, 176–7 and 211–19; G. E. M. Anscombe, 'Aristotle' in G. E. M. Anscombe and P. T. Geach, *Three Philosophers*; Hugh R. King, 'Aristotle without *prima materia*', *J.Hist.Ideas*, XVII (1956), 370–89; Friedrich Solmsen, 'Aristotle and prime matter, a reply to Hugh R. King', *J.Hist.Ideas*, XIX (1958), 243–52; W. Charlton, 'Prime matter: a rejoinder', *Phronesis*, XXVIII (1983), 197–211. In what follows I rely heavily on Joseph Owens, 'Matter and predication in Aristotle', in *Aristotle*, ed. J. M. E. Moravcsik. This is a Thomist interpretation of Aristotle but I make no apologies for that because it contains the basis of the views that Boyle found in current Aristotelianism and attacked.
10 Aristotle, *Physics*, I.7, especially 190a32–191a22. Owens in Moravcsik (ed.), p. 204.
11 Aristotle, *Physics*, III.1 and V.i; *de Gen. et Corrup.*, I.2 and 4.
12 Owens in Moravcsik (ed.), p. 206.
13 Aristotle, *de Gen. et Corrup.*, I.4; Anscombe in Anscombe and Geach, *Three Philosophers*, pp. 52–4; W. D. Ross, *Aristotle*, p. 105.
14 Ross, *Aristotle*, pp. 21–5. The source for this doctrine is Aristotle, *Categories*.
15 Owens in Moravcsik (ed.), pp. 198–201.
16 Owens in Moravcsik (ed.), pp. 199–200.
17 Aristotle shows an awareness of the importance of this distinction at *de Gen. et Corrup.*, I.2.
18 G. W. Leibniz, *Discourse on Metaphysics* (1686). References are to the edition by P. G. Lucas and L. Grint. Leroy E. Loemker, in 'Boyle and Leibniz', *J.Hist.Ideas*, XVI, shows that Leibniz was influenced by Boyle in this matter. He had read *O.F.Q.* when he talked with Boyle in London in 1673.
19 G. W. Leibniz, Letter to Arnauld, 14 July 1686 in Leibniz, *Philosophical Writings*, ed. G. H. R. Parkinson, p. 63.
20 J. C. Scaliger (1484–1558), polymath, chemist and medical practitioner, wrote commentaries on Hippocrates, Aristotle and Cardan.

3 Boyle's Corpuscular Philosophy

1 But see: *A Disquisition about the Final Causes of Natural Things* (1688) in *Works*, vol. V, pp. 392–452; *A Discourse of Things Above Reason* (1681) in *Works*, vol. IV, pp. 406–47; and Stewart, *Philosophical Papers*, pp. 208–42.
2 For an opposed view see M. D. Wilson, 'Superadded properties: the limits of mechanism in Locke', *A.P.Q.*, XVI (1979), 143–50.
3 Boyle, *About the Excellency and Grounds of the Mechanical Hypothesis*

(1647), in *Works*, vol. IV, pp. 67–8. Quotations are taken from this edition, indicated by *E.G.* followed by a page number. References also given to Stewart, *Philosophical Papers*, indicated by *S* followed by a page number.

4 By, among others, E. M. Curley in 'Locke, Boyle and the distinction between primary and secondary qualities', *Phil.Rev.*, LXXXI (1972), 438–64. See my reply 'Curley on Locke and Boyle', *Phil.Rev.*, LXXXIII (1974), 229–37.

5 Tubal-cain, son of Lamesh, was the first blacksmith. Genesis 4.22.

6 The word 'decomposition' meant the opposite of what we mean by it. 'Composition' meant 'putting together simple parts'; 'decomposition' meant 'putting together parts already composite'. See *O.E.D.*

7 See Boyle, *Works*, vol. I, p. 2.

8 See, e.g., Newton, *Opticks* (1704), Book III, Pt. 1.

9 Boyle, *Mechanical Origin*, *Works*, vol. IV, pp. 314–15. Quoted in Boas Hall, *Robert Boyle*, pp. 247–8.

10 F. J. O'Toole in his 'Qualities and powers in the corpuscular philosophy of Robert Boyle', published in the same year as my 'Locke and Boyle on primary and secondary qualities', took a view of Boyle similar to mine until he came to secondary qualities. I have not changed my mind about this but have elaborated my original account. Further reasons are given in Chap. 7 below.

4 Ideas

1 See, e.g., G. J. Warnock, *Berkeley*; D. J. O'Connor, *John Locke*; D. M. Armstrong (ed.), *Berkeley's Philosphical Writings.*

2 J. W. Yolton, *John Locke and the Way of Ideas*, Chap. III, sect. 2.

3 Newton's use of 'phantasm' in this way is to be found in his letter to Locke dated 30 June 1691, *The Correspondence of John Locke*, ed. E. S. de Beer, vol IV, pp. 288–90.

4 See *O.E.D.*, defs. 'Phantasm', 'Notion', 'Species'.

5 Armstrong (ed.), *Berkeley*, p. 8.

6 N. Kemp Smith, *New Studies in the Philosophy of Descartes*, pp. 52–3

7 Aristotle, *de Anima*, 434a17 and 425b23.

8 D. W. Hamlyn, *Sensation and Perception*, p. 21.

9 Epicurus *Letters, Principal Doctrines, Vatican Sayings*, ed. R. M. Geer, pp. 14–17.

10 *Works of Thomas Reid*, ed. Sir William Hamilton, vol. II, note M. pp. 951–60.

11 F. C. Copleston, *History of Philosophy*, vol. III, pp. 248–50.

12 Leibniz, *Monadology*, ed. R. Latta, pp. 219–20, fn. 10.

13 Descartes, *Dioptiques* in Descartes, *Discourse on Method, Optics, Geometry*, ed. P. J. Olscamp, p. 68.

14 See also Descartes, *Principles of Philosophy*, in Descartes, *Philosophical Writings*, ed. Elizabeth Anscombe and P. T. Geach, I, lxix, p. 195: 'It is true that, upon seeing a body, we are no less certain of its existence *qua* that which appears coloured than *qua* that which appears with a shape...' Tables never appear as painful.

15 Yolton, *Compass*, p. 121.
16 Locke sometimes refers to them as 'faculties' and sometimes as 'operations'. At II.ix.15, for example, he refers to perceptions as 'the first Operation of all our intellectual Faculties'. There is probably no important distinction being made. The mind is said to be often active in memory at II.x.7.
17 It is worth noting that the 'Port-Royal Logic', i.e. A. Arnauld, *The Art of Thinking* (1662), does not make this distinction but includes discerning in abstracting. See Part I, Chap. 5.

5 Qualities

1 See, e.g., Jonathan Bennett, *Locke, Berkeley, Hume*, p. 28. Bennett's discussion is complicated by his use of a corrupt text.
2 About Locke's revisions after reading Newton, see below Chap. 7.
3 See, e.g., J. L. Mackie, *Problems from Locke*; Reginald Jackson, 'Locke's distinction between primary and secondary qualities' in *Locke and Berkeley*, ed. C. B. Martin and D. M. Armstrong, pp. 53–77.
4 See almost any commentary written before the 1960s and Warnock, *Berkeley*, p. 94.
5 Manna is a laxative prepared from the solidified gum of the manna ash. It produces griping pains and nausea.
6 O'Connor, *Locke*, p. 66; J. L. Mackie, *Problems*, pp. 9–11, 20–1.
7 See Chapter 11, n. 1.

6 Which Qualities are Primary?

1 See, e.g., Bennett, *Locke, Berkeley, Hume, passim*; Curley, *Phil.Rev.*, LXXXI; D. H. Sanford, 'Does Locke think hardness is a primary quality?' *Locke Newsletter*, 1 (1970), 17–29.
2 In my article 'Curley on Locke and Boyle', *Phil.Rev.*, LXXXIII, p. 235 I made a foolish and mystifying mistake. I said 'Size is intended to include bulk, which Locke calls "solidity"...' That is, of course, quite wrong. Locke uses 'bulk' for 'size' just as Boyle does.
3 Boyle, *O.F.Q.*, p. 22.
4 See above pp. 135–6.
5 Hilary Putnam, 'On properties' (1970), in Putnam, *Philosophical Papers*, vol. I, pp. 305–22.
6 There is an interesting discussion of this matter in G. C. Nerlich's critical notice of *Reduction, Time and Reality*, ed. Healey, *Phil. Q.*, XXXII (1982), pp. 272–9.

7 Powers

1 See, e.g., Bennett, *Locke, Berkeley, Hume*; Jackson, 'Locke's distinction between primary and secondary qualities' in Martin and Armstrong (eds.), pp. 53–77; Mackie, *Problems*; also Robert Cummins, 'Two troublesome

claims about qualities in Locke's *Essay*', *Phil.Rev.*, LXXXIV (1975), 401–18.
2 See Jackson in *Locke and Berkeley*, ed. Martin and Armstrong.
3 That Locke was aware of this point can be seen in his discussion of essences. See esp. III.vi.5.
4 Mackie, *Problems*, pp. 11–12.
5 For problems associated with Locke's view of necessary connections see R. S. Woolhouse, *Locke's Philosophy of Science and Knowledge*, pp. 141–3 and Chap. IX.
6 See Nidditch's introduction to his edition of Locke's *Essay*, p. xv.
7 See Nidditch's edition of the *Essay*, pp. 135–6. In the last line of the quotation 'other' was omitted from editions 2 and 3 and 'than by' was replaced by 'without'.
8 Rogers, 'Locke's *Essay* and Newton's *Principia*', *J.Hist.Ideas*, XXXVII.
9 *The Philosophical Works of Descartes*, ed. E. S. Haldane and G. R. T. Ross, vol. I, p. 269.

8 What are Secondary Qualities?

1 See, e.g., E. M. Curley in *Phil.Rev.*, LXXXI (1972), p. 440; Bennet, *Locke, Berkeley, Hume*, passim.
2 N. Kretzmann, 'The main thesis of Locke's semantic theory', *Phil.Rev.*, LXXVII (1968), 175–96, reprinted in *Locke on Human Understanding*, ed. Tipton, pp. 123–40. References are to the reprint.
R. J. Butler, 'Substance un-Locked', *P.A.S.*, LXXIV (1973/4), 131–60.
3 Boyle, *An Introduction to the History of Particular Qualities* (1671), *Works*, vol. III, pp. 292–305. This quotation from p. 292. Also in *Philosophical Papers*, ed. Stewart, p. 7.

9 Observability

1 Yolton, *Compass*, pp. 46–81.
2 Maurice Mandelbaum, *Philosophy, Science and Sense Perception*, p. 18.
3 See e.g. Bennett, *Locke, Berkeley, Hume*, pp. 68–79; Mackie, *Problems*, pp. 51–6.
4 Yolton, *Compass*, pp. 127–37, 208–23.
5 I. C. Tipton, *Berkeley*, pp. 18–26 and Chap. 6.
6 Tipton, *Berkeley*, pp. 23–6.

10 Patterns and Resemblance

1 G. Berkeley, *Philosophical Commentaries* (1707–8) in Berkeley, *Philosophical Works*, ed. M. R. Ayers, p. 297, note 484. References will be given to Ayers' edition of Berkeley's works either, for the *Dialogues*, by page number or, for other works by section and page number. See also Tipton, *Berkeley*, p. 33.

2 Berkeley, *Three Dialogues between Hylas and Philonous*, in Berkeley, ed. Ayers, p. 163.
3 Berkeley, *Principles of Human Knowledge*, in Berkeley, ed. Ayers, I.8, p. 79.
4 Curley in *Phil.Rev.*, LXXXI (1972), 451–4.
5 Peter Alexander, 'Boyle and Locke on primary and secondary qualities', *Ratio*, XVI (1974), reprinted in Tipton (ed.), pp. 77–104.

11 Substance-in-General

1 Locke's correspondence with Stillingfleet started as a result of Stillingfleet's remarks about the *Essay* in his *A Discourse in Vindication of the Doctrine of the Trinity* (1696) and comprises three letters from Locke and two replies by Stillingfleet. Locke's side of the correspondence consisted of:

I. *A Letter to the Right Reverend Edward, Bishop of Worcester* ... (1697)
II. *Mr. Locke's Reply to the Right Reverend the Lord Bishop of Worcester's Answer to his Letter* ... (1697)
III. *Mr. Locke's Reply to the Right Reverend the Lord Bishop of Worcester's Answer to his Second Letter* ... (1698)

As is customary I refer to these as the first, second and third letters respectively and quote them from Locke, *Works*, 1823, vol. IV, indicated by *L. S.* followed by a page number.
2 In Stillingfleet's *Discourse*.
3 Mandelbaum, *Philosophy, Science*, pp. 31–46.
M. B. Bolton, 'Substances, substrata and names of substances in Locke's *Essay*', *Phil.Rev.*, LXXXV (1976), 488–513.
4 M. R. Ayers, 'The ideas of power and substance in Locke's philosophy', *Phil.Q.*, XXV (1975), 1–27, reprinted in *Locke on Human Understanding*, ed. Tipton, pp. 77–104. References are to the reprint.
5 Ayers in Tipton (ed.), p. 91.
6 Yolton, *Compass*, p. 45; Ayers in Tipton (ed.), p. 81.
7 Boyle, *A Free Inquiry into the Vulgarly Received Notion of Nature* (1686) in *Works*, vol. V, pp. 158–254. Part of this work is in *Philosophical Papers*, ed. Stewart, pp. 176–91.
8 See footnotes to p. 297 in Nidditch's edition.
9 See footnote to p. 298 in Nidditch's edition.
10 The third edition was published in 1695, the fourth in 1700; the letters to Stillingfleet are dated 1696–7, 1697 and 1698.
11 The relevant definition of this word in *O.E.D.* is 'mental separation of idea or fact'.
12 Locke's argument in this passage of the *Essay* would seem to be that people think of God as a thinking material being because that is how they think of themselves, but once we see that matter cannot *originate* thought we see that God, at least, must be immaterial even if he can make matter think. Once this

is admissible we can as well regard that which thinks in us as immaterial. However, my main point here is that Locke thinks that he has shown, with high probability, that this is so.

13 *The Bishop of Worcester's Answer to Mr. Locke's Letter* ... (1697), pp. 78–9.

14 The details are clear in Nidditch's edition, footnote to p. 284.

15 David Berman, 'Anthony Collins: aspects of his thought and writing', *Hermathena*, CXIX (1975), 54. I am grateful to David Berman for help with this and other points.

16 Anthony Collins, *An Answer to Mr. Clark's Third Defence of His Letter to Mr. Dodwell* (1708).

17 The reference is to the first letter; see Locke, *Works*, p. 10.

12 Language and Meaning

1 Berkeley, *Principles*, in Ayers (ed.), Introduction, Sections 22–5, pp. 83–4.

2 Putnam, 'Language and philosophy', in his *Philosophical Papers*, vol. II, pp. 1–32.

3 W. P. Alston, *Philosophy of Language*, Chap. 1.

4 For example, four of the abridged editions I have looked at omit the whole of the important chapter 'Of Particles'. They are those edited by Pringle-Pattison, Woozley, Wilburn and Cranston. Yolton is one honourable exception.

5 Leibniz, *New Essays on Human Understanding*, trans. and ed. Peter Remnant and Jonathan Bennett, p. 52.

6 See, e.g., J. A. Fodor, *The Language of Thought*, esp. Chap. 2.

7 Kretzmann in Tipton (ed.). I am inclined to think, however, that Locke needs less revision than Kretzmann suggests.

8 Note, also, that even the flavour of pineapple appears to be in the pineapple, on the tongue; I point into my mouth, not to any other part of my head, and not nowhere.

9 This passage is not further relevant to my problem here: it involves accepting the hypothesis that the parrot is a rational language-user, in all such respects like a human being, and asking the question whether we would, even then, call it a man. The answer is: no, because it has a parrot's body.

10 That, incidentally, is one reason why Berkeley's criticism of Locke on the abstract idea of a triangle does not touch Locke's view.

11 Berman, *Hermathena*, CXIX, p. 55.

12 Anthony Collins, *An Essay Concerning the Use of Reason in Propositions* (1707), pp. 8–9.

13 Kretzmann in Tipton (ed.), pp. 126–7 argues that Locke is mainly concerned with nouns and adjectives.

14 Berkeley, *Principles*, in Ayers (ed.), Introduction, sect. 13, p. 70.

15 Berkeley, *Principles*, in Ayers (ed.), Introduction, sect. 16, p. 72.

16 Berkeley, *Principles*, in Ayers (ed.), Introduction, sect. 16, pp. 71–2.

17 Berkeley, *Principles*, in Ayers (ed.), sect. 12, p. 70.
18 Berkeley, *Principles*, in Ayers (ed.), sect. 18, p. 73.
19 Berkeley, *An Essay Towards a New Theory of Vision*, in Ayers (ed.), sect. 123, p. 44.
20 Berkeley, *Principles*, in Ayers (ed.), Introduction, sect. 13, p. 70.
21 Criticisms of Locke's view from a different standpoint may be found in Peter Geach, *Mental Acts*.

13 Essences, Species and Kinds

1 For various views on this matter see: R. S. Woolhouse, *Locke's Philosophy of Science*, Chap. VI; John W. Yolton *Compass*, pp. 28–35, 80–6; Maurice Mandelbaum, *Philosophy, Science and Sense Perception*, pp. 41–6; R. M. Yost, Jr.. 'Locke's rejection of hypotheses about sub-microscopic events', *J.Hist.Ideas*, XII (1951), 111–30.
2 Saturday, 19 September 1676. Reprinted in *An Early Draft of Locke's Essay*, ed. R. I. Aaron and Jocelyn Gibb, p. 83. See also p. 98.
3 Mackie, *Problems*, p. 88.
4 Mackie, *Problems*, p. 94.
5 Mackie, *Problems*, p. 87.
6 I am grateful to David Hirschmann of the University of Bristol for help with this and other points.

14 Knowledge

1 M. D. Wilson in *A.P.Q.* XVI, 143–50 and reply by M. R. Ayers, 'Mechanism, superaddition and the proof of God's existence in Locke's *Essay*', *Phil.Rev.*, XC (1981), 210–51 and M. D. Wilson 'Superadded properties: a reply to M. R. Ayers', *Phil.Rev.*, XCI (1982), 247–52.
2 For a discussion of this passage from Boyle see above, Chap. 3.
3 Tipton, *Berkeley*, pp. 18–26.
4 Boyle, *Some Considerations Touching the Usefulness of Experimental Natural Philosophy* (1663) in *Works*, vol. II, pp. 1–246. This quotation from p. 46.

BIBLIOGRAPHY

AARON, R. I., *John Locke*, Oxford, Oxford University Press, 1937. 2nd edition, 1955; 3rd edition, 1971.

AARON, R. I., and GIBB, Jocelyn (eds.), *An Early Draft of Locke's Essay*, Oxford, Oxford University Press, 1936.

AARSLEFF, Hans, 'Leibniz on Locke and language', *American Philosophical Quarterly*, I, No. 3, 1964.

ALEXANDER, Peter, 'Boyle & Locke on primary and secondary qualities', *Ratio*, XVI (1974), 51–67, reprinted in Tipton (ed.), pp. 77–104.

'Curley on Locke & Boyle', *Philosophical Review*, LXXXIII (1974), 229–37.

'The names of secondary qualities', *Proceedings of the Aristotelian Society*, LXXVII (1977), 203–20.

'Locke on substance-in-general' I and II, *Ratio*, XXII (1980), 91–105 and XXIII (1981), 1–19.

'The case of the lonely corpuscle: reductive explanation and primitive expressions', in *Reduction, Time and Reality*, ed. Richard Healey, Cambridge, Cambridge University Press, 1981, pp. 17–35.

ALSTON, W. P., *Philosophy of Language*, Englewood Cliffs, Prentice-Hall, 1964.

ANSCOMBE, G. E. M. and GEACH, P. T., *Three Philosophers*, Oxford, Blackwell, 1963 Art. 'Aristotle' by G. E. M. Anscombe.

ARISTOTLE, *de Generatione et Corruptione*, trans. E. S. Foster, London, Heinemann, 1955.

ed. C. J. F. Williams, Oxford, Clarendon Press, 1982.

Physics I, II, ed. W. Charlton, Oxford, Clarendon Press, 1970.

Physics, ed. P. H. Wicksteed and F. M. Cornford, London, Heinemann, 1929.

de Anima, trans. W. S. Hett, London, Heinemann, 1975.

Categories, trans. H. P. Cook, London, Heinemann, 1938.

ARMSTRONG, D. M. (ed.), *Berkeley's Philosophical Writings*, London, Collier-Macmillan, 1965.

ARNAULD, Antoine and NICOLE, P., *The Art of Thinking (Part-Royal Logic)* (1662), London, Taylor, 1702.

ASHWORTH, E. J., '"Do words signify ideas or things?" The scholastic sources of Locke's theory of language', *Journal of the History of Philosophy*, XIX (1981), 299–326.

ASPELIN, G., '"Idea" and "Perception" in Locke's *Essay*', a reply to Greenlee (q.v.) in Tipton (ed.), pp. 47–51.

ATTFIELD, Robin, 'Clarke, Collins and compounds', *Journal of the History of Philosophy*, XV (1977), 45–54.

AVERILL, E. W., 'The primary-secondary quality distinction', *Philosophical Review*, XCI (1982), 343–61.

AYERS, M. R., 'The ideas of power and substance in Locke's philosophy', *Philosophical Quarterly*, XXV (1975), 1–27. Reprinted in I. C. Tipton (ed.), pp. 77–104.

'Locke's doctrine of abstraction: some aspects of its historical and philosophical significance', in R. Brandt (ed.), *John Locke Symposium, Wolfenbüttel (1979)*, Berlin, de Gruyter, 1981, 5–24.

'Mechanism, superaddition and the proof of God's existence in Locke's *Essay*', *Philosophical Review*, XC (1981), 210–51.

'Locke versus Aristotle on natural kinds', *Journal of Philosophy*, LXXVIII (1981), 247–71.

AYERS, M. R. (ed.), Berkeley, *Philosophical Works*, London, Dent, 1975. See also RÉE, Jonathan.

BACON, Francis, *Novum Organum* (1600), trans. William Wood, in *The Physical and Metaphysical Works of Lord Bacon*, ed. J. Devey, London, Bell, 1901.

BARGER, Bill, *Locke on Substance* together with Boyle's *Origin of Forms and Qualities (The Theorical Part)*, Manhattan Beach, California, Sheffield Press, 1976.

BENNETT, Jonathan, 'Substance, reality & primary qualities', *American Philosophical Quarterly*, II (1965), reprinted in C. B. Martin & D. M. Armstrong (eds.), pp. 86–124.

Locke, Berkeley, Hume, Oxford, Oxford University Press, 1971.

BERKELEY, G., ed. D. M. Armstrong, *Berkeley's Philosophical Writings*, London, Collier-Macmillan, 1965.

ed. M. R. Ayers, Berkeley, *Philosophical Works*, London, Dent, 1975.

BERMAN, David, 'On missing the wrong target', *Hermathena*, CXIII (1972), 54–67.

'Anthony Collins: aspects of his thought and writings', *Hermathena*, CXIX (1975), 49–70.

BOAS, Marie, 'The establishment of the mechanical philosophy', *Osiris*, X (1953), 413–541.

Robert Boyle and Seventeenth Century Chemistry, Cambridge, Cambridge University Press, 1958.

see also HALL, Marie Boas.

BOLTON, Martha Brant, 'Substances, substrata and names of substances in Locke's *Essay*', *Philosophical Review*, LXXXV (1976), 488–513.

'The origins of Locke's doctrine of primary and secondary qualities', *Philosophical Quarterly*, XXVI (1976), 305–16.

'A defence of Locke and the representative theory of perception' in C. E. Jarrett et al., *New Essays on Rationalism & Empiricism (Canadian Journal of Philosophy*, Supp. 4 (1978)), 101–120.

BOYLE, Robert, *Works*, 6 vols., London, Rivington etc., 1772.

The Sceptical Chymist (1661), ed. M. M. Pattison Muir, London, Dent, N. D.

Experiments and Considerations Touching Colours (1664), ed. Marie Boas Hall, New York, Johnson Reprint Corporation, 1964.

BRACKEN, H. M., *The Early Reception of Berkeley's Immaterialism*, The Hague, 1959.

BRANDT, R., 'Historical observations on the genesis of the three-dimensional optical picture', *Ratio*, XVII (1975), 176–90.

BRETT, G. S., *The Philosophy of Gassendi*, London, Macmillan, 1908.

BURTHOGGE, Richard, *The Philosophical Writings of Richard Burthogge*, ed. Margaret W. Landes, Chicago, Open Court, 1921.

BUTLER, R. J., 'Substance un-Locked', *Proceedings of the Aristotelian Society*, LXXIV (1973–4), 131–60.

CAMPBELL, John, 'Locke on qualities', *Canadian Journal of Philosophy*, X (1980), 567–85.

CAMPBELL, K., Primary and secondary qualities', *Canadian Journal of Philosophy*, II (1972). 219–32.

CARRÉ, M. H., 'Gassendi and the new philosophy', *Philosophy*, XXXIII (1958), 112–20.

CHARLTON, W., 'Prime matter: a rejoinder', *Phronesis*, XXVIII (1983), 197–211.

COIRAULT, G., 'Gassendi et non Locke créatur de la doctrine sensualiste moderne sur la génération des idées', *Actes du Congrès du Tricentenaire de Pierre Gassendi*, Digne, 1955, pp. 71–94.

COLLINS, Anthony, *An Essay Concerning the Use of Reason in Propositions*, London, 1707

An Answer to Mr. Clarke's Third Defence of his Letter to Mr. Dodwell, London, Baldwin, 1708.

COPLESTON, F. C., *History of Philosophy*, London, Burns, Oates & Washbourne, 8 vols., 1946–66.

CRANSTON, Maurice, *John Locke, a Biography*, London, Longmans, 1957.

CROMBIE, A. C., *Medieval and Early Modern Science*, New York, Doubleday, 1959.

CUMMINS, Phillip D., 'Locke's anticipation of Hume's use of "Impression"', *Modern Schoolman*, I (1973), 297–301.

CUMMINS, Robert, 'Two troublesome claims about qualities in Locke's *Essay*', *Philosophical Review*, LXXXIV (1975), 401–18.

CURLEY, E. M., 'Locke, Boyle and the distinction between primary and secondary qualities', *Philosophical Review*, LXXXI (1972), 438–64.

DEBUS, A. G., *The English Paracelsians*, London, Oldbourne, 1965.

The Chemical Philosophy, 2 vols., New York, Science History Publications, 1977.

DESCARTES, René, *The Philosophical Works of Descartes*, eds. Elizabeth S. Haldane and G. R. T. Ross, New York, Dover, 1959.

Discourse on Method, Optics, Geometry, ed. P. J. Olscamp, Indianapolis, Bobbs-Merrill, 1965.

Philosophical Writings, selections translated Elizabeth Anscombe and Peter Thomas Geach, London, Nelson, 1970.

DEWHURST, K., *John Locke (1632–1704), Physician and Philosopher*, London, 1963.

DRISCOLL, E. A., 'The influence of Gassendi on Locke's hedonism', *International Philosophical Quarterly*, XII (1972), 87–110.

ENGLEBRETSEN, George, 'Locke's language of proper names', *The Locke Newsletter*, No. 4 (1973), 25–31.

EPICURUS, *Letters, Principal Doctrines, Vatican Sayings*, trans. Russel M. Geer, Indianapolis, Bobbs-Merrill, 1978.

ERASTUS, Thomas, *Disputationes de Medicina nova Paracelsi*, Basel, 1572.

FODOR, J. A., *The Language of Thought*, Hassocks, Harvester Press, 1976.

GABRIEL, M. L. and FOGEL, S. (eds.), *Great Experiments in Biology*, Englewood Cliffs, Prentice-Hall, 1955.

GEACH, Peter, *Mental Acts*, London, Routledge, 1957.

GIBSON, J., *Locke's Theory of Knowledge*, Cambridge, Cambridge University Press, 1917.

GIVNER, David A., 'Scientific preconceptions in Locke's philosophy of language', *Journal of the History of Ideas*, XXIII (1962), 340–54.

GREENLEE, D., 'Locke's idea of "Idea"', *Theoria*, XXXIII (1967), 98–106, reprinted in Tipton (ed.), pp. 41–7.

'Idea and object in the *Essay*', a reply to Aspelin (q.v.), in Tipton (ed.), pp. 51–4.

HACKER, P. M. S., 'Locke and the meaning of colour words', in G. Vesey (ed.), *Impressions of Empiricism (Royal Institute of Philosophy Lectures*, IX, 1974–5), London, Macmillan, 1976, pp. 23–46.

HACKING, Ian M., *Why Does Language Matter to Philosophy*?, Cambridge, Cambridge University Press, 1975.

HALL, A. R., *The Scientific Revolution 1500–1800*, London, Longmans, 1954.

HALL, Marie Boas (ed.), Robert Boyle, *Experiments and Considerations Touching Colours*, (1664) in facsimile edition, New York, Johnson Reprint Corporation, 1964.

Robert Boyle on Natural Philosophy, Bloomington, Indiana University Press, 1965.

HALL, Roland (ed.), *The Locke Newsletter* Nos. 1–14, York, Roland Hall, 1970–1983.

HALL, Roland and WOOLHOUSE, Roger, *Eighty Years of Locke Scholarship: A Bibliographical Guide*, Edinburgh, Edinburgh University Press, 1963.

HAMILTON, Sir William (ed.), *Works of Thomas Reid* (q.v.).

HAMLYN, D. W., *Sensation and Perception*, London, Routledge, 1961.

HARRISON, John and LASLETT, Peter (eds.), *The Library of John Locke*, Oxford, Oxford University Press, 1971

van HELMONT, J. B., *Ortus Medicina*, Amsterdam, 1748.

HENZE, D. F., 'Locke on "Particles"', *Journal of the History of Philosophy*, IX (1971), 222–6.

HOFFMAN, J., 'Locke on whether a thing can have two beginnings of existence', *Ratio*, XXII (1980), 106–11.

JACKSON, Reginald, 'Locke's distinction between primary and secondary qualities', *Mind*, XXXVIII (1929), reprinted in C. B. Martin and D. M. Armstrong (eds.) *Locke and Berkeley*, pp. 53–77.

'Locke's version of the doctrine of representative perception', *Mind*, XXXIX (1930), reprinted in Martin and Armstrong (eds.), pp. 125–54.

KARGON, R. H., *Atomism in England from Hariot to Newton*, Oxford, Oxford University Press, 1966.

KELEMEN, J., 'Locke's theory of language and semiotics', *Language Sciences*, XL (1976), 16–24.

KEMP SMITH, N., *New Studies in the Philosophy of Descartes*, London, Macmillan, 1953.

KING, Hugh R., 'Aristotle without *prima materia*', *Journal of the History of Ideas*, XVII (1956), 370–89.

See also SOLMSEN.

KRETZMANN, Norman, 'The main thesis of Locke's semantic theory', *Philosophical Review*, LXXVII (1968), 175–96, reprinted in I. Tipton (ed.), pp. 123–40.

LANDESMAN, C., 'Locke's theory of meaning', *Journal of the History of Philosophy*, XXIV (1976), 23–35.

LATTA, Robert (ed.), Leibniz: *The Monadology and Other Philosophical Writings*, Oxford, Oxford University Press, 1898, 2nd impression 1925.

LAUDAN, Laurens, 'The nature and sources of Locke's views on hypotheses', *Journal of the History of Ideas*, XXVIII (1967), 211–23, reprinted in Tipton (ed.), pp. 149–62

'The clock metaphor and probabilism', *Annals of Science*, XXII (1966), 73–104.

LEFEBVRE, Nicholas, *A Compleat Body of Chymistry* (1660), trans. P. D. C. Landen, T. Ratcliffe, 1664.

LEIBNIZ, G. W., *Discourse on Metaphysics* (1686) trans. Peter G. Lucas and Leslie Grint, Manchester, University Press, 2nd edition, 1953.

Letter to Arnauld IX, July 14th, 1686, in Parkinson (ed.)

New Essays on Human Understanding (1704), ed. Peter Remnant and Jonathan Bennett, Cambridge, Cambridge University Press, 1981.

See also LATTA, Robert.

LOCKE, John, *An Essay Concerning Human Understanding* (1690) ed. P. H. Nidditch, Oxford, Oxford University Press, 1975.

Essay, ed. John W. Yolton, 2 vols., Dent, London, 1961.

Essay, ed. A. C. Fraser, 2 vols., Oxford, Oxford University Press, 1894.

An Essay Concerning Human Understanding (Abridged) ed. Maurice Cranston, New York, Collier Books, 1965.

ed. Pringle-Pattison, Oxford, Oxford University Press, 1924.

ed. Raymond Wilburn, London, Dent, 1947.

ed. A. D. Woozley, London, Collins, 1964.

ed. John W. Yolton, London, Dent, 1977.

Draft A of Locke's Essay, ed. P. H. Nidditch, University of Sheffield, 1980.
Draft B of Locke's Essay, ed. P. H. Nidditch, University of Sheffield, 1982.
Works in 10 volumes, Tegg, etc., London, 1823.
Letters to Edward Stillingfleet, Bishop of Worcester
I. 1696–7, *Works*, Vol. IV, 3–96.
II. 1697, *Works*, Vol. IV, 99–189.
III. 1698, *Works*, Vol. IV, 193–498.
The Correspondence of John Locke, ed. E. S. de Beer, Oxford, Oxford University Press, 1976–.
LOEMKER, Leroy E., 'Boyle and Leibniz', *Journal of the History of Ideas*, XVI (1955), 22–43.
LUCRETIUS, *De Rerum Natura*, trans. W. H. D. Rouse, London, Heinemann, 1975.
MABBOTT, J. D., *John Locke*, London, Macmillan, 1973.
McGUIRE, J. E., 'Transmutation and immutability: Newton's doctrine of physical qualities', *Ambix*, XIV (1967), 69–95.
MACINTOSH, J. J., 'Primary and secondary qualities', *Studia Leibnitiana*, VIII (1976), 88–104.
MACKIE, J. L., *Problems from Locke*, Oxford, Oxford University Press, 1976.
MANDELBAUM, Maurice, *Philosophy, Science and Sense Perception*, Baltimore, The Johns Hopkins Press, 1964, 2nd edition, 1977.
MARTIN, C. B., 'Substance substantiated', *Australasian Journal of Philosophy*, LVIII (1980), 3–10.
MARTIN, C. B. and ARMSTRONG, D. M. (eds.), *Locke and Berkeley*, London, Macmillan, N. D. [1968].
MATTERN, Ruth, 'Locke: "Our knowledge, which all consists in propositions"', *Canadian Journal of Philosophy*, VIII (1978), 677–95.
'Locke on power and causation: excerpts from the 1685 draft of the *Essay*', *Philosophical Research Archives*, VII (1981), No. 1357.
MATTHEWS, H. E., 'Locke, Malebranche & the representative theory', *The Locke Newsletter*, No. 2 (1971), 12–21, reprinted in Tipton (ed.), pp. 55–61.
van MELSEN, Andrew G., *From Atomos to Atom*, Duquesne, Duquesne University Press, 1952, reprinted N.Y., Harper and Brothers, 1960.
MORAVCSIK, J. M. E. (ed.), *Aristotle*, London, Macmillan, 1968.
MULTHAUF, R. P., *The Origins of Chemistry*, London, Oldbourne, 1966.
NERLICH, G. C., Review of *Reduction, Time and Reality*, ed. Richard Healey, Cambridge, Cambridge University Press, 1981, in *Philosophical Quarterly*, XXXII (1982), 272–9.
NEWTON, Isaac, *Principia Mathematica* (1687), ed. Cajori, 2 vols., Berkeley, University of California Press, 1962.
Opticks (1704), reprint of the 4th edition (1730), New York, Dover, 1952.
O'CONNOR, D. J., *John Locke*, Harmondsworth, Penguin, 1952.
ODEGARD, Douglas, 'Locke and the signification of words', *The Locke Newsletter* No. 1 (1970). 11–17.

'Locke and substance', *Dialogue*, VIII (1969), 243–55.

'Locke, Geach and individual essences', *Philosophical Studies*, XXII (Minneapolis, 1971), 70–3.

OSLER, M. J., 'Providence and the divine will: the theological background to Gassendi's views on scientific knowledge', *Journal of the History of Ideas*, XLIV (1983), 549–60.

'John Locke and the changing ideal of scientific knowledge', *Journal of the History of Ideas*, XXXI (1970), 3–16.

O'TOOLE, F. J., 'Qualities and powers in the corpuscular philosophy of Robert Boyle', *Journal of the History of Philosophy*, XII (1974), 295–315.

OWENS, Joseph, 'Matter and predication in Aristotle', in *The Concept of Matter in Greek and Medieval Philosophy*, ed. E. McMullen, Notre Dame, 1963. Reprinted in *Aristotle*, ed. J. M. E. Moravcsik, London, Macmillan, 1968, pp. 191–214.

PALMER, David, 'Boyle's corpuscular hypothesis and Locke's primary–secondary quality distinction', *Philosophical Studies*, XXIX (Dordrecht, 1976), 181–9.

PARACELSUS, *The Hermetic and Alchemical Writings*, English translation by A. E. Waite (1894) reprinted by University Books, New York, 1967.

PARKINSON, G. H. R. (ed.), Leibniz, *Philosophical Writings*, London, Dent, 1973.

PARTINGTON, J. R., *A History of Chemistry*, 4 vols., London, Macmillan, 1958–1970.

PELLETIER, F. J., 'Locke's doctrines of substance' in C. E. Jarrett et al., *New Essays on Rationalism and Empiricism*, 121–40, *Canadian Journal of Philosophy*, supp. 4 (1978).

PERRY, David L., 'Locke on mixed modes, relations & knowledge', *Journal of the History of Philosophy*, V (1967), 219–35.

POPPER, K. R., *Conjectures and Refutations*, London, Routledge, 1963.

PUTNAM, H., 'On properties' (1970) in *Philosophical Papers*, Vol. I, Cambridge, Cambridge University Press, 1975, pp. 305–22.

'Language & Philosophy' in *Philosophical Papers*, Vol. II, Cambridge, Cambridge University Press, 1975, pp. 1–32.

RÉE, Jonathan, AYERS, Michael and WESTOBY, Adam, *Philosophy and its Past*, Hassocks, Harvester, 1978.

REID, Thomas, *Works of Thomas Reid*, ed. Sir William Hamilton, 2 vols., Edinburgh, Maclachlan and Stewart, 1846.

ROGERS, G. A. J., 'The systems of Locke and Newton', in Z. Beckler (ed.), *Contemporary Newtonian Research*, Bloomington, Indiana University Press, 1965, pp. 215–38.

'Boyle, Locke and reason', *Journal of the History of Ideas*, XXVII (1966), 205–16.

'Locke's *Essay* and Newton's *Principia*', *Journal of the History of Ideas*, XXXVII (1978), 217–32.

'Descartes and the method of English science', *Annals of Science*, XXIX (1972), 237–55.

'The empiricism of Locke & Newton', in S. C. Brown (ed.), *Philosophers of the Enlightenment* (R. I. P. Lectures, vol. 12), Sussex, Harvester Press, 1979, pp. 1–30.

ROSS, W. D., *Aristotle*, London, Methuen, 1937.

RYLE, Gilbert, 'John Locke on the Human Understanding', *Tercentenary Addresses on John Locke*, ed., J. L. Stocks, Oxford, Oxford University Press, 1933, reprinted in C. B. Martin & D. M. Armstrong (eds.), pp. 14–39.

SANFORD, David H., 'Does Locke think hardness is a primary quality?' *The Locke Newsletter*, No. 1, York, Roland Hall, 1970.

SCHANKULA, H. A. S., 'Locke, Descartes and the science of nature', *J Hist. Ideas*, XLI (1980), 459–77.

SCHOFIELD, Robert E., *Mechanism and Materialism*, Princeton, Princeton University Press, 1970.

SCOTT, Wilson L., *The Conflict between Atomism and Conservation Theory, 1644 to 1860*, London, Macdonald, 1970.

SENNERT, Daniel, *de Chymicorum cum Aristotelicis et Galenicis Consensu ac Dissensu Liber*, Wittenburg, 1619.

SOLMSEN, Friedrich, 'Aristotle and prime matter, a reply to Hugh R. King', *Journal of the History of Ideas*, XIX (1958), 243–52.

SPECHT, R., 'Uber empiricistische ansätze Lockes', *Allgemeine Zeitschrift für Philosophie*, III (1977), 1–35.

SPRAT, Thomas, *The History of the Royal Society*, London, 1667.

Facsimile edition J. I. Cope and H. W. Jones (eds.), London, Routledge & Kegan Paul, 1959.

SQUADRITO, K., *Locke's theory of sensitive knowledge*, Washington, University Press of America, 1978.

STEWART, M. A. (ed.), *Selected Philosophical Papers of Robert Boyle*, Manchester, University Press, 1979.

STEWART, M. A., 'Locke's mental atomism and the classification of Ideas', I, *The Locke Newsletter*, No. 10 (1979), 53–82; II, *The Locke Newsletter*, No. 11 (1980), 25–62.

'Locke's professional contacts with Robert Boyle', *The Locke Newsletter*, No. 12 (1981), 19–44.

'The authenticity of Robert Boyle's anonymous writings on Reason', *Bodleian Library Record*, 10 (1981), 280–9.

STILLINGFLEET, Edward (Bishop of Worcester), *A Discourse in Vindication of the Doctrine of the Trinity*, second edition, London, Mortlock, 1697.

Letters to John Locke

I London, Mortlock, 1697.

II London, Mortlock, 1698.

STROUD, Barry, *Hume*, London, Routledge, 1977.

'Berkeley v. Locke on primary qualities', *Philosophy*, LV (1980), 149–66.

TIPTON, I. C., *Berkeley*, London, Methuen, 1974.

(ed.) *Locke on Human Understanding*, Oxford, Oxford University Press, 1977.

TROYER, John, 'Locke on the names of substances', *The Locke Newsletter*, No. 6 (1975), 27–39.

WADE, Ira O., *The Intellectual Origins of the French Enlightenment*, Princeton, Princeton University Press, 1971.

WAITE, A. E. (ed.), *The Hermetic and Alchemical Writings of Paracelsus* (1894), facsimile reprinted by University Books, New York, 2 vols., 1967.

WARNOCK, G. J., *Berkeley*, Harmondsworth, Penguin, 1953.

WELBOURNE, M., 'The community of knowledge', *Philosophical Quarterly*, XXXI (1981), 302–14.

WESTOBY, Adam, See RÉE, Jonathan.

WILLIAMS, C. J. F., 'Are primary qualities qualities?', *Philosophical Quarterly*, XIX (1969), 310–23.

WILSON, Margaret D., 'Leibniz and Locke on "First Truths"', *Journal of the History of Ideas*, XXVIII (1967), 347–66.

'Superadded properties. The limits of mechanism in Locke', *American Philosophical Quarterly*, XVI (1979), 143–50.

'Superadded properties: a reply to M. R. Ayers', *Philosophical Review*, XCI (1982), 247–52.

See also AYERS, M. R.

WOLFE, David, 'Sydenham and Locke on the limits of anatomy', *Bulletin of the History of Medicine*, XXXV (1961), 193–220.

WOOD, N., 'The Baconian character of Locke's *Essay*', *Studies in the History and Philosophy of Science*, VI (1975), 43–84.

WOOLHOUSE, R. S., 'Substance & substances in Locke's *Essay*', *Theoria*, XXXV (1969), 153–67.

Locke's Philosophy of Science & Knowledge, Oxford, Basil Blackwell, 1971.

'Locke on modes, substances and knowledge', *Journal of the History of Philosophy*, X (1972), 417–24.

'Locke, Geach & individuals' essences', *Philosophical Studies*, XXIV (1973), 204–7.

'Locke, Leibniz and the reality of Ideas', *John Locke Symposium, Wolfenbüttel (1979)*, ed. Reinhard Brandt, Berlin, de Gruyter, 1981, 193–206.

WOOZLEY, A. D., 'Some remarks on Locke's account of knowledge', *The Locke Newsletter*, No. 3 (1972), 7–17, reprinted in Tipton (ed.), pp. 141–8.

YOLTON, J. W. (ed), *Locke: Problems & Perspectives*, Cambridge, Cambridge University Press, 1969.

YOLTON, J. W., *John Locke and the Way of Ideas*, Oxford, Oxford University Press, 1956.

Locke and the Compass of Human Understanding, Cambridge, Cambridge University Press, 1970.

YOST, R. M. Jr., 'Sydenham's philosophy of science', *Osiris*, IX (1950), 84–105.

'Locke's rejection of hypotheses about sub-microscopic events', *Journal of the History of Ideas*, XII (1951), 111–30.

INDEX

THE PHILOSOPHICAL WRITINGS OF *DESCARTES*

VOLUME I, VOLUME II

Translated by
J. G. COTTINGHAM
Department of Philosophy, University of Reading

ROBERT STOOTHOFF
Department of Philosophy and Religious Studies,
University of Canterbury, New Zealand

and DUGALD MURDOCH
Department of Philosophy and Religious Studies,
University of Canterbury, New Zealand

These two volumes provide a completely new translation of the philosophical works of Descartes, based on the best available Latin and French texts. They are intended to replace the only reasonably comprehensive selection of his works in English, by Haldane and Ross, first published in 1911. All the works included in that edition are translated here, together with a number of additional texts crucial for an understanding of Cartesian philosophy, including important material from Descartes' scientific writings. The result should meet the widespread demand for an accurate and authoritative edition of Descartes' philosophical writings in clear and readable modern English.

CONTENTS

VOLUME ONE

VOLUME TWO

G. W. LEIBNIZ

NEW ESSAYS ON HUMAN UNDERSTANDING

Translated and edited by
Peter Remnant and Jonathan Bennet

The *New Essays on Human Understanding* is a philosophical classic. It discloses much of Leibniz's philosophy, and is the most important single confrontation between the philosophical traditions of empiricism and rationalism and between the intellectual temperaments which are associated with them. In it Leibniz argues chapter by chapter with John Locke's *Essay Concerning Human Understanding*, providing a critique which ranges over the main topics where the two traditions diverge. Leibniz expounds his thought here with a freedom and panache absent from most of his better-known writings, and the force of his intellect can be felt on almost every page. The references to his contemporaries and the discussions of the ideas and institutions of the age make this a rich and fascinating document in the history of ideas; the editors display this aspect of the work in an extensive glossary.

Written in French in 1703–5, the *Nouveaux essais* was not published until 1765, one of its first effects being to acquaint Kant directly with Leibniz's thought. The only previous English translation, published in 1896, was seriously deficient and based on a corrupt text. The version by Peter Remnant and Jonathan Bennett, based on the sole reliable edition of the French text (published in 1962), is accurate and readable, and will become the standard English translation. It is indispensable for serious students of Leibniz's thought and of that period in the history of philosophy and the history of ideas.

'The appearance of Remnant and Bennett's translation has been eagerly awaited by Leibniz scholars, and they have not been disappointed in their expectations... [Remnant and Bennett] have produced a superb translation which is everywhere clear, fluent and readable ... what [they] have accomplished might be compared to the cleaning of a venerable Old Master; layers of centuries-old dirt and varnish have been removed, and we now confront the original as it would have appeared to its contemporaries ... it seems fairly clear that the *New Essays* of Remnant and Bennett will take its place among the great translations of philosophical classics: it will surely come to rank with, and indeed possibly surpass, Kemp Smith's *Critique of Pure Reason*.'

Philosophical Books

Published in hard covers and as a paperback

G. W. LEIBNIZ

NEW ESSAYS ON HUMAN UNDERSTANDING

Translated and edited by
Peter Remnant and Jonathan Bennett

Abridged paperback edition

This is an abridgement of the complete translation of the *New Essays*, first published in 1981, designed for use as a study text. The material extraneous to philosophy – more than a third of the original – and the glossary of notes have been cut, and a new philosophical introduction and bibliography of recent work on Leibniz have been provided by the translators. The marginal pagination has been retained for ease of cross-reference to the full edition.

JOHN W. YOLTON

THE LOCKE READER

Selections from the works of John Locke with a General Introduction and Commentary

John Yolton seeks to allow readers of Locke to have accessible in one volume selections from a wide range of Locke's books, structured so that some of the interconnections of his thought can be seen and traced. Although Locke did not write from a system of philosophy, he did have in mind an overall division of human knowledge. The readings begin with Locke's essay on Hermeneutics and the portions of his *Essay Concerning Human Understanding* on how to read a text. The rest of the selections are organized around Locke's division of human knowledge into natural science, ethics, and the theory of signs. Extensive selections are taken from the *Essay, Two Treatises of Government, A Letter Concerning Toleration, Some Thoughts Concerning Education, The Reasonableness of Christianity, the Conduct of Understanding,* and *An Examination of P. Malebranche's Opinion of Seeing All Things in God.* Yolton's introduction and commentary explicate Locke's doctrines and provide the reader with the general background knowledge of other seventeenth-century writers and their works necessary to an understanding of Locke and his time.

Published in hard covers and as a paperback

J. DUNN

THE POLITICAL THOUGHT OF JOHN LOCKE

An historical account of the argument of the 'Two Treatises of Government'

'Tart, trenchant, sticking closely to the texts yet full of sudden forays and insights, it is the product of tense thinking and wide reading . . . gives a new and original twist to Locke's thought and character.' *The New Statesman*

'Interesting: fresh and lively, and full of promise for the future.'
The Times Literary Supplement

Published in hard covers and as a paperback.